T0292642

Parallel Programming

Parallel Programming

Concepts and Practice

Bertil Schmidt
Institut für Informatik
Staudingerweg 9
55128 Mainz
Germany

Jorge González-Domínguez
Computer Architecture Group
University of A Coruña
Edificio área científica (Office 3.08), Campus de Elviña
15071, A Coruña
Spain

Christian Hundt
Institut für Informatik
Staudingerweg 9
55128 Mainz
Germany

Moritz Schlarb
Data Center
Johannes Gutenberg-University Mainz
Germany

Anselm-Franz-von-Bentzel-Weg 12
55128 Mainz
Germany

Morgan Kaufmann is an imprint of Elsevier
50 Hampshire Street, 5th Floor, Cambridge, MA 02139, United States

Library of Congress Cataloging-in-Publication Data
A catalog record for this book is available from the Library of Congress

British Library Cataloguing-in-Publication Data
A catalogue record for this book is available from the British Library

ISBN: 978-0-12-849890-3

For information on all Morgan Kaufmann publications
visit our website at https://www.elsevier.com/books-and-journals

Publisher: Katey Birtcher
Acquisition Editor: Steve Merken
Developmental Editor: Nate McFadden
Production Project Manager: Sreejith Viswanathan
Designer: Christian J. Bilbow

Typeset by VTeX

Contents

Preface

Parallelism abounds. Nowadays, any modern CPU contains at least two cores, whereas some CPUs feature more than 50 processing units. An even higher degree of parallelism is available on larger systems containing multiple CPUs such as server nodes, clusters, and supercomputers. Thus, the ability to program these types of systems efficiently and effectively is an essential aspiration for scientists, engineers, and programmers. The subject of this book is a comprehensive introduction to the area of parallel programming that addresses this need. Our book teaches practical parallel programming for shared memory and distributed memory architectures based on the C++11 threading API, Open Multiprocessing (OpenMP), Compute Unified Device Architecture (CUDA), Message Passing Interface (MPI), and Unified Parallel C++ (UPC++), as well as necessary theoretical background. We have included a large number of programming examples based on the recent C++11 and C++14 dialects of the C++ programming language.

This book targets participants of "Parallel Programming" or "High Performance Computing" courses which are taught at most universities at senior undergraduate level or graduate level in computer science or computer engineering. Moreover, it serves as suitable literature for undergraduates in other disciplines with a computer science minor or professionals from related fields such as research scientists, data analysts, or R&D engineers. Prerequisites for being able to understand the contents of our book include some experience with writing sequential code in C/C++ and basic mathematical knowledge.

In good tradition with the historic symbiosis of High Performance Computing and natural science, we introduce parallel concepts based on real-life applications ranging from basic linear algebra routines over machine learning algorithms and physical simulations but also traditional algorithms from computer science. The writing of correct yet efficient code is a key skill for every programmer. Hence, we focus on the actual implementation and performance evaluation of algorithms. Nevertheless, the theoretical properties of algorithms are discussed in depth, too. Each chapter features a collection of additional programming exercises that can be solved within a web framework that is distributed with this book. The **System for Automated Code Evaluation (SAUCE)** provides a web-based testing environment for the submission of solutions and their subsequent evaluation in a classroom setting: the only prerequisite is an HTML5 compatible web browser allowing for the embedding of interactive programming exercise in lectures. SAUCE is distributed as docker image and can be downloaded at

https://parallelprogrammingbook.org

This website serves as hub for related content such as installation instructions, a list of errata, and supplementary material (such as lecture slides and solutions to selected exercises for instructors).

If you are a student or professional that aims to learn a certain programming technique, we advise to initially read the first three chapters on the fundamentals of parallel programming, theoretical models, and hardware architectures. Subsequently, you can dive into one of the introductory chapters on C++11 Multithreading, OpenMP, CUDA, or MPI which are mostly self-contained. The chapters on Advanced C++11 Multithreading, Advanced CUDA, and UPC++ build upon the techniques of their preceding chapter and thus should not be read in isolation.

If you are a lecturer, we propose a curriculum consisting of 14 lectures mainly covering applications from the introductory chapters. You could start with a lecture discussing the fundamentals from the first chapter including parallel summation using a hypercube and its analysis, the definition of basic measures such as speedup, parallelization efficiency and cost, and a discussion of ranking metrics. The second lecture could cover an introduction to PRAM, network topologies, weak and strong scaling. You can spend more time on PRAM if you aim to later discuss CUDA in more detail or emphasize hardware architectures if you focus on CPUs. Two to three lectures could be spent on teaching the basics of the C++11 threading API, CUDA, and MPI, respectively. OpenMP can be discussed within a span of one to two lectures. The remaining lectures can be used to either discuss the content in the advanced chapters on multithreading, CUDA, or the PGAS-based UPC++ language.

An alternative approach is splitting the content into two courses with a focus on pair-programming within the lecture. You could start with a course on CPU-based parallel programming covering selected topics from the first three chapters. Hence, C++11 threads, OpenMP, and MPI could be taught in full detail. The second course would focus on advanced parallel approaches covering extensive CUDA programming in combination with (CUDA-aware) MPI and/or the PGAS-based UPC++.

We wish you a great time with the book. Be creative and investigate the code! Finally, we would be happy to hear any feedback from you so that we could improve any of our provided material.

Acknowledgments

This book would not have been possible without the contributions of many people.

Initially, we would like to thank the anonymous and few non-anonymous reviewers who commented on our book proposal and the final draft: Eduardo Cesar Galobardes, Ahmad Al-Khasawneh, and Mohammad Olaimat.

Moreover, we would like to thank our colleagues who thoroughly peer-reviewed the chapters and provided essential feedback: André Müller for his valuable advise on C++ programming, Robin Kobus for being a tough code reviewer, Felix Kallenborn for his steady proofreading sessions, Daniel Jünger for constantly complaining about the CUDA chapter, as well as Stefan Endler and Elmar Schömer for their suggestions.

Additionally, we would like to thank the staff of Morgan Kaufman and Elsevier who coordinated the making of this book. In particular we would like to mention Nate McFadden.

Finally, we would like to thank our spouses and children for their ongoing support and patience during the countless hours we could not spend with them.

CHAPTER

INTRODUCTION

1

Abstract

In the recent past, teaching and learning of parallel programming has become increasingly important due to the ubiquity of parallel processors in portable devices, workstations, and compute clusters. Stagnating single-threaded performance of modern CPUs requires future computer scientists and engineers to write highly parallelized code in order to fully utilize the compute capabilities of current hardware architectures. The design of parallel algorithms, however, can be challenging especially for inexperienced students due to common pitfalls such as race conditions when concurrently accessing shared resources, defective communication patterns causing deadlocks, or the non-trivial task of efficiently scaling an application over the whole number of available compute units. Hence, acquiring parallel programming skills is nowadays an important part of many undergraduate and graduate curricula. More importantly, education of concurrent concepts is not limited to the field of *High Performance Computing* (HPC). The emergence of deep learning and big data lectures requires teachers and students to adopt HPC as an integral part of their knowledge domain. An understanding of basic concepts is indispensable for acquiring a deep understanding of fundamental parallelization techniques.

The goal of this chapter is to provide an overview of introductory concepts and terminologies in parallel computing. We start with learning about speedup, efficiency, cost, scalability, and the computation-to-communication ratio by analyzing a simple yet instructive example for summing up numbers using a varying number of processors. We get to know about the two most important parallel architectures: distributed memory systems and shared memory systems. Designing efficient parallel programs requires a lot of experience and we will study a number of typical considerations for this process such as problem partitioning strategies, communication patterns, synchronization, and load balancing. We end this chapter with learning about current and past supercomputers and their historical and upcoming architectural trends.

Keywords

Parallelism, Speedup, Parallelization, Efficiency, Scalability, Reduction, Computation-to-communication ratio, Distributed memory, Shared memory, Partitioning, Communication, Synchronization, Load balancing, Task parallelism, Prefix sum, Deep learning, Top500

CONTENTS

Parallel Programming. DOI: 10.1016/B978-0-12-849890-3.00001-0

1.1 MOTIVATIONAL EXAMPLE AND ITS ANALYSIS

In this section we learn about some basic concepts and terminologies. They are important for analyzing parallel algorithms or programs in order to understand their behavior. We use a simple example for summing up numbers using an increasing number of processors in order to explain and apply the following concepts:

- **Speedup.** You have designed a parallel algorithm or written a parallel code. Now you want to know how much faster it is than your sequential approach; i.e., you want to know the speedup. The speedup (S) is usually measured or calculated for almost every parallel code or algorithm and is simply defined as the quotient of the time taken using a single processor ($T(1)$) over the time measured using p processors ($T(p)$) (see Eq. (1.1)).

$$S = \frac{T(1)}{T(p)} \tag{1.1}$$

- **Efficiency and cost.** The best speedup you can usually expect is a *linear speedup*; i.e., the maximal speedup you can achieve with p processors or cores is p (although there are exceptions to this, which are referred to as *super-linear speedups*). Thus, you want to relate the speedup to the number of utilized processors or cores. The *Efficiency* E measures exactly that by dividing S by P (see Eq. (1.2)); i.e., linear speedup would then be expressed by a value close to 100%. The cost C is similar but relates the runtime $T(p)$ (instead of the speedup) to the number of utilized processors (or cores) by multiplying $T(p)$ and p (see Eq. (1.3)).

$$E = \frac{S}{p} = \frac{T(1)}{T(p) \times p} \tag{1.2}$$

$$C = T(p) \times p \tag{1.3}$$

- **Scalability.** Often we do not only want to measure the efficiency for one particular number of processors or cores but for a varying number; e.g. $P = 1, 2, 4, 8, 16, 32, 64, 128$, etc. This is called *scalability analysis* and indicates the behavior of a parallel program when the number of processors increases. Besides varying the number of processors, the input data size is another parameter that you might want to vary when executing your code. Thus, there are two types of scalability: *strong scalability* and *weak scalability*. In the case of strong scalability we measure efficiencies for a varying number of processors and keep the input data size fixed. In contrast, weak scalability shows the behavior of our parallel code for varying both the number of processors and the input data size; i.e. when doubling the number of processors we also double the input data size.
- **Computation-to-communication ratio.** This is an important metric influencing the achievable scalability of a parallel implementation. It can be defined as the time spent calculating divided by the time spent communicating messages between processors. A higher ratio often leads to improved speedups and efficiencies.

The example we now want to look at is a simple *summation*; i.e., given an array A of n numbers we want to compute $\sum_{i=0}^{n-1} A[i]$. We parallelize this problem using an array of *processing elements* (PEs). We make the following (not necessarily realistic) assumptions:

- **Computation.** Each PE can add two numbers stored in its local memory in one time unit.
- **Communication.** A PE can send data from its local memory to the local memory of any other PE in three time units (independent of the size of the data).
- **Input and output.** At the beginning of the program the whole input array A is stored in PE #0. At the end the result should be gathered in PE #0.
- **Synchronization.** All PEs operate in lock-step manner; i.e. they can either compute, communicate, or be idle. Thus, it is not possible to overlap computation and communication on this architecture.

Speedup is relative. Therefore, we need to establish the runtime of a sequential program first. The sequential program simply uses a single processor (e.g. PE #0) and adds the n numbers using $n-1$ additions in $n-1$ time units; i.e. $T(1,n) = n-1$. In the following we illustrate our parallel algorithm for varying p, where p denotes the number of utilized PEs. We further assume that n is a power of 2; i.e., $n = 2^k$ for a positive integer k.

- $p = 2$**.** PE #0 sends half of its array to PE #1 (takes three time units). Both PEs then compute the sum of their respective $n/2$ numbers (takes time $n/2-1$). PE #1 sends its partial sum back to PE #0 (takes time 3). PE #0 adds the two partial sums (takes time 1). The overall required runtime is $T(2,n) = 3 + n/2 - 1 + 3 + 1$. Fig. 1.1 illustrates the computation for $n = 1024 = 2^{10}$, which has a runtime of $T(2, 1024) = 3 + 511 + 3 + 1 = 518$. This is significantly faster than the sequential runtime. We can calculate the *speedup* for this case as $T(1, 1024)/T(2, 1024) = 1023/518 = 1.975$. This is very close to the optimum of 2 and corresponds to an *efficiency* of 98.75% (calculated dividing the speedup by the number of utilized PEs; i.e. 1.975/2).
- $p = 4$**.** PE #0 sends half of the input data to PE #1 (takes time 3). Afterwards PE #0 and PE #1 each send a quarter of the input data to PE #2 and PE #3 respectively (takes time 3). All four PEs then compute the sum of their respective $n/4$ numbers in parallel (takes time $n/4-1$). PE #2 and PE #3 send their partial sums to PE #0 and PE #1, respectively (takes time 3). PE #0 and PE #1 add their respective partial sums (takes time 1). PE #1 then sends its partial sum to PE #0 (takes time 3). Finally, PE #0 adds the two partial sums (takes time 1). The overall required runtime is $T(4,n) = 3+3+n/4-1+3+1+3+1$. Fig. 1.2 illustrates the computation for $n = 1024 = 2^{10}$, which has a runtime of $T(4, 1024) = 3+3+255+3+1+3+1 = 269$. We can again calculate the *speedup* for this case as $T(1, 1024)/T(4, 1024) = 1023/269 = 3.803$ resulting in an *efficiency* of 95.07%. Even though this value is also close to 100%, it is slightly reduced in comparison to $p = 2$. The reduction is caused by the additional communication overhead required for the larger number of processors.
- $p = 8$**.** PE #0 sends half of its array to PE #1 (takes time 3). PE #0 and PE #1 then each send a quarter of the input data to PE #2 and PE #3 (takes time 3). Afterwards, PE #0, PE #1, PE #2, and PE #3 each send a 1/8 of the input data to PE #5, PE #6, PE #7, and PE #8 (takes again time 3). Fig. 1.3 illustrates the three initial data distribution steps for $n = 1024 = 2^{10}$. All eight PEs then compute the sum of their respective $n/8$ numbers (takes time $n/8-1$). PE #5, PE #6, PE #7, and PE #8 send their partial sums to PE #0, PE #1, PE #2, and PE #3, respectively (takes time 3).

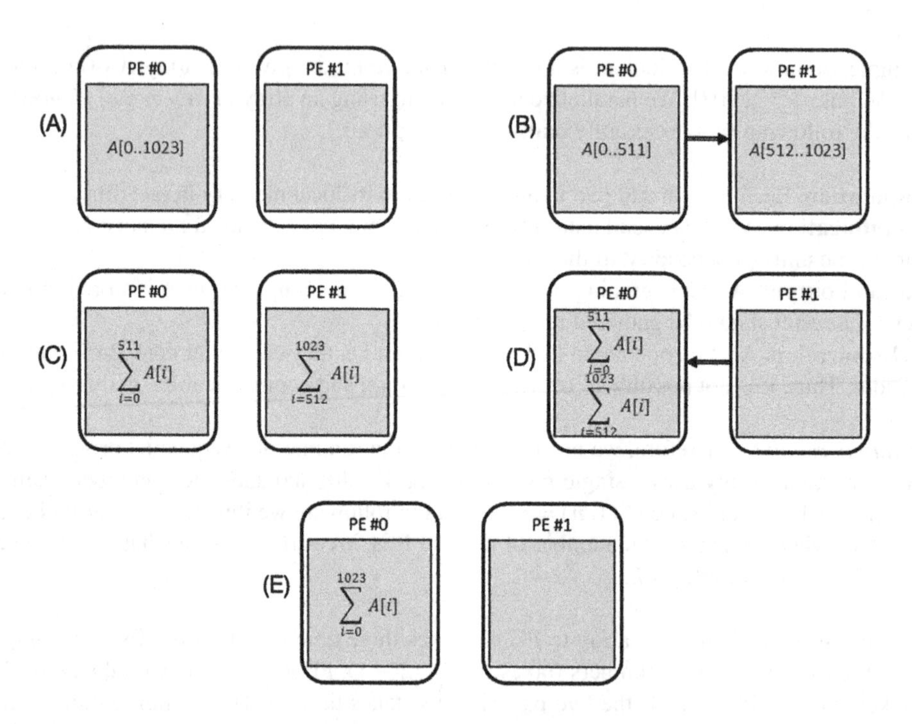

FIGURE 1.1

Summation of $n = 1024$ numbers on $p = 2$ PEs: (A) initially PE #0 stores the whole input data locally; (B) PE #0 sends half of the input to PE #1 (takes time 3); (C) Each PE sums up its 512 numbers (takes time 511); (D) PE #1 sends its partial sum back to PE #0 (takes time 3); (E) To finalize the computation, PE #0 adds the two partial sums (takes time 1). Thus, the total runtime is $T(2, 1024) = 3 + 511 + 3 + 1 = 518$.

Subsequently, PE #0, PE #1, PE #2, and PE #3 add their respective partial sums (takes time 1). PE #2 and PE #3 then send their partial sums to PE #0 and PE #1, respectively (takes time 3). PE #0 and PE #1 add their respective partial sums (takes time 1). PE #1 then sends its partial sum to PE #0 (takes time 3). Finally, PE #0 adds the two partial sums (takes time 1). The overall required runtime is $T(8, n) = 3 + 3 + 3 + n/8 - 1 + 3 + 1 + 3 + 1 + 3 + 1$. The computation for $n = 1024 = 2^{10}$ thus has a runtime of $T(8, 1024) = 3 + 3 + 3 + 127 + 3 + 1 + 3 + 1 + 3 + 1 = 148$. The *speedup* for this case is $T(1, 1024)/T(8, 1024) = 1023/148 = 6.91$ resulting in an *efficiency* of 86%. The decreasing efficiency is again caused by the additional communication overhead required for the larger number of processors.

We are now able to analyze the runtime of our parallel summation algorithm in a more general way using $p = 2^q$ PEs and $n = 2^k$ input numbers:

- **Data distribution time:** $3 \times q$.
- **Computing local sums:** $n/p - 1 = 2^{k-q} - 1$.
- **Collecting partial results:** $3 \times q$.
- **Adding partial results:** q.

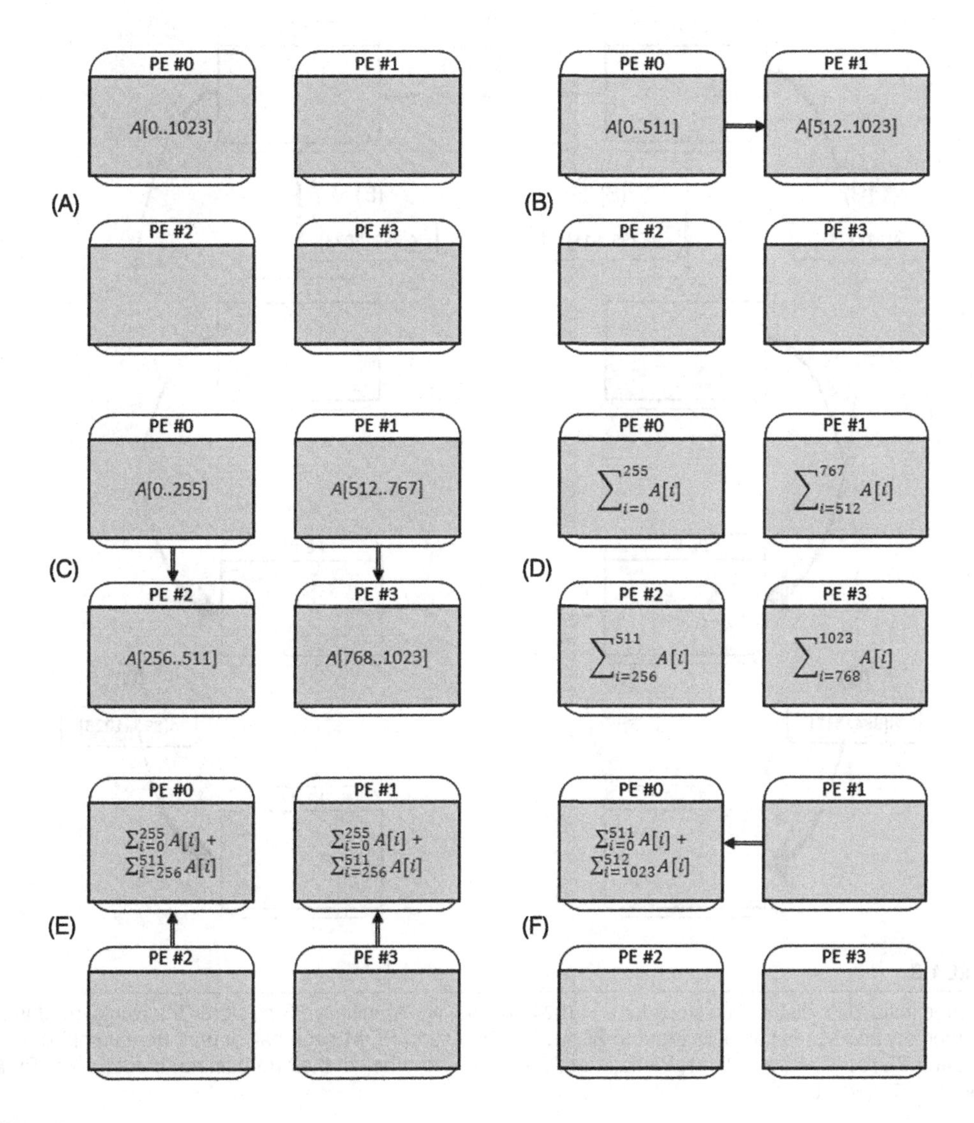

FIGURE 1.2

Summation of $n = 1024$ numbers on $p = 4$ PEs: (A) initially PE #0 stores the whole input in its local memory; (B) PE #0 sends half of its input to PE #1 (takes time 3); (C) PE #0 and PE #1 send half of their data to PE #2 and PE #3 (takes time 3); (D) Each PE adds its 256 numbers (takes time 255); (E) PE #2 and PE #3 send their partial sums to PE #0 and PE #1, respectively (takes time 3). Subsequently, PE #0 and PE #1 add their respective partial sums (takes time 1); (F) PE #1 sends its partial sum to PE #0 (takes time 3), which then finalizes the computation by adding them (takes time 1). Thus, the total runtime is $T(4, 1024) = 3 + 3 + 511 + 3 + 1 + 3 + 1 = 269$.

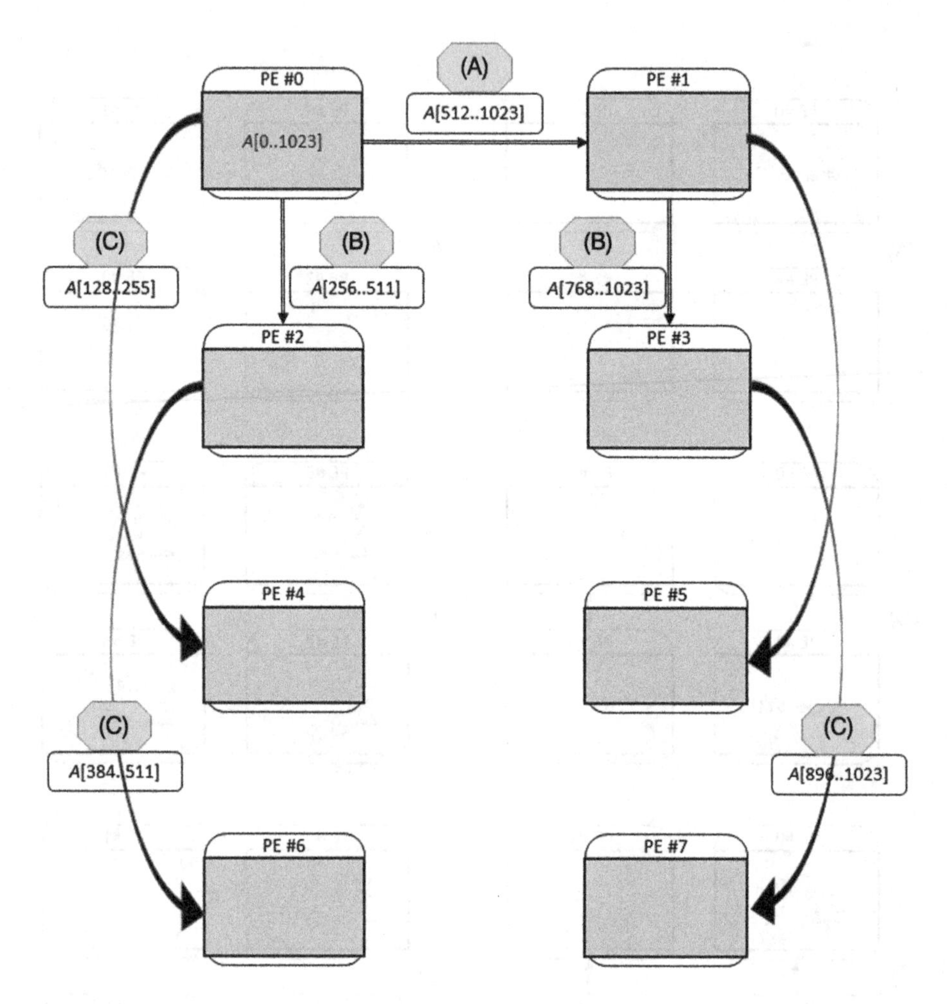

FIGURE 1.3

The three initial data distribution steps for $n = 1024$ and $p = 8$: (A) Initially PE #0 stores the whole input in its local memory and sends half of its input to PE #1; (B) PE #0 and PE #1 send half of their (remaining) data to PE #2 and PE #3; (C) PE #0, PE #1, PE #2, and PE #3 each send half of their (remaining) input data to PE #5, PE #6, PE #7, and PE #8.

Thus, we get the following formula for the runtime:

$$T(p,n) = T(2^q, 2^k) = 3q + 2^{k-q} - 1 + 3q + q = 2^{k-q} - 1 + 7q. \quad (1.4)$$

Fig. 1.4 shows the runtime, speedup, cost, and efficiency of our parallel algorithm for $n = 1024$ and p ranging from 1 to 512. This type of runtime analysis (where the input size is kept constant and the number of PEs is scaled) is called *strong scalability analysis*. We can see that the efficiency

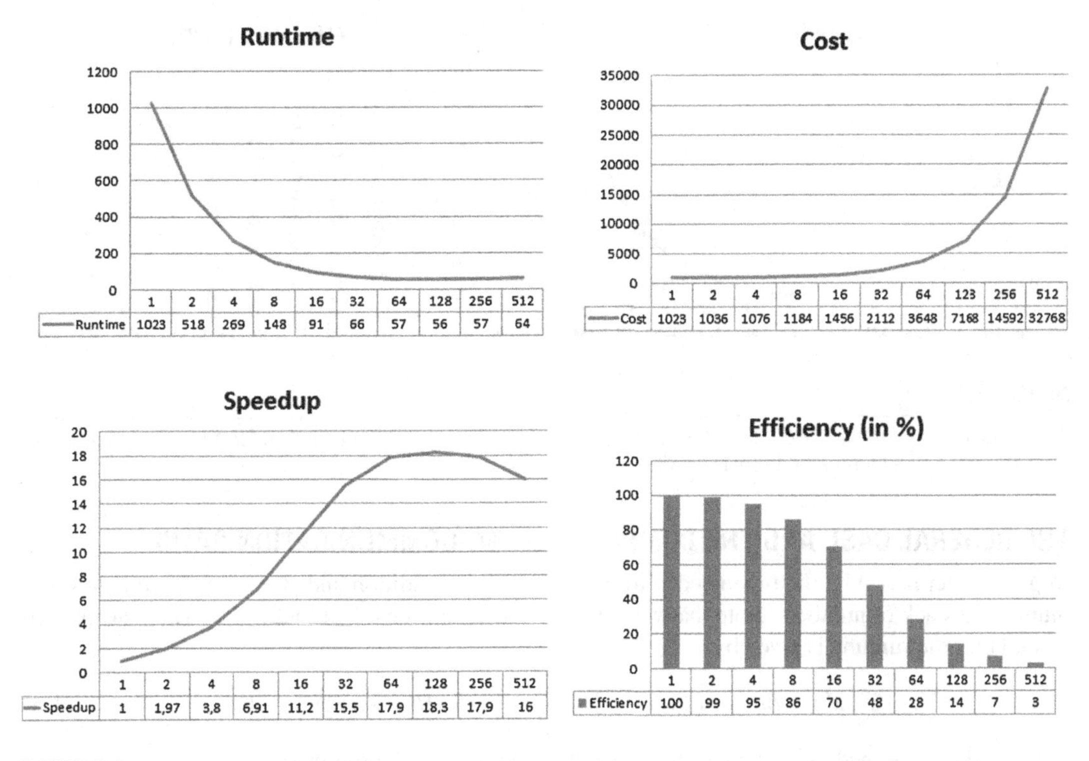

FIGURE 1.4

Strong scalability analysis: runtime, speedup, cost, and efficiency of our parallel summation algorithm for adding $n = 1024$ numbers on a varying number of PEs (ranging from 1 to 512).

is high for a small number of PEs (i.e. $p \ll n$), but is low for a large number of PEs (i.e. $p \approx n$). This behavior can also be deduced from Eq. (1.4): for the case $p \ll n$, holds $2^{k-q} \gg 7q$ (i.e., the term for computation time dominates), while it holds $2^{k-q} \ll 7q$ for the case $p \approx n$ (i.e., the term for communication time dominates). Thus, we can conclude that our algorithm is not strongly scalable.

Now, we want to change our analysis a bit by not only increasing the number of PEs but additionally increasing the input data size at the same time. This is known as *weak scalability analysis*. Fig. 1.5 shows the speedup and efficiency of our algorithm for n ranging from 1024 to 524,288 and p ranging from 1 to 512. We can see that the efficiency is kept high (close to 100%) even for a large number of PEs. This behavior can again be deduced from Eq. (1.4): since both n and p are scaled at the same rate, the term relating to the computation time is constant for varying number of PEs (i.e. $2^{k-q} = 1024$ in Fig. 1.5), while the term for the communication time ($7q = 7 \times \log(p)$) only grows at a logarithmic rate. Thus, we can conclude that our algorithm is weakly scalable.

The terms weak and strong scalability are also related to two well-known laws in parallel computing: *Amdahl's law* and *Gustafsson's law*, which we will discuss in more detail in Chapter 2.

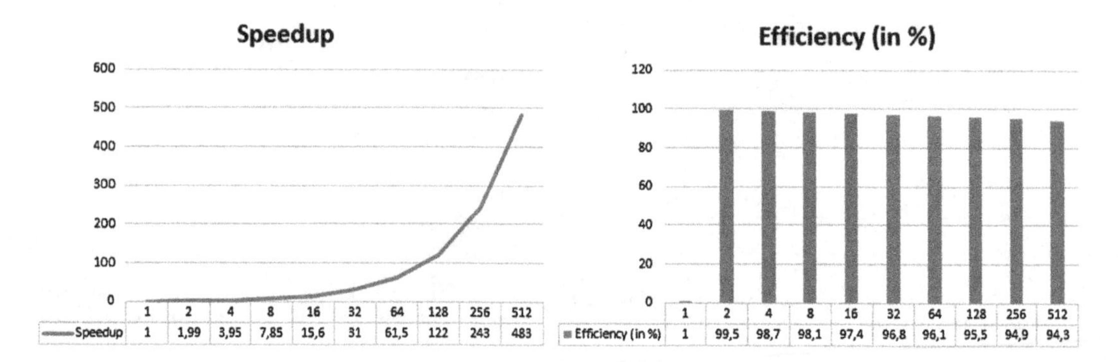

FIGURE 1.5

Weak scalability analysis: speedup and efficiency of our parallel summation algorithm for adding $n = 1024 \times p$ numbers on p PEs (p ranging from 1 to 512).

THE GENERAL CASE AND THE COMPUTATION-TO-COMMUNICATION RATIO

In general, let $\alpha > 0$ be the time needed to perform a single addition and $\beta > 0$ be the time to communicate a stack of numbers. Note that we have previously chosen $\alpha = 1$ and $\beta = 3$. Then the general formula for the runtime is given by

$$T_{\alpha,\beta}(2^q, 2^k) = \beta q + \alpha(2^{k-q} - 1) + \beta q + \alpha q = 2\beta q + \alpha\left(2^{k-q} - 1 + q\right) \quad . \tag{1.5}$$

The speedup is defined as quotient of the sequential and the parallel runtime:

$$S_{\alpha,\beta}(2^q, 2^k) = \frac{T_{\alpha,\beta}(2^0, 2^k)}{T_{\alpha,\beta}(2^q, 2^k)} = \frac{\alpha(2^k - 1)}{2\beta q + \alpha\left(2^{k-q} - 1 + q\right)} \quad . \tag{1.6}$$

For our example we define the *computation-to-communication ratio* as $\gamma = \frac{\alpha}{\beta}$. The speedup then tends to zero if we compute the limit $\gamma \to 0$ for $q > 0$:

$$S_\gamma(2^q, 2^k) = \frac{\gamma(2^k - 1)}{2q + \gamma\left(2^{k-q} - 1 + q\right)} \quad \text{and} \quad \lim_{\gamma \to 0} S_\gamma(2^q, 2^k) = 0 \quad . \tag{1.7}$$

The first derivative of $S_\gamma(2^q, 2^k)$ with respect to γ for fixed q and k is always positive, i.e. the speedup is monotonically decreasing if we increase the communication time (reduce the value of γ). Let $k > q > 0$, $A(k) = 2^k - 1 > 0$, and $B(q,k) = 2^{k-q} - 1 + q > 0$; then we can simply apply the quotient rule:

$$\frac{\mathrm{d}}{\mathrm{d}\gamma} S_\gamma(2^q, 2^k) = \frac{\mathrm{d}}{\mathrm{d}\gamma} \frac{\gamma A(k)}{2q + \gamma B(q,k)} = \frac{2q A(k)}{\left(2q + \gamma B(q,k)\right)^2} > 0 \quad . \tag{1.8}$$

As a result, decreasing the computation-to-communication ratio is decreasing the speedup independent of the number of used compute units $p = 2^q > 1$ – an observation that is true for the majority of parallel algorithms. The speedup $S_\gamma(2^q, 2^k)$ interpreted as function of q exhibits a local maximum at

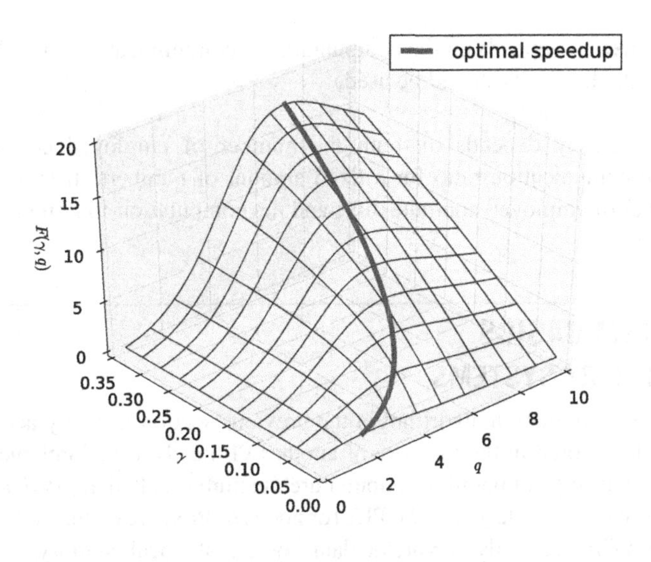

FIGURE 1.6

Functional dependency $F(\gamma, q) = S_\gamma(p, n)$ of the speedup for a varying number of compute units $p = 2^q$, varying computation-to-communication ratios γ, and a fixed amount $n = 2^{10}$ of processed numbers (strong scaling). The thick line represents the points of optimal speedups $S_\gamma(p(\gamma), n)$ where $p(\gamma) = \frac{\gamma \ln 2}{\gamma + 2} n$.

$p = \frac{\gamma \ln 2}{2+\gamma} n$ since

$$\begin{aligned}\frac{\mathrm{d}}{\mathrm{d}q} S_\gamma(2^q, 2^k) &= \frac{\mathrm{d}}{\mathrm{d}q} \frac{\gamma A(k)}{2q + \gamma(2^{k-q} - 1 + q)} \\ &= -\frac{\gamma A(k)(2 - \gamma 2^{k-q} \ln 2 + \gamma)}{\left(2q + \gamma(2^{k-q} - 1 + q)\right)^2} \overset{!}{=} 0 \end{aligned} \tag{1.9}$$

and thus $2 + \gamma - \gamma 2^{k-q} \ln 2 \overset{!}{=} 0 \Leftrightarrow 2^q = \frac{\gamma \ln 2}{2+\gamma} 2^k$.

For $\gamma = \frac{1}{3}$ and $n = 1024$, as in our toy model, we obtain roughly $p \approx 100$ compute units for an optimal speedup. Furthermore, we observe that the longer the communication takes the less compute units should be used. The functional dependency for the speedup $F(\gamma, q) := S_\gamma(2^q, 2^{10})$ for a fixed amount of $n = 2^{10} = 1024$ numbers is plotted in Fig. 1.6.

Concluding, let us derive some take-home-messages from what we have seen in our general analysis:

1. Speedup depends on both the number of employed compute units and the computation-to-communication ratio for a fixed amount of numbers.

   - Speedup usually increases the more compute units are employed up to a local maximum. However, it will decrease if we use too many units.

- The optimal speedup depends on the computation-to-communication ratio. The longer communication takes the less units should be used.

2. Parallelization efficiency depends on both the number of employed compute units and the computation-to-communication ratio for a fixed amount of numbers. It is a monotonic function in both the number of employed compute units and the computation-to-communication ratio.

1.2 PARALLELISM BASICS

DISTRIBUTED MEMORY SYSTEMS

Each PE in our parallel summation algorithm in the previous section has only access to its own local memory. Access to data stored in the memory of another PE needs to be implemented by an explicit communication step. This type of parallel computer architecture is called a *distributed memory system*. Fig. 1.7(A) illustrates its general design. All CPUs (or nodes or PEs) are connected through an *interconnection network*. Each CPU can only operate on data stored in its local memory. Remote data accesses need to be explicitly implemented through message passing over the interconnection network. For example a *point-to-point communication* for sending data from CPU 1 to CPU 2 can be implemented by CPU 1 calling a function to send data stored in its local memory to CPU 2 and CPU 2 calling a function to receive data from CPU 1 and store it in its local memory. In a *collective communication* operation all CPUs in a group can participate. Examples include broadcasting of data from one CPU to all other CPUs or computing the global sum (or another type of associative reduction operation such as minimum or product) of a variable stored in every CPU.

The interconnection network is an important architectural factor of a distributed memory system. It is typically implemented with point-to-point links or a switching network. Standard network protocols (such as Infiniband or Ethernet) are often used to implement the communication tasks. The *network topology* determines the scalability of the architecture for many applications. In Chapter 3 we will study some typical topologies and discuss their qualities in terms of graph theoretic concepts (such as degree, bisection width, and diameter). Prominent examples of distributed memory systems are compute clusters and network-on-chip (NOC) architectures.

You will learn about programming languages for distributed memory systems in detail in Chapter 9 (MPI) and Chapter 10 (UPC++). The *message passing interface* (MPI) is arguably the most popular language for parallel programming on distributed memory systems. MPI creates a (fixed) number processes at start-up time (e.g. one process per compute node or CPU). Each process can only access its local memory. Data exchanges between two processes can be implemented using (versions of) `MPI_Send` and `MPI_Recv` commands while data communication between groups of processes can be implemented by collective communication functions such as `MPI_Bcast`, `MPI_Reduce`, `MPI_Gather`, or `MPI_Scatter`.

We can already deduce from this basic description that *data partitioning*, i.e. the distribution of data between processes, is a key issue in programming distributed memory systems. Fig. 1.7(B) shows a partitioning of an 8×8 matrix onto four processes: each process stores a 4×4 submatrix. Let us now assume we want to implement a stencil code on this matrix where each array element needs to be updated by accessing its left, upper, lower, and right neighbor (i.e. a 5-point stencil). In this case

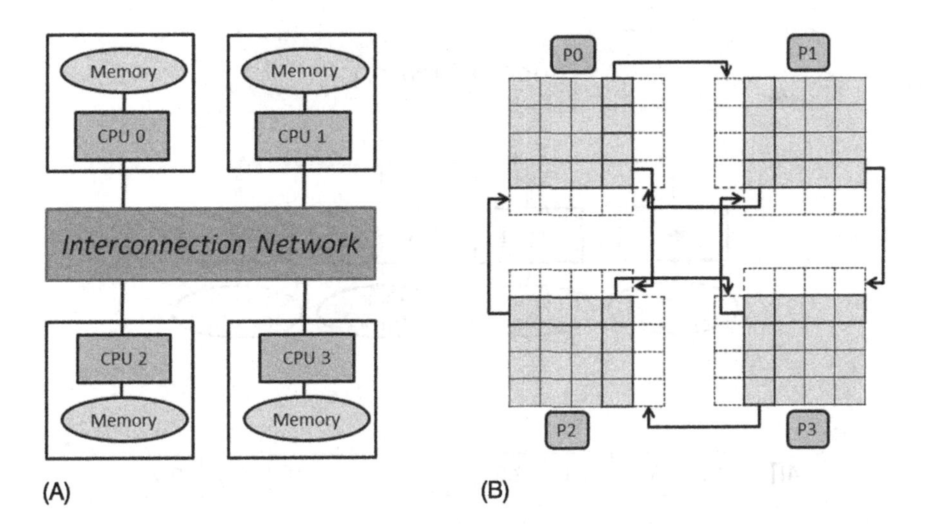

FIGURE 1.7

(A) General design of a distributed memory system; (B) Distributed memory partitioning of an 8×8 matrix onto four processes (P0, P1, P2, P3) for the implementation of a 5-point stencil code. Access to the neighboring cells requires the sending and receiving of data between pairs of processes.

each process needs to allocate additional memory in order to store an additional row and an additional column to be received from another process. Furthermore, it needs to send one row and one column to another process. We will learn about a number of such typical data distributions and associated communication patterns in more detail in Chapter 9.

Partitioned Global Address Space (PGAS) is another popular approach to develop programs for distributed memory systems. It combines distributed memory programming with shared memory concepts by using a global address space that is logically partitioned. A portion of it is local to each process. Thus, portions of the shared memory space have affinity to a particular process, thereby exploiting locality of reference. The PGAS model is the basis of UPC++ which we will study in Chapter 10.

SHARED MEMORY SYSTEMS

This brings us to *shared memory systems*, the second important type of parallel computer architecture. Fig. 1.8 illustrates the general design. All CPUs (or cores) can access a common memory space through a shared bus or crossbar switch. Prominent examples of such systems are modern multi-core CPU-based workstations in which all cores share the same main memory. In addition to the shared main memory each core typically also contains a smaller local memory (e.g. Level 1 cache) in order to reduce expensive accesses to main memory (known as the von Neumann bottleneck). In order to guarantee correctness, values stored in (writable) local caches must be coherent with the values stored in shared memory. This is known as *cache coherence* and is explained in more detail in Chapter 3. Modern multi-core systems support cache coherence and are often also referred to as *cache coherent non-uniform access architectures* (ccNUMA).

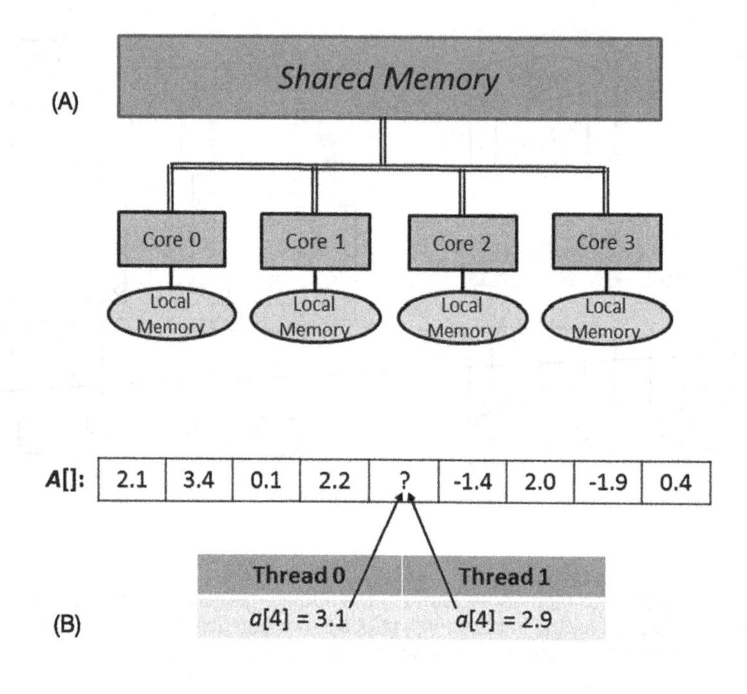

FIGURE 1.8

(A) General design of a shared memory system; (B) Two threads are writing to the same location in a shared array *A* resulting in a race conditions.

Programming of shared memory systems will be studied in detail in Chapter 4 (C++ multi-threading), Chapter 6 (OpenMP), and Chapter 7 (CUDA). Parallelism is typically created by starting threads running concurrently on the system. Exchange of data is usually implemented by threads reading from and writing to shared memory locations. Thus, multiple threads work on the same data simultaneously and programmers need to implement the required coordination among threads wisely. In particular, *race conditions* should be avoided. A race condition can occur when two threads access a shared variable simultaneously (without any locking or synchronization), which could lead to unexpected results (see Fig. 1.8). A number of programming techniques (such as *mutexes*, *condition variables*, *atomics*), which can be used to avoid race conditions, will be discussed in Chapter 4.

You will learn about the implementation of multi-threaded programs on multi-core CPUs using C++11 threads in Chapter 4. A program typically starts with one process running a single thread. This master thread creates a number of slave threads which later join the master thread in order to terminate. Each thread can define its own local variables but has also access to shared variables. Thread creation is much more lightweight and faster compared to process creation. Therefore threads are often dynamically created and terminated during program execution. Table 1.1 shows that the time difference in initialization overhead between a thread and a process on a typical Intel CPU can be more than two orders of magnitude.

OpenMP is another approach to multi-threaded programming (Chapter 6) based on semiautomatic parallelization. OpenMP provides an *application programming interface* (API) in order to simplify

Table 1.1 Difference in initialization overhead between the creation of a thread and the creation of process on an Intel i5 CPU using Visual Studio.

Function call	Time
CreateProcess(..)	12.76 ms
CreateThread(..)	0.037 ms

multi-threaded programming based on *pragmas*. Pragmas are preprocessor directives that a compiler can use to generate multi-threaded code. Thus, when parallelizing a sequential code with OpenMP, a programmer often only needs to annotate the code with the suitable pragmas. Nevertheless, achieving a highly efficient and scalable implementation can still require in-depth knowledge.

The utilized number of threads in a program can range from a small number (e.g., using one or two threads per core on a multi-core CPU) to thousands or even millions. This type of massive multi-threading is used on modern accelerator architectures. We will study the *CUDA* programming language in Chapter 7 for writing efficient massively parallel code for GPUs.

CONSIDERATIONS WHEN DESIGNING PARALLEL PROGRAMS

Assume you are given a problem or a sequential code that needs to be parallelized. Independent of the particular architecture or programming language that you may use, there are a few typical considerations that need to be taken into account when designing a parallel solution:

- **Partitioning:** The given problem needs to be decomposed into pieces. There are different ways how to do this. Important examples of partitioning schemes are data parallelism, task parallelism, and model parallelism.
- **Communication:** The chosen partitioning scheme determines the amount and types of required communication between processes or threads.
- **Synchronization:** In order to cooperate in an appropriate way, threads or processes may need to be synchronized.
- **Load balancing:** The amount of work needs to be equally divided among threads or processes in order to balance the load and minimize idle times.

One of the first thoughts is usually about the potential sources of parallelism. For example, you are given a sequential code that contains a for-loop where the result of iteration step i depends on iteration step $i - 1$. Finding parallelism in case of such a *loop-carried data dependency* appears to be difficult but is not impossible.

Consider the case of a *prefix sum* consisting of the following loop:

```
for (i=1; i<n; i++) A[i] = A[i] + A[i-1]
```

A possible approach to parallelize a prefix sum computation first performs a data partitioning where the input array A is equally divided among p cores. Each core then computes the prefix sum of its local array in parallel. Subsequently, the rightmost value of each local array is taken to form an array B of size p for which yet another prefix sum is computed. This can be done in parallel in $\log_2(p)$ steps. Each

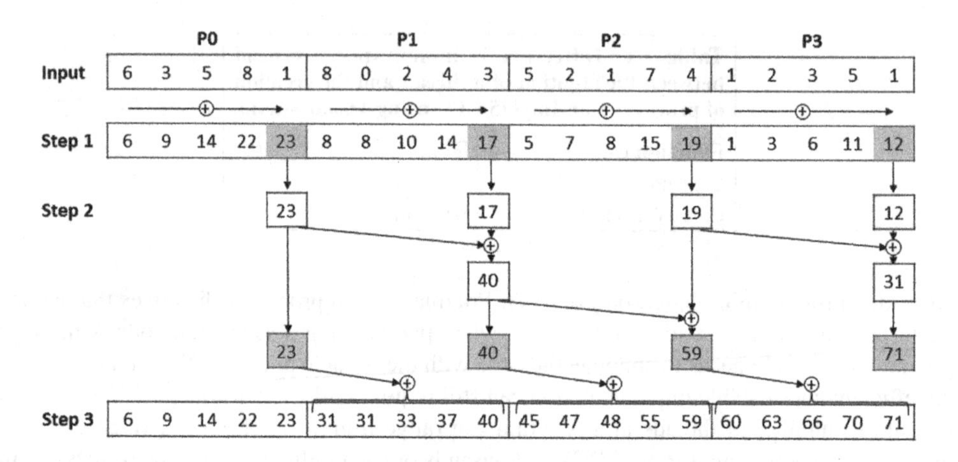

FIGURE 1.9

Parallel prefix computation using four processors where each processor is assigned five elements of the input array. **Step 1:** Local summation within each processor; **Step 2:** Prefix sum computation using only the rightmost value of each local array; **Step 3:** Addition of the value computed in Step 2 from the left neighbor to each local array element.

core then adds the corresponding value of B to each element of its local array in parallel to calculate the overall prefix sum. This concept is illustrated in Fig. 1.9. Indeed, parallel prefix computations (not only for summation but also other binary associative operators) are important building blocks when designing parallel algorithms and we will analyze their theoretical efficiency in more detail in Chapter 2.

As is evident from the examples discussed so far, your choice of *partitioning strategy* is crucial for parallel algorithm design. *Data parallelism* distributes the data across different processors or cores which can then operate on their assigned data. An example is the domain decomposition of a matrix for implementing a stencil code on a distributed memory system as illustrated in Fig. 1.7. The chosen partitioning scheme also determines what communication is required between tasks. Some data parallel algorithms are even *embarrassingly parallel* and can operate independently on their assigned data; e.g., in an image classification task different images can be assigned to different processors which can then classify each image independently in parallel (see Fig. 1.10). Other partitioning schemes are more complex and require significant inter-task communication. Consider again the stencil code shown in Fig. 1.7. This partitioning requires a communication scheme between pairs of processors where a whole column or row of the assigned submatrix needs to be send to another process.

For the implementation of a data parallel algorithm you sometimes need to perform *synchronization* between processes or threads. For example, the implementation of the described parallel prefix computation using multiple threads may require a barrier synchronization between the different stages of the algorithm to guarantee that the required data is available for the subsequent stage. An implementation of the stencil code using MPI may require a synchronization step to guarantee that the required row and columns have been received from neighboring processes before the border values of the assigned submatrix can be updated.

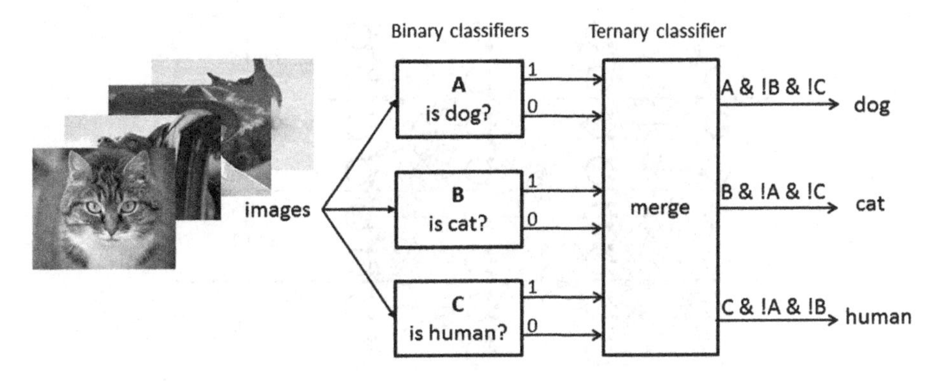

FIGURE 1.10

Design of a ternary classifier where each input image is classified to contain either a cat, a dog, or a human. A data-parallel approach runs the full classifier on each processor or core on different images. A task-parallel approach runs a different binary classifier on each processor or core on all images and then merges the results for each image.

Task parallelism (or functional decomposition), on the other hand, assigns different operations to processor or cores which are then performed on the (same) data. Consider the ternary image classification task shown in Fig. 1.10, where each input image is classified to contain either a cat, a dog, or a human using three corresponding binary classifiers. In a task-parallel approach, a different binary classifier (dog, cat, human) is assigned to a different process (let's say P0, P1, and P2). Every process then classifies each image using the assigned classifier. At the end of the computation the three binary classification results for each image are sent to a single processor (e.g. P0) and merged. Note that in this example the amount of tasks that can run in parallel is limited to three. Thus, in order to scale towards a large number of processors, task parallelism may need to be combined with data parallelism.

The equal distribution of work to processors or cores is called *load balancing*. Let us assume that the binary classifier for humans is more complex than the ones for dog and cat. In this case, in the task-parallel approach Process P2 (assigned with the human classifier) takes longer time than the other two processes. As a result the merge operation at the end of the computation needs to wait for the completion of P2 while P0 and P1 are running idle resulting in a *load imbalance* which limits the achievable speedup. *Dynamic scheduling* policies can be applied to achieve better load balancing. For example, in the data-parallel approach, the input images could be divided into a number of batches. Once a process has completed the classification of its assigned batch, the scheduler dynamically assigns a new batch.

By training neural networks with a large number of layers (known as *deep learning*), computers are now able to outperform humans for a large number of image classification tasks (superhuman performance). However, training such types of neural network models is highly compute-intensive since they need to be trained with a large set of images. Thus, compute clusters with a large number massively parallel GPUs are frequently used for this task in order to reduce associated runtimes. However, the size of a complex neural network can often exceed the main memory of a single GPU. Therefore, a data-parallel approach where the same model is used on every GPU but trained with different images

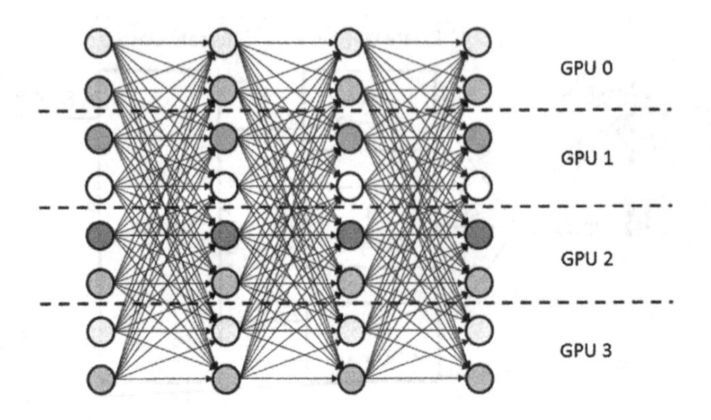

FIGURE 1.11

Model parallelism: A fully connected neural network with four layers is partitioned among four GPUs.

does not always work. As a consequence, a different partitioning strategy called *model parallelism* can be used for implementing deep learning on a number of GPUs. This approach splits the weights of the neural network equally among GPUs (see Fig. 1.11). Each GPU then only needs to store and work on a part of the model. However, all GPUs need to cooperate on the training of this model with a given set of images. The distributed output vector generated after each layer needs to be gathered in each GPU before the computation of the next layer can proceed (i.e. both communication and synchronization are required).

1.3 HPC TRENDS AND RANKINGS

Which are currently the fastest computers in the world? This is always a question of great interest and there are a number of projects that regularly create rankings of supercomputers. Such rankings provide valuable sources for identifying historical trends and new developments in HPC. Arguably the most well-known one is the TOP500 (top500.org) project, which is released bi-annually since 1993. In this list supercomputers are ranked according to their maximally achieved LINPACK performance. This benchmark measures the floating-point performance of an HPC system for solving a dense system of linear equations ($A \cdot x = b$) in terms of *billion floating-point operations per second* (GFlop/s).

Fig. 1.12 shows the historical performance of the top ranked system, the system ranked 500, and the sum and the mean of the performance of all 500 systems in the list for each release since 1993. Performance has been increasing at an exponential rate: The fastest system in 1993 – the Connection Machine CM5/1024 – had a LINPACK performance of 59.7 GFlop/s while the fastest machine in 2016 – the Sunway Taihu Light – can achieve 93,014,600 GFlop/s! This corresponds to a performance improvement of over six orders of magnitude.

In earlier years performance improvements could often be attributed to the increased single-threaded performance of newer processors. However, since the first decade of the twenty-first century

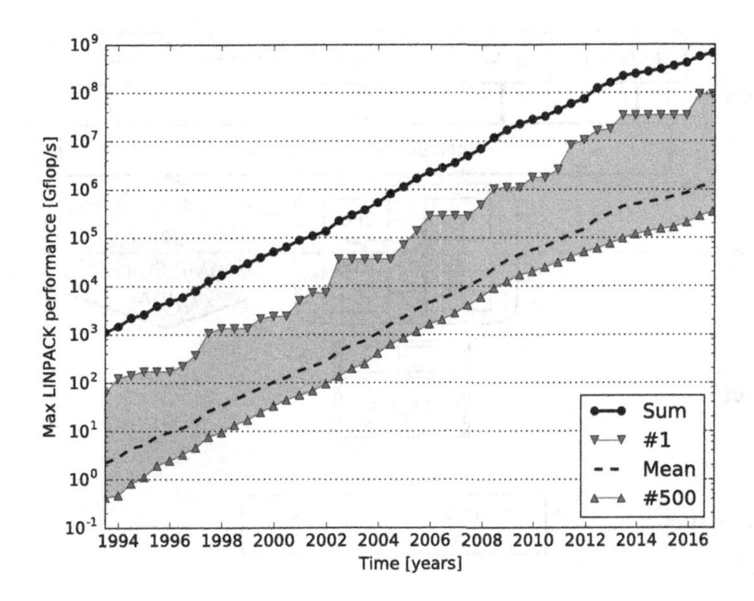

FIGURE 1.12

Growth of supercomputer performance in the TOP500 since 1993 in terms of maximally achieved LINPACK performance in GFlop/s.

single-threaded CPU performance has stagnated (e.g. in 2004 Intel canceled its latest single-core development efforts and moved instead to a multi-core design). Thus, subsequent performance improvements can mainly be attributed to a massive increase in parallelism as evidenced by the total number of cores: the Connection Machine CM5/1024 contained only 1024 cores while the Sunway Taihu Light contains a total of over 10 million cores.

However, growing performance at any cost is not a sustainable solution since supercomputers are not only expensive but also consume large amounts of electrical power; e.g., the Sunway Taihu Light consumes over 15 MW. The Green500 list therefore considers the *power-efficiency* of an HPC system by ranking supercomputer performance in terms of Flop/s-per-Watt of achieved LINPACK performance. The usage of accelerator architectures (also called *heterogeneous computing*) has become an important trend to achieve the best energy-efficiency. The majority of the ten best systems in the Green500 list as of November 2016 are using either CUDA-enabled GPUs or Xeon Phis as accelerators within each node. Yet another trend that may arise in the near future is the increased usage of *neo-heterogeneous architectures*. This approach uses different types of cores integrated on the same chip – the SW26010 chip employed in the nodes of Sunway Taihu Light is an example of such an architecture.

Writing code that can scale up to the huge number of cores available on a current supercomputer is very challenging. Current systems ranked in the TOP500 typically contain several levels of parallelism. Thus, code has usually to be written using a hybrid combination of various languages (see Fig. 1.13):

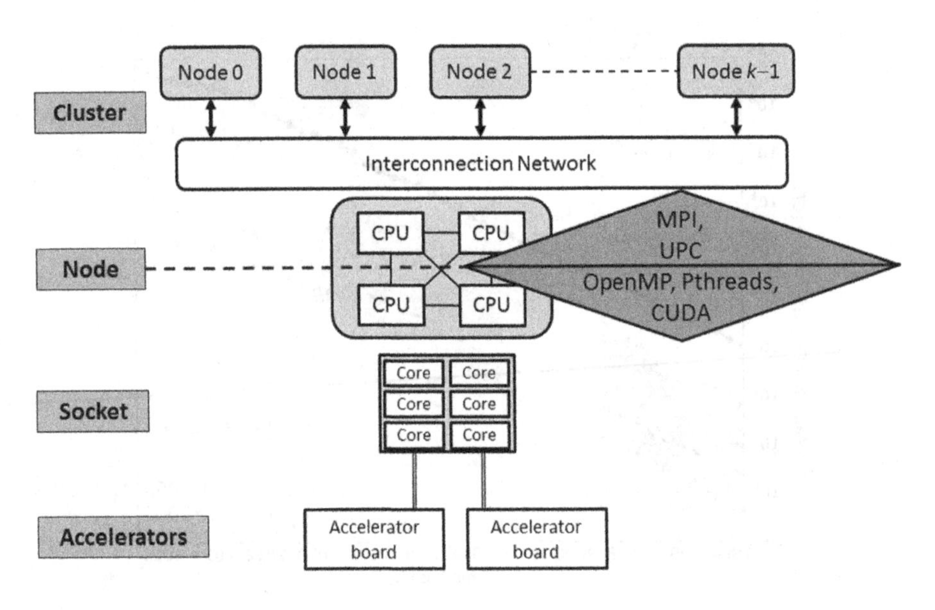

FIGURE 1.13

Example of a heterogeneous HPC system and associated parallel programming languages.

- **Node-level parallelization:** requires the implementation of algorithms for a distributed memory machine model using for example MPI (studied in detail in Chapter 9) or UPC++ (studied in detail in Chapter 10).
- **Intra-node parallelization:** is usually based on languages for shared memory systems (multi-core CPUs), such as C++ multi-threading (studied in detail in Chapter 4) or OpenMP (studied in detail in Chapter 6).
- **Accelerator-level parallelization:** offloads some of the computation to an accelerator such as a massively parallel GPU using language such as CUDA (studied in detail in Chapter 7).

1.4 ADDITIONAL EXERCISES

1. Analyze the speedup and the efficiency of the parallel summation algorithm presented in Section 1.1 using $n = 2048$ numbers assuming that each PE can add two numbers in one millisecond and each PE can send m numbers in $2 + m/1024$ milliseconds to another PE. Vary the number of PEs from 1 to 1024 using powers of 2.
2. Consider the parallel prefix computation algorithm presented in Fig. 1.9. Describe how this parallel algorithm works in general on a shared memory machine using an input array of size $n = 2^k$ and $n/4$ cores. How high is the achieved speedup?
3. The computation of a histogram is a frequent operation in image processing. The histogram simply counts the number of occurrences of each tonal value in the given image. Consider a 2-dimensional input gray-scale image I of size $n \times n$. Its histogram (initialized with all zeros) can be sequentially

computed as follows:

```
for (i=0; i<n; i++)
   for (j=0; j<n; j++)
       histogram[I[i,j]]++
```

Discuss the advantages and disadvantages of the two following partitioning strategies for computing an image histogram in parallel:

a. The histogram slots are partitioned between processors.
b. The input image is partitioned between processors.

4. Another well-known HPC ranking is the Graph500 project. Study the website www.graph500.org and answer the following questions:
   a. What is the utilized benchmark and how is performance measured?
   b. What is the measured performance and the specification of the top-ranked system?
   c. What are some of the advantages and disadvantages compared to the aforementioned TOP500 project?
5. Write down all the prime numbers between 1 and 1000 on the board. At each step, you are allowed to erase two numbers on the board (say, x and y) and in place of the two erased numbers write the number $x + y + x \cdot y$. Repeat this process over and over until only a single number remains (call it Q). Over all possible combination of numbers, what is the smallest value of Q? Assume we have already precomputed the n prime numbers between 1 and 1000. Moreover, we use a third-party library for arbitrary long integers[1] and thus do not need to worry about potential overflow of integers.
   (i) Prove that the operation $x \odot y := x + y + x \cdot y$ is commutative and associative.
   (ii) Using the result of (i), how can we efficiently parallelize this algorithm?
   (iii) Investigate the runtime of the algorithm on p processor. Discuss the result.

[1] Alternatively, we could use Python natively providing big integer support.

CHAPTER

THEORETICAL BACKGROUND 2

Abstract

Our approach to teaching and learning of parallel programming in this book is based on practical examples. Nevertheless, it is important to initially study a number of important theoretical concepts in this chapter before starting with actual programming. We begin with the PRAM model, an abstract shared memory machine model. It can be viewed as an idealized model of computation and is frequently used to design parallel algorithms. These theoretical designs often have an influence on actual parallel implementations (e.g. efficient PRAM algorithms are often the most efficient algorithms at CUDA thread block level). We analyze a few popular PRAM algorithms in terms of their cost and study the design of some cost-optimal PRAM algorithms. Subsequently, we learn about some typical topologies of distributed memory systems and network architectures. We compare their qualities in terms of the graph theoretic concepts degree, bisection width, and diameter. Utilized topologies have an influence on the efficiency of implementing the communication between processors; e.g. collective communication operations in MPI. Amdahl's law and Gustafson's law are arguably the two most famous laws in parallel computing. They can be used to derive an upper bound on the achievable speedup of parallel program. We study both laws as special cases of a more general scaled speedup formulation. This chapter ends with Foster's methodology for parallel algorithm design. It is particularly useful for exploring and comparing possible parallelization approaches for distributed memory architecture.

Keywords

PRAM, Parallel reduction, Prefix scan, Network topology, Degree, Diameter, Bisection-width, Linear array, Mesh, Binary tree, Hypercube, Amdahl's law, Gustafson's law, Strong scaling, Weak scaling, Scaled speedup, Iso-efficiency analysis, Foster's methodology, Parallel algorithm design, Jacobi iteration

CONTENTS

Parallel Programming. DOI: 10.1016/B978-0-12-849890-3.00002-2

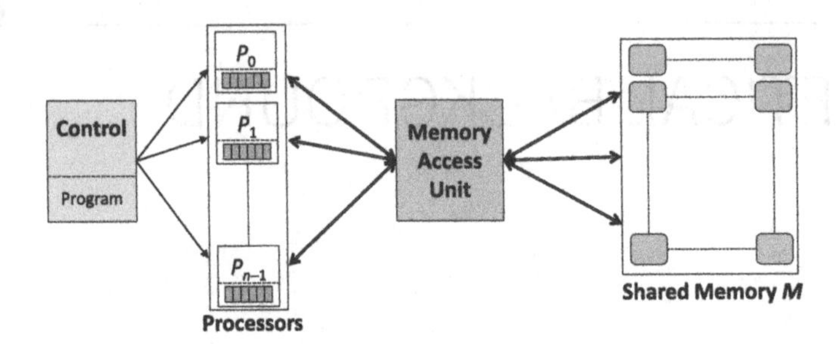

FIGURE 2.1

Important features of a PRAM: n processors $P_0, \dots, P_{n-1}$ are connected to a global shared memory M, whereby any memory location is uniformly accessible from any processor in constant time. Communication between processors can be implemented by reading and writing to the globally accessible shared memory.

2.1 PRAM

Instead of implementing a (complicated) algorithm on a specific parallel architecture, it is sometimes better to take a step back first and explore possible sources and limitations of the algorithm independently of a specific architecture or programming language. For such explorations the usage of theoretical computer models is beneficial. One of the most popular models in this context is the *Parallel Random Access Machine* (PRAM) model. The PRAM can be viewed as an idealized shared memory architecture that does not consider many characteristics of real computer systems such as slow and irregular memory access times, synchronization overheads, caches, etc. Thus, when designing PRAM algorithms we can focus on the best possible parallel algorithm design rather than on how to avoid certain specific technological limitations. Asymptotic runtimes of optimal PRAM algorithms can often be taken as lower bounds for implementations on actual machines. Furthermore, we can often transfer the techniques used for our PRAM design to practical parallel implementations. For example, the first implementation of the merge-sort algorithm on massively-parallel CUDA-enabled GPUs [6], used key insights from parallel merging algorithms on the PRAM.

Fig. 2.1 displays some general features of a PRAM. It consists of n identical processors P_i, $i = 0, \dots, n-1$, operating in lock-step. In every step each processor executes an instruction cycle in three phases:

- **Read phase:** Each processor can simultaneously read a single data item from a (distinct) shared memory cell and store it in a local register.
- **Compute phase:** Each processor can perform a fundamental operation on its local data and subsequently stores the result in a register.
- **Write phase:** Each processor can simultaneously write a data item to a shared memory cell, whereby the exclusive write PRAM variant allows writing only distinct cells while concurrent write PRAM variant also allows processors to write to the same location (race conditions).

Three-phase PRAM instructions are executed synchronously. We should note that communication in the PRAM needs to be implemented in terms of reading and writing to shared memory. This type of memory can be accessed in a uniform way; i.e., each processor has access to any memory location

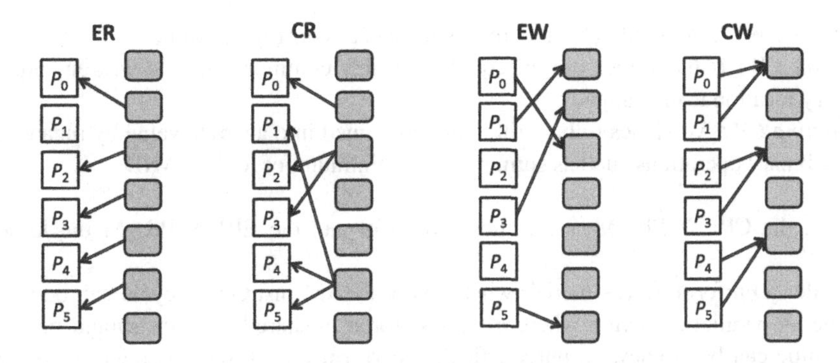

FIGURE 2.2

The four different variants of reading/writing from/to shared memory on a PRAM.

in unit (constant) time. This makes it more powerful than real shared memory machines in which accesses to (a large) shared memory is usually non-uniform and much slower compared to performing computation on registers. The PRAM can therefore be regarded as an idealized model of a shared memory machine; e.g. we cannot expect that a solution for a real parallel machine is more efficient than an optimal PRAM algorithm.

PRAM VARIANTS

You might already have noticed that conflicts can occur when processors read or write to the same shared memory cell during the same instruction cycle. In order to resolve such conflicts, several types of PRAM variants have been defined which differ in the way how data in shared memory can be read/written: only exclusively or also concurrently. Fig. 2.2 illustrates the four possible combinations: ER (exclusive read), CR (concurrent read), EW (exclusive write), and CW (concurrent write).

We now describe the three most popular variants the EREW PRAM, the CREW PRAM, and the CRCW PRAM.

- **Exclusive Read Exclusive Write (EREW):** No two processors are allowed to read or write to the same shared memory cell during any cycle.
- **Concurrent Read Exclusive Write (CREW):** Several processors may read data from the same shared memory cell simultaneously. Still, different processors are not allowed to write to the same shared memory cell.
- **Concurrent Read Concurrent Write (CRCW):** Both simultaneous reads and writes to the same shared memory cell are allowed in this variant. In case of a simultaneous write (analogous to a race condition) we need to further specify which value will actually be stored. Four common approaches to deal with the situation where two (or more) processors attempt to write to the same memory location during the same clock cycle are:

  1. *Priority CW*: Processors have been assigned distinct priorities and the one with the highest priority succeeds writing its value.

2. *Arbitrary CW*: A randomly chosen processor succeeds writing its value.
3. *Common CW*: If the values are all equal, then this common value is written; otherwise, the memory location is unchanged.
4. *Combining CW:* All values to be written are combined into a single value by means of an associative binary operations such as sum, product, minimum, or logical AND.

Obviously, the CRCW PRAM is the most powerful and the EREW PRAM the least powerful variant.

We consider a uniform-access model, where every PRAM instruction cycle can be performed in constant time. A parallel reduction where n values stored in shared memory should be accumulated in a single value can be (somewhat unrealistically) performed with only a single instruction (i.e. in constant time $\mathcal{O}(1)$) on a Combining CRCW PRAM using n processors. However, on a EREW or CREW PRAM this would require $\lceil \log_2(n) \rceil + 1$ instructions. Furthermore, broadcasting a single value stored in a register of a single processor to all n processors requires only constant time ($\mathcal{O}(1)$) on a CREW or CRCW PRAM while it requires logarithmic time ($\mathcal{O}(\log(n))$) on an EREW PRAM.

PARALLEL PREFIX COMPUTATION ON A PRAM

We now want to design and analyze an algorithm for a PRAM with exclusive write access (i.e. an EREW or a CREW PRAM) for computing a prefix sum (or scan) of a given array of n numbers. Parallel prefix sums are an important building block with many applications. We already looked at it briefly in Section 1.2. A sequential algorithm consists of simple loop:

```
for (i=1; i<n; i++) A[i] = A[i] + A[i-1];
```

The corresponding computational complexity of this problem is obviously linear or $\mathcal{O}(n)$ when expressed in terms of asymptotic notation. Our goal is to design a *cost-optimal* PRAM algorithm for prefix summation; i.e., an algorithm where the cost $C(n) = T(n, p) \times p$ is linear in n. $T(n, p)$ denotes the time required for n input numbers and p processors.

Our first approach uses $p = n$ processors and is based on a recursive doubling technique as illustrated in Fig. 2.3. From the pseudo-code given in Listing 2.1, we can easily see that $\lceil \log_2(n) \rceil$ iterations are required. This leads to a cost of $C(n) = T(n, p) \times p = \mathcal{O}(\log(n)) \times n = \mathcal{O}(n \times \log(n))$.

```
1  // each processor copies an array entry to a local register
2  for (j=0; j<n; j++) do_in_parallel
3      reg_j = A[j];
4
5  // sequential outer loop
6  for (i=0; i<ceil(log(n)); i++) do
7      // parallel inner loop performed by Processor j
8      for (j = pow(2,i); j<n; j++) do_in_parallel {
9          reg_j += A[j-pow(2,i)];  // perform computation
10         A[j] = reg_j; // write result to shared memory
11     }
```

Listing 2.1: Parallel prefix summation of an array A of size n stored in shared memory on an EREW PRAM with n processors.

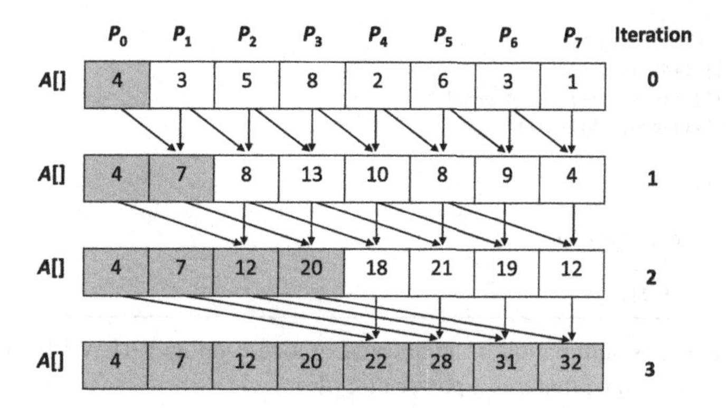

FIGURE 2.3

Parallel prefix summation of an array A of size 8 on a PRAM with eight processors in three iteration steps based on recursive doubling.

Our first approach is therefore not cost-optimal. How can we improve it? In order to reduce the cost from log-linear to linear, we either need to (asymptotically) lower the runtime or the number processors. Since reducing the former is difficult, we simply reduce the number of utilized processors from $p = n$ to $p = \frac{n}{\log_2(n)}$. With this reduced number of processors we now design a PRAM algorithm with logarithmic number of steps in three stages as follows:

1. Partition the n input values into chunks of size $\log_2(n)$. Each processor computes local prefix sums of the values in one chunk in parallel (takes time $\mathcal{O}(\log(n))$).
2. Perform the old non-cost-optimal prefix sum algorithm on the $\frac{n}{\log_2(n)}$ partial results (takes time $\mathcal{O}(\log(n/\log(n)))$).
3. Each processor adds the value computed in Stage 2 by its left neighbor to all values of its chunk (takes time $\mathcal{O}(\log(n))$).

This approach has actually already been discussed in Section 1.2 and is illustrated in Fig. 1.9. We further provide the algorithmic details in Listing 2.2. For the sake of simplicity we assume in Listing 2.2 that n is a power of 2 ($n = 2^k$) and therefore $p = \frac{n}{k}$. We have further removed the explicit copying to local registers (as in Listing 2.1), which we now implicitly assume in each step. Since each of the three stages in Listing 2.2 can be performed in logarithmic time, our second approach leads to a cost of $C(n) = T(n, p) \times p = \mathcal{O}(\log(n)) \times \mathcal{O}(\frac{n}{\log(n)}) = \mathcal{O}(n)$, which is cost-optimal.

```
1  //Stage 1: each Processor i computes a local
2  //prefix sum of a subarray of size n/p = log(n) = k
3  for (i=0; i<n/k; i++) do_in_parallel
4      for (j=1; j<k; j++)
5          A[i*k+j] += A[i*k+j-1];
6
7  //Stage 2: Prefix sum computation using only the rightmost value
```

```
 8  // of each subarray which takes O(log(n/k)) steps
 9  for (i=0; i<log(n/k); i++) do
10      for (j = pow(2,i); j<n/k; j++) do_in_parallel
11          A[j*k-1] += A[(j-pow(2,i))*k-1];
12
13  //Step 3: each Processor i adds the value computed in Step 2 by
14  //Processor i-1 to each subarray element except for the last one
15  for (i=1; i<n/k; i++) do_in_parallel
16      for (j=0; j<k-1; j++)
17          A[i*k+j] += A[i*k-1];
```

Listing 2.2: Parallel prefix summation of an array A of size n on an EREW PRAM. The number of processors is $\frac{n}{k}$ and we assume that n is a power of 2 (2^k).

SPARSE ARRAY COMPACTION ON A PRAM

Parallel prefix computations can be used as a primitive for the efficient implementation of a variety of applications. We are discussing one such example now: array compaction. Assume you have a one-dimensional array A where the majority of entries are zero. In this case we can represent the array in a more memory-efficient way by only storing the values of the non-zero entries (in an array V) and their corresponding coordinates (in an array C). An example is shown in Fig. 2.4.

A sequential algorithm can simply iterate over the n elements of A from left to right and incrementally build V and C in time linear in n. We can now build a cost-optimal PRAM algorithm using $p = \frac{n}{\log_2(n)}$ processors by using our parallel prefix approach as follows:

1. We generate a temporary array (tmp) with $tmp[i] = 1$ if $A[i] \neq 0$ and $tmp[i] = 0$ otherwise. We then perform a parallel prefix summation on tmp. For each non-zero element of A, the respective value stored in tmp now contains the destination address for that element in V.
2. We write the non-zero elements of A to V using the addresses generated by the parallel prefix summation. The respective coordinates can be written to C in a similar way.

Fig. 2.5 illustrates this process for creating V using an example. Again, using $p = \frac{n}{\log_2(n)}$ processors each step be accomplished in logarithmic time leading to a cost-optimal solution: $C(n) = T(n, p) \times p = \theta(\log(n)) \times \theta(\frac{n}{\log(n)}) = \theta(n)$.

In summary, we can conclude that the PRAM can be used as a theoretical model to explore potential sources of parallelism for a variety of algorithms and compare their efficiency. Many of the techniques explored on the PRAM have also high relevance for practical implementations. For example, Satish et al. [6] have followed prior work on merging pairs of sorted sequences in parallel for the PRAM model for designing an efficient merge-sort implementation for massively parallel GPUs using CUDA. We have also included the merging of sorted sequences on a PRAM as an exercise in Section 2.5. Furthermore, a vast amount of various PRAM algorithms can be found in literature; see e.g. in [5]. Besides the PRAM many other models for parallel computations have been proposed, such as the BSP (*Bulk-Synchronous Parallel*) model [7].

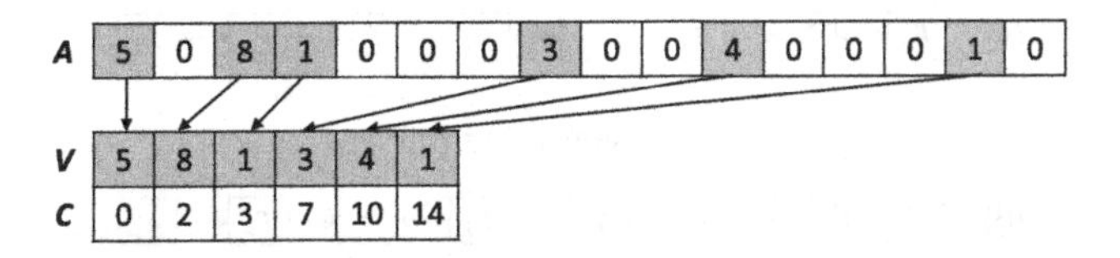

FIGURE 2.4

Example of compacting a sparse array A of size 16 into two smaller arrays: V (values) and C (coordinates).

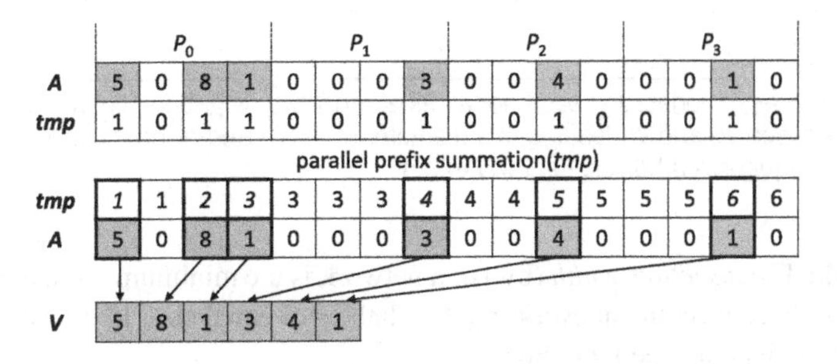

FIGURE 2.5

Example of compacting the values of a sparse array A of size 16 into the array V on a PRAM with four processors.

2.2 NETWORK TOPOLOGIES

Interconnection networks are important architectural factors of parallel computer systems. The two main network types are *shared* and *switched.* A prominent example of a shared network is a bus (such as traditional Ethernet), which can communicate at most one message at a time. This usually limits its scalability in comparison to switched networks, which are able to simultaneously transfer several messages between different pairs of nodes. Interconnection networks in high-performance distributed memory architectures are therefore typically implemented as switching networks allowing for fast point-to-point communication between processors. The specific *network topology* is a key factor for determining the scalability and performance of a parallel computer architecture.

We will now study some typical switch network topologies and compare their qualities in terms of graph theoretical concepts. We represent a network as a connected graph whose vertices represent nodes (these could be switches or processors) and the edges represent communication links. They can be classified into *direct* and *indirect* networks. In a direct network all nodes have a processor attached, i.e. there are direct connections between processors. On the other hand, indirect networks also incorporate intermediate routing-only nodes.

We will now use the following three features to compare the qualities of different network topologies:

- **Degree:** The *degree* (deg) of a network is the maximum number of neighbors of any node.
- **Diameter:** The *diameter* (diam) of a network is the length of the longest of all shortest paths between any two nodes.

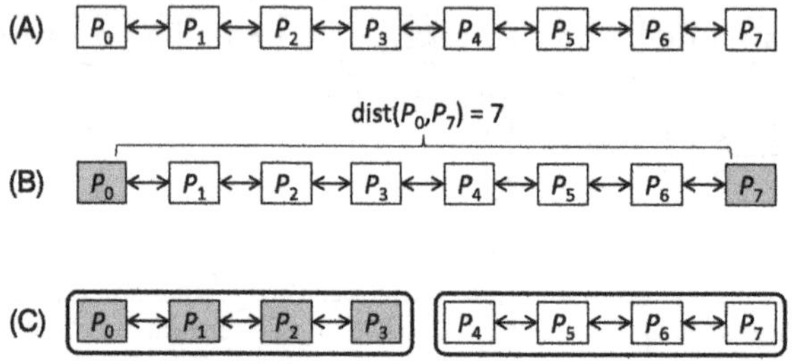

FIGURE 2.6

(A) The linear array with eight nodes; L_8: each node has at most two neighbors; i.e. $\deg(L_8) = 2$. (B) The longest distance is between P_0 and P_7 leading to a diameter of 7. (C) Removing the link between P_3 and P_4 disconnects L_8 in two equal-sized halves; i.e. $\text{bw}(L_8) = 1$.

- **Bisection-width:** The *bisection-width* (bw) of a network is the minimum number of edges (or links) to be removed to disconnect the network into two halves of equal size. In case of an odd number of nodes, one half can include one more node.

The design of an interconnection network is usually a trade-off between a number of *contradictory requirements*. Using the three features we have just defined, we can formulate the following three desirable properties.

- **Constant degree:** The degree of a network should be constant; i.e., it should be independent of the network size. This property would allow a network to scale to a large number of nodes without the need of adding an excessive number of connections.
- **Low diameter:** In order to support efficient communication between any pair of processors, the diameter should be minimized.
- **High bisection-width:** The bisection-width identifies a potential bottleneck of a network and has an implication on its internal bandwidth. A low bisection-width can slow down many collective communication operations and thus can severely limit the performance of applications. However, achieving a high bisection width may require non-constant network degree.

The first topology we consider is the **linear array** with n nodes $P_0, \ldots, P_{n-1}$, denoted as L_n. Node P_i $(0 < i < n-1)$ is connected to its left neighbor P_{i-1} and its right neighbor P_{i+1} and thus the degree is 2; i.e. $\deg(L_n) = 2$. The longest distance between any two nodes is between the leftmost node P_0 and the rightmost node P_{n-1}. Data communicated between them needs to traverse $n - 1$ links and thus $\text{diam}(L_n) = n - 1$. Finally, the bisection-width is 1 ($\text{bw}(L_n) = 1$) since only the link between $P_{\lfloor (n-1)/2 \rfloor}$ and $P_{\lceil n/2 \rceil}$ needs to be removed in order to split L_n into two disconnected halves. Fig. 2.6 illustrates L_8 as an example.

By increasing the dimensionality of the linear array it is possible to improve both bisection-width and diameter while keeping a constant degree. In a **2D mesh** the n nodes are arranged in a (usually square) grid of size $n = k \times k$, denoted as $M_{k,k}$. For the sake of simplicity let us further assume that k is even. We illustrate this topology for $k = 4$ in Fig. 2.7. It can be seen that each node is connected to

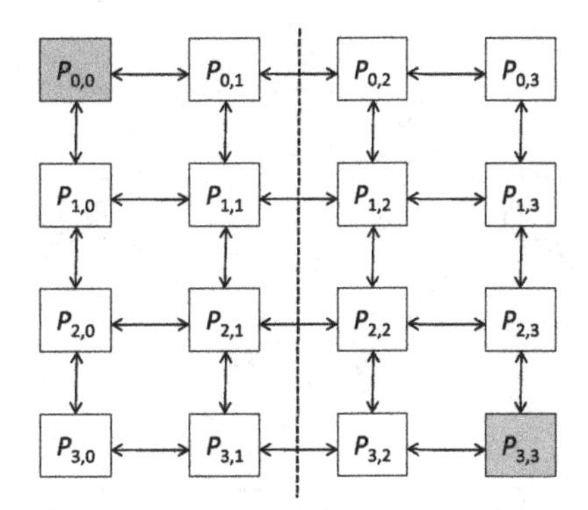

FIGURE 2.7

A 2D mesh of size 4 × 4 ($M_{4,4}$). Each node has at most four neighbors; i.e. $\text{deg}(M_{4,4}) = 4$. The longest distance is for example between $P_{0,0}$ and $P_{3,3}$, leading to a diameter of 6. Removing all links between the second and the third column (or the second and the third row) of nodes disconnects $M_{4,4}$ into two equal-sized halves; i.e. $\text{bw}(M_{4,4}) = 4$.

at most four other nodes; i.e. $\text{deg}(M_{k,k}) = 4$. Thus, the degree is independent of the actual mesh size – one of our formulated desirable network properties. The longest distance between a pair of nodes occurs when traveling between the upper left and the lower right or the upper right and the lower left node. This requires traversing $2(k-1)$ edges, leading to $\text{diam}(M_{k,k}) = 2(k-1) = 2(\sqrt{n}-1)$ – a significant improvement compared to the linear array. In order to split $M_{k,k}$ into two equal-sized unconnected halves, we need to remove at least k edges; e.g., all edges connecting the two middle rows or connecting the two middle columns. Thus, it holds that $\text{bw}(M_{k,k}) = k = \sqrt{n}$, which is significantly larger than the bisection-width of the linear array.

A frequently used extension of the mesh is the torus. For example, a **2D torus** $T_{k,k}$ extends $M_{k,k}$ by adding *wrap-around edges* to directly connect the left node of each row and the right node of each row as well the bottom node of each column and the top node of each column. This reduces the diameter and the bisection width by a factor of 2 compared to the 2D mesh of the same size while keeping the degree constant at 4. A **3D mesh** $M_{k,k,k}$ extends the 2D mesh by another dimension. It has a degree of 6, a diameter of $3(k-1) = 3(\sqrt[3]{n}-1)$, and a bisection width of $k^2 = n^{2/3}$. A **3D torus** further extends a 3D mesh in a similar way as in the 2D case. A 3D tours has many desirable features such as a constant degree, relatively low diameter, and a relatively high bisection width. Thus, it has been used as an interconnection network in a number of Top500 supercomputers including IBM's Blue Gene/L and Blue Gene/P, and the Cray XT3.

If we want to further reduce the network diameter, a tree-based structure can be used. For example, in a **binary tree** of height of depth d, denoted as BT_d, the $n = 2^d - 1$ nodes are arranged in a complete binary tree of depth d as illustrated in Fig. 2.8 for $d = 3$. Each node (except the root and the leaves) is connected to its parent and two children, leading to a degree of 3; i.e. $\text{deg}(BT_d) = 3$. The longest distance in BT_d occurs when traveling between a leaf node on the left half of the tree to one on the right half which requires going up to the root ($d-1$ links) and then down again ($d-1$ links); i.e. $\text{diam}(BT_k) = 2(d-1) = 2\log_2(n+1)$. Thus, the degree is constant and the diameter is low (logarithmic in the number of nodes) – two desirable properties. However, the disadvantage of the binary tree

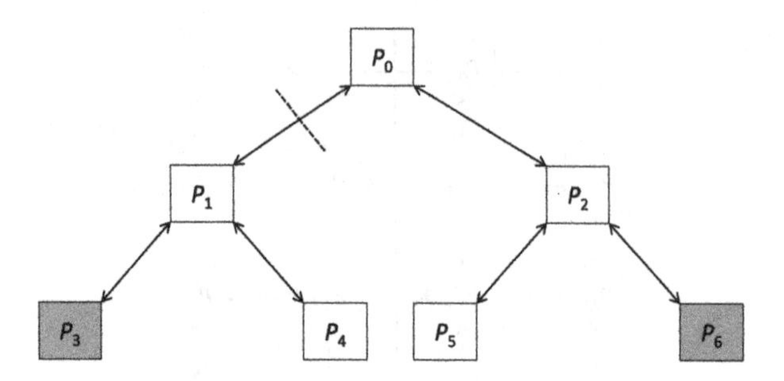

FIGURE 2.8

A binary tree of depth 3 (BT_3). Each node has at most three neighbors; i.e. $\text{deg}(BT_3) = 3$. The longest distance is for example between P_3 and P_6, leading to a diameter of 4. Removing a single link adjacent to the root disconnects the tree into two (almost) equal-sized halves; i.e. $\text{bw}(BT_3) = 1$.

topology is its extremely low bisection-width: by removing only a single link incident with the root, we can split the network into two disconnected components differing by node; i.e. $\text{bw}(BT_k) = 1$. This drawback is addressed in **fat tree** network topologies by introducing more links nearer the top of the hierarchy; i.e., those inks are *fatter* than links further down the hierarchy, thereby improving bisection-width. A variation of the fat tree is the **hypertree network** which will be discussed in an exercise in Section 2.5. Fat tree interconnection networks are used in a number of supercomputers including Tianhe-2 and the Earth Simulator.

The **hypercube** network of dimension d ($d \geq 1$), denoted by Q_d, can be represented as a graph with $n = 2^d$ nodes that has each vertex labeled with a distinct bit-string of length d. Furthermore, two vertices are connected if and only if the associated bit-strings differ in *exactly one* bit. Fig. 2.9 shows Q_4 as an example. Obviously, each node is connected to exactly d other nodes since each of the d bits of the associated bit-string can be flipped; i.e. $\text{deg}(Q_d) = d = \log_2(n)$. The longest distance between two nodes occurs when the two associated bit-strings differ in every position (i.e. two bit-strings with Hamming distance d). In this case d bit-flips are required in order to transform one bit-string into the other which corresponds to traversing d links in the network; thus $\text{diam}(Q_d) = d = \log_2(n)$. Removing all links between the nodes starting with label 0 and all nodes starting with 1 disconnects H_d in two equal-sized halves. Since each node labeled $0x_1 \ldots x_{n-1}$ is connected to exactly one other node staring with label 1 ($1\overline{x_1} \ldots \overline{x_{n-1}}$), it holds that $\text{bw}(Q_d) = 2^{d-1} = n/2$. Overall, the hypercube has the highest bisection width of the networks we have studied so far (linear in the number of nodes). Furthermore, the diameter is very low (logarithmic in the number of nodes). Its disadvantage is the non-constant degree. The number of required links per node is a (logarithmic) function of network size, making it difficult to scale up to large number of nodes. As a consequence, hypercube topologies have been used in some early message-passing machines but are not seen in current state-of-the-art supercomputers.

We summarize the properties of the discussed network topologies in Table 2.1 in terms of the number of nodes using asymptotic notation. Besides these topologies, there are many more and some of them are discussed in the exercises in Section 2.5. Furthermore, additional features and properties may be used to characterize network topologies such as maximum edge (link) length, average distance between nodes, or fault tolerance. To conclude, the design of an interconnection network is often a

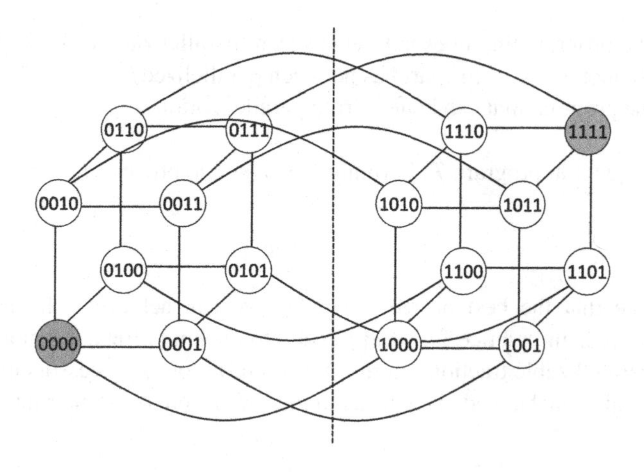

FIGURE 2.9

A 4-dimensional hypercube (Q_4). Each node has exactly four neighbors; i.e. $\deg(Q_4) = 4$. The longest distance is, for example, between the node labeled 0000 and the one labeled 1111, leading to a diameter of 4. Removing all links between the nodes starting with label 0 and all nodes starting with 1 disconnects H_4 into two equal-sized halves; i.e. $\mathrm{bw}(Q_4) = 8$.

Table 2.1 Degree, diameter, and bisection-width of the discussed interconnection network topologies in terms of the number of nodes (n) using asymptotic notation.

Topology	Degree	Diameter	Bisection-width
Linear Array	$\mathcal{O}(1)$	$\mathcal{O}(n)$	$\mathcal{O}(1)$
2D Mesh/Torus	$\mathcal{O}(1)$	$\mathcal{O}(\sqrt{n})$	$\mathcal{O}(\sqrt{n})$
3D Mesh/Torus	$\mathcal{O}(1)$	$\mathcal{O}(\sqrt[3]{n})$	$\mathcal{O}(n^{2/3})$
Binary Tree	$\mathcal{O}(1)$	$\mathcal{O}(\log(n))$	$\mathcal{O}(1)$
Hypercube	$\mathcal{O}(\log(n))$	$\mathcal{O}(\log(n))$	$\mathcal{O}(n)$

trade-off between various contradictory requirements (such as a high bisection width and a constant degree). Thus, there is no single ideal network topology but the different topologies have their advantages and disadvantages.

2.3 AMDAHL'S AND GUSTAFSON'S LAWS

We will now learn about a theoretical method to derive an upper bound on the achievable speedup when parallelizing a given sequential program. It can be used in advance to any actual parallelization in order to analyze whether the effort might be worth it. The method requires that the execution time of the program could be split into two parts:

- T_{ser}: the part of the program that does not benefit from parallelization (think of this part as either being inherently sequential or as the part has not been parallelized).
- T_{par}: the part of the program that can benefit from parallelization.

The runtime of the sequential program $T(1)$ running on a single processor is simply the sum of these two parts:

$$T(1) = T_{ser} + T_{par} \,. \tag{2.1}$$

We further assume that the best possible speedup we can achieve is linear (i.e., super-linear speedups that might occur in practice for example due to caching effects are not considered by this method). Thus, the parallelizable fraction can run p times faster on p processors in the best case while the serial fraction remains unchanged. This leads to the lower bound for the runtime on p processors $T(p)$:

$$T(p) \geq T_{ser} + \frac{T_{par}}{p} \,. \tag{2.2}$$

By diving $T(1)$ by $T(p)$ this then results in an upper bound for the achievable speedup $S(p)$ using p processors:

$$S(p) = \frac{T(1)}{T(p)} \leq \frac{T_{ser} + T_{par}}{T_{ser} + \frac{T_{par}}{p}} \,. \tag{2.3}$$

Instead of using the absolute runtimes (T_{ser} and T_{par}), we will now use their fraction. Let f denote the fraction of T_{ser} relative to T_1; i.e. $T_{ser} = f \cdot T(1)$. Then $1 - f$ is the fraction of T_{par} relative to T_1; i.e. $T_{par} = (1 - f) \cdot T(1)$. Obviously, f is a number between 0 and 1 ($0 \leq f \leq 1$).

$$T_{ser} = f \cdot T(1) \text{ and } T_{par} = (1 - f) \cdot T(1) \qquad (0 \leq f \leq 1)$$

Substituting this into Eq. (2.3) results in an upper bound for the speedup that only depends on f and p:

$$S(p) = \frac{T(1)}{T(p)} \leq \frac{T_{ser} + T_{par}}{T_{ser} + \frac{T_{par}}{p}} = \frac{f \cdot T(1) + (1 - f) \cdot T(1)}{f \cdot T(1) + \frac{(1-f)\cdot T(1)}{p}} = \frac{f + (1 - f)}{f + \frac{(1-f)}{p}} = \frac{1}{f + \frac{(1-f)}{p}} \,. \tag{2.4}$$

Eq. (2.4) is known as **Amdahl's law** [1]. By knowing f, we can use it to predict the theoretically achievable speedup when using multiple processors. This situation is further illustrated in Fig. 2.10. Here are two typical examples for applying Amdahl's law:

- **Example 1:** 95% of a program's execution time occurs inside a loop that we want to parallelize. What is the maximum speedup we can expect from a parallel version of our program executed on six processors?

$$S(6) \leq \frac{1}{0.05 + \frac{(0.95)}{6}} = 4.8$$

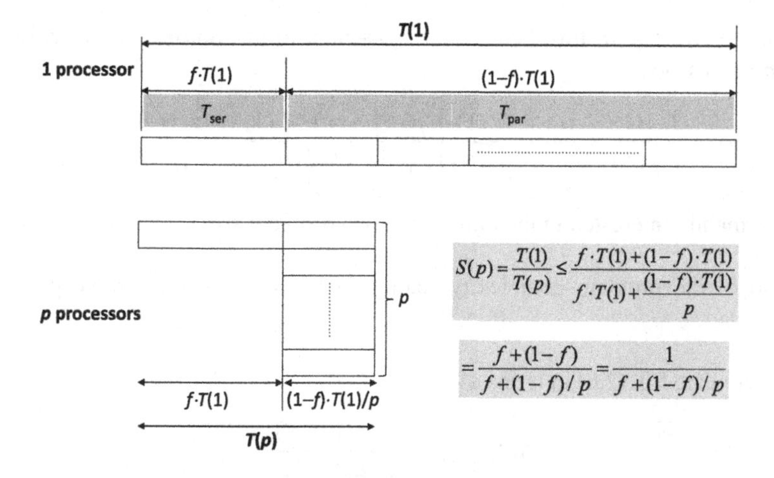

FIGURE 2.10

Illustration of Amdahl's law for the establishing an upper bound for the speedup with constant problem size.

- **Example 2:** 10% of a program's execution time is spent within inherently sequential code. What is the limit to the speedup achievable by a parallel version of the program?

$$S(\infty) \leq \lim_{p \to \infty} \frac{1}{0.1 + \frac{(0.9)}{p}} = 10$$

A well-known limitation of Amdahl's law is that it only applies in situation where the problem size is **constant** and the number of processors varies (*strong scalability*– a concept we already discussed in Section 1.1). However, when using more processors, we may also use larger problem sizes (similar to the concept of *weak scalability*– also discussed in Section 1.1). In this case the time spent in the parallelizable part (T_{par}) may grow faster in comparison to the non-parallelized part T_{ser}. In order to also incorporate such scenarios in the calculation of the achievable speedup (also called **scaled speedup**), we now derive a more general law which allows for scaling of these two parts with respect to the problem's complexity:

- α: scaling function of the part of the program that does not benefit from parallelization with respect to the complexity of the problem size.
- β: scaling function of the part of the program that benefits from parallelization with respect to the complexity of the problem size.

Using these two scaling functions, we decompose the sequential runtime under consideration of scaling over the problem size complexity:

$$T_{\alpha\beta}(1) = \alpha \cdot T_{\text{ser}} + \beta \cdot T_{\text{par}} = \alpha \cdot f \cdot T(1) + \beta \cdot (1 - f) \cdot T(1) . \tag{2.5}$$

By diving $T_{\alpha\beta}(1)$ by $T_{\alpha\beta}(p)$ this then results in a **scaled upper bound** for the achievable speedup $S_{\alpha\beta}(p)$ using p processors:

$$S_{\alpha\beta}(p) = \frac{T_{\alpha\beta}(1)}{T_{\alpha\beta}(p)} \leq \frac{\alpha \cdot f \cdot T(1) + \beta \cdot (1-f) \cdot T(1)}{\alpha \cdot f \cdot T(1) + \frac{\beta \cdot (1-f) \cdot T(1)}{p}} = \frac{\alpha \cdot f + \beta \cdot (1-f)}{\alpha \cdot f + \frac{\beta \cdot (1-f)}{p}}. \tag{2.6}$$

Since we are mainly interested in the ratio of the two problem size scaling functions, we define:

- $\gamma = \frac{\beta}{\alpha}$: ratio of the problem complexity scaling between the parallelizable part and the non-parallelizable part.

We now reformulate Eq. (2.6) in terms of γ:

$$S_\gamma(p) \leq \frac{f + \gamma \cdot (1-f)}{f + \frac{\gamma \cdot (1-f)}{p}}. \tag{2.7}$$

By using different functions for γ in terms of the number of p yields the following special cases:

1. $\gamma = 1$ (i.e. $\alpha = \beta$): The ratio is constant and thus this special case is exactly **Amdahl's law** (see Eq. (2.4)).
2. $\gamma = p$ (e.g. $\alpha = 1$; $\beta = p$): The parallelizable grows linear in p while the non-parallelizable part remains constant. This special case is known as **Gustafson's law** [4] and is shown here:

$$S(p) \leq f + p \cdot (1-f) = p + f \cdot (1-p). \tag{2.8}$$

3. γ is any other function depending on p.

Gustafson's law is further illustrated in Fig. 2.11. By knowing f, we can use it to predict the theoretically achievable speedup using multiple processors when the parallelizable part scales linearly with the problem size while the serial part remains constant.

Here are two typical examples for applying our derived generalized scaling law.

- **Example 1:** Suppose we have a parallel program that is 15% serial and 85% linearly parallelizable for a given problem size. Assume that the (absolute) serial time does not grow as the problem size is scaled.

  (*i*) How much speedup can we achieve if we use 50 processors without scaling the problem?

$$S_{\gamma=1}(50) \leq \frac{f + \gamma \cdot (1-f)}{f + \frac{\gamma \cdot (1-f)}{p}} = \frac{1}{0.15 + \frac{(0.85)}{50}} = 5.99$$

  (*ii*) Suppose we scale up the problem size by a factor of 100. How much speedup could we achieve with 50 processors?

$$S_{\gamma=100}(50) \leq \frac{f + \gamma \cdot (1-f)}{f + \frac{\gamma \cdot (1-f)}{p}} = \frac{0.15 + 100 \cdot 0.85}{0.15 + \frac{100 \cdot 0.85}{50}} = 46.03$$

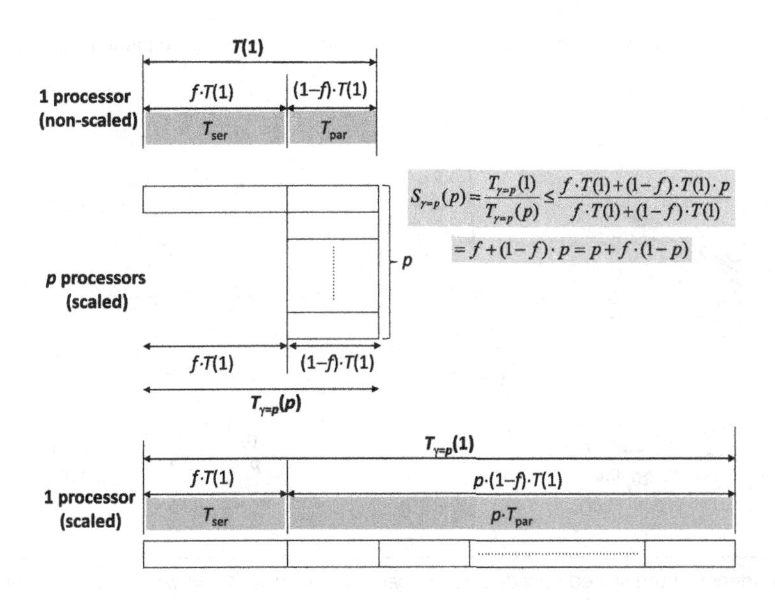

FIGURE 2.11

Illustration of Gustafson's law for establishing an upper bound for the scaled speedup.

- **Example 2:** Assume that you want to write a program that should achieve a speedup of 100 on 128 processors.
  (*i*) What is the maximum sequential fraction of the program when this speedup should be achieved under the assumption of strong scalability?
  We start with Amdahl's law and then isolate f as follows:

$$100 = \frac{1}{f + \frac{(1-f)}{128}} = \frac{128}{128 \cdot f + 1 - f} = \frac{128}{127 \cdot f + 1} \Longrightarrow f = \frac{0.28}{127} = 0.0022 .$$

  Thus, only less than 1% of your program can be serial in the strong scaling scenario!
  (*ii*) What is the maximum sequential fraction of the program when this speedup should be achieved under the assumption of weak scalability whereby the ratio γ scales linearly?
  We now start with Gustafson's law and then isolate f as follows:

$$100 = 128 + f \cdot (1 - 128) = 128 - 127 \cdot f \Longrightarrow f = \frac{28}{127} = 0.22 .$$

  Thus, in this weak scaling scenario a significantly higher fraction can be serial!

Finally, we analyze how the upper bound of the scaled speedup $S_\gamma(p)$ behaves with respect to p and γ for different values of f: Since $\gamma = 1$ refers to Amdahl's law and $\gamma = p$ to Gustafson's law, we choose a suitable parameterization $\gamma(p, \delta) = p^\delta$ to model the dependency of γ on the degree of p. The choice $\delta = 0$ now refers to $\gamma = 1$ and $\delta = 1$ to $\gamma = p$. The resulting expressions for the scaled speedup

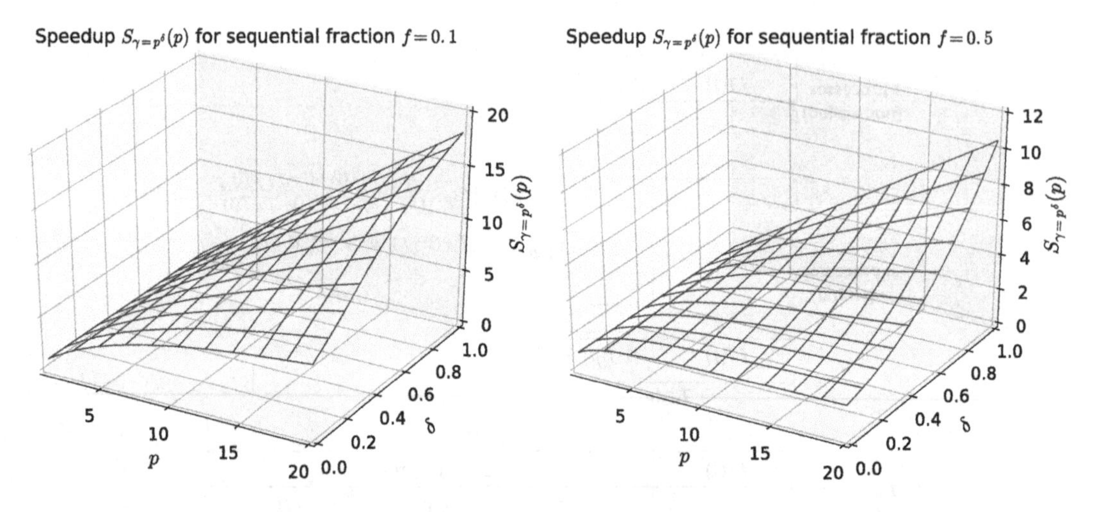

FIGURE 2.12

Functional dependency of the scaled speedup $S_\gamma(p)$ parameterized with $\gamma = p^\delta$. The parameter δ is sampled from the interval $[0, 1]$ referring to Amdahl' law for $\delta = 0$ and Gustafson's law for $\delta = 1$. The serial fraction is fixed at $f = 0.1$ (upper panel) and $f = 0.5$ (lower panel). Obviously, a smaller value of f implies better scalability.

and efficiency read as follows:

$$S_{\gamma=p^\delta}(p) = \frac{f + (1-f) \cdot p^\delta}{f + (1-f) \cdot p^{\delta-1}} \quad , \quad E_{\gamma=p^\delta}(p) = \frac{f + (1-f) \cdot p^\delta}{p \cdot f + (1-f) \cdot p^\delta} \ . \tag{2.9}$$

As we can see, the scaled speedup is bound by either 1 or $1/f$ for degrees $\delta \leq 0$ since

$$\lim_{p\to\infty} \left. \frac{f + (1-f) \cdot p^\delta}{f + (1-f) \cdot p^{\delta-1}} \right|_{\delta<0} = \frac{f}{f} \quad \text{and} \quad \lim_{p\to\infty} \left. \frac{f + (1-f) \cdot p^\delta}{f + (1-f) \cdot p^{\delta-1}} \right|_{\delta=0} = \frac{1}{f} \ . \tag{2.10}$$

In contrast, $S_{\gamma=p^\delta}(p)$ is unbound for any $\delta > 0$, i.e. we can produce virtually any speedup if we increase the number of processing units as long as the input scaling function $\gamma(p)$ has a monotonous dependency on p. Fig. 2.12 illustrates the described behavior for the different serial fractions $f = 0.1$ and $f = 0.5$. Moreover, the **parallel efficiency** $E_{\gamma=p^\delta}(p)$ tends to $(1 - f)$ in the limit $p \to \infty$ for the Gustafson case ($\delta = 1$) and vanishes completely for Amdahl's law ($\delta = 0$) as illustrated in Fig. 2.13.

In general, one might be interested in the dependency of the amount of used processing units p on the remaining parameters of the efficiency function such as γ or δ in our case. In detail, it is of high interest how to choose suitable scaling schemes for a given algorithm with known time complexity in order to preserve high parallel efficiency while increasing the value of p. Linear parameterizations of the arguments (p and δ in our case) that guarantee constant efficiency are called **iso-efficiency lines**. Mathematically speaking, we are interested in all tuples (p, δ) that result in the same efficiency. Unfortunately, it is not straightforward to compute analytic solutions in the general case. Numeric solutions can be obtained by locally applying the implicit function theorem. Fig. 2.13 shows the corresponding

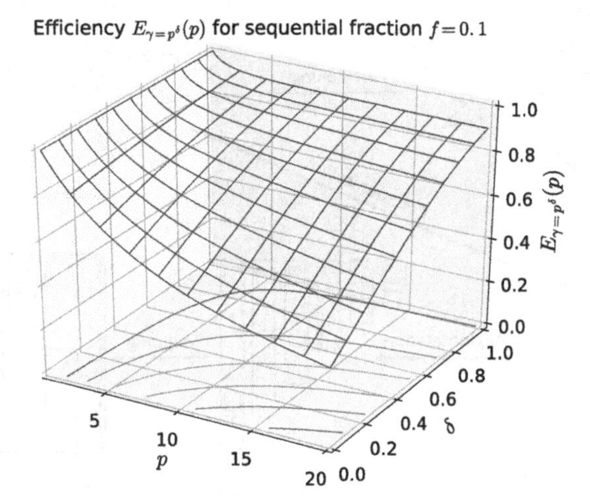

FIGURE 2.13

Functional dependency of the scaled efficiency $E_\gamma(p) = S_\gamma(p)/p$ parameterized with $\gamma = p^\delta$. The parameter δ is sampled from the interval $[0, 1]$ referring to Amdahl' law for $\delta = 0$ and Gustafson's law for $\delta = 1$. The six curves in the p–δ plane are projected iso-efficiency lines of $E_{\gamma=p^\delta}(p)$. Obviously, we have to significantly increase the degree δ of the functional dependency of the scaling ratio $\gamma = p^\delta$ in order to preserve efficiency when increasing the number of processing units p.

iso-efficiency lines in the p–δ plane of the plot for our generalized scaling law. Obviously, we always have to increase the parameter δ in order to preserve parallelization efficiency for increasing values of p. Loosely speaking, the parallelizable part of our problem has to grow if we want to keep the same efficiency while using more processing units.

Both Amdahl's law and Gustafson's law neglect the impact of communication on the runtime of our program. An interested reader might refer to Grama et al. [3] which proposes a more general model for the analysis of iso-efficiency. We have included an exercise in Section 2.5 where you are asked to calculate the **iso-efficiency function** (the scaling of the problem size to keep the efficiency fixed) of a given parallel system (Jacobi iteration) considering both the computation and the communication time.

2.4 FOSTER'S PARALLEL ALGORITHM DESIGN METHODOLOGY

Let us assume you have been given an interesting problem or a sequential program that you are asked to parallelize. This is a challenging task since there is no single known recipe how to convert any problem or sequential program into an efficient parallel program. To make things worse there are often several different possible parallel solutions. In order to approach the parallelization problem, we have already discussed a few guidelines in Section 1.2, that we can consider when designing a parallel solution:

- Partitioning
- Communication

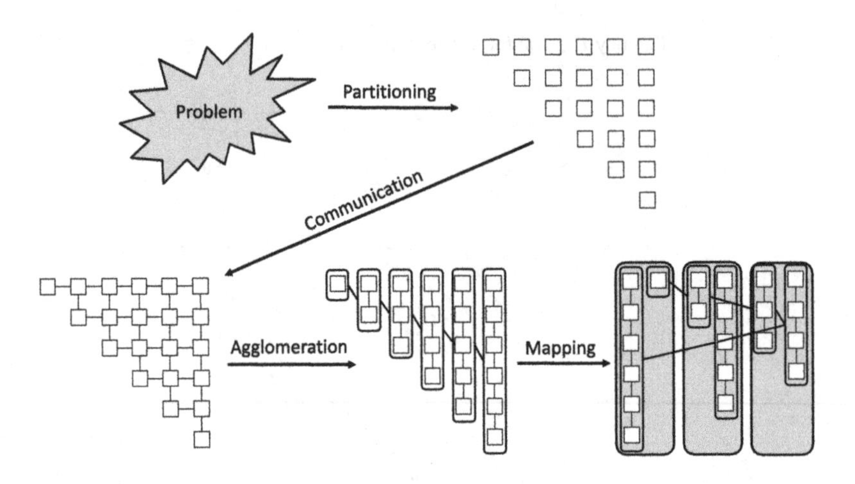

FIGURE 2.14

Illustration of Foster's parallel algorithm design methodology. (*i*) A given problem is first partitioned into large number of small tasks. (*ii*) Communication between the tasks is then specified by linking the respective tasks. (*iii*) Small tasks are agglomerated into large tasks. In this case all tasks within the same column are combined thereby reducing the communication overhead to the left and right neighbor. (*iv*) Lastly, tasks are mapped onto processors in order to reduce the overall execution time. In this example, the six tasks are mapped onto three processors in order to achieve a good (static) load balance.

- Synchronization
- Load balancing

But how should we exactly apply these guidelines and in which order? In order to explore possible parallel solutions in a systematic fashion, Ian Foster proposed a **parallel algorithm design methodology** [2]. It consists of four stages that can be described by the acronym **PCAM** (Partitioning, Communication, Agglomeration, Mapping), as illustrated in Fig. 2.14. The first two stages aim at the discovery of (fine-grained) parallelism, while the remaining two stages focus on maximizing data locality and dividing the workload between processors.

- **Stage 1: Partitioning.** In this first stage we want to identify potential sources of parallelism by partitioning (or decomposing) the problem into small tasks. We already learned about two alternative partitioning strategies in Section 1.2: *domain decomposition* and *functional decomposition*. In a domain decomposition approach, we first determine an appropriate partitioning scheme for the data and then deal with the associated computation. In a functional decomposition approach this order is reversed: we first decompose the computation and then deal with the associated data. By applying alternative partitioning strategies to the same problem, we generally obtain different parallel solutions. Since we want to identify as much parallelism as possible in this stage, we usually aim at maximizing the number of identified tasks (called *fine-grained parallelism*). Note that data parallelism (domain decomposition) often has a finer granularity than task parallelism (functional decomposition).

- **Stage 2: Communication.** In this stage we want to determine the required communication between the tasks identified in Stage 1. We specify the data that must be transferred between two tasks in terms as a channel linking these tasks. On a channel, one task can send messages and the other can receive. In practice, we encounter many different communication patterns, such as local or global, structured or unstructured, static or dynamic, and synchronous or asynchronous. For example, in a domain decomposition with local communication each task needs to communicate with a small set of neighboring tasks in order to perform computation, while in a domain decomposition with global communication each task needs to communicate with all other tasks.
- **Stage 3: Agglomeration.** In the third stage we want to increase the granularity of our design from Stages 1 and 2 by combining a number of small tasks into larger tasks. Our fine-grained parallel design developed in Stages 1 and 2 might actually not be highly efficient when directly implementing it on an actual machine (such as a distributed memory architecture). A reason for that is that executing a large number of small tasks in parallel using different processes (or threads) can be highly inefficient due to the communication overhead. To reduce such overheads, it could be beneficial to agglomerate several small tasks into a single larger task within the same processor. This often improves *data locality* and therefore reduces the amount of data to be communicated among tasks.
- **Stage 4: Mapping.** In the fourth stage we want to map tasks to processors for execution (e.g. through processes scheduled onto processors of a compute cluster). The mapping procedure generally aims at: (*i*) minimizing communication between processors by assigning tasks with frequent interactions to the same processor, (*ii*) enabling concurrency by assigning tasks to different processors that can execute in parallel, and (*iii*) balance the workload between processors. The mapping can sometimes be straightforward. For example, in cases of algorithms with a fixed number of equally-balanced tasks and structured local/global communication, tasks can be mapped so that inter-processor communication is minimized. However, in more complex scenarios such as algorithms with variable amounts of work per task, unstructured communication patterns, or dynamic task the creation of efficient mapping strategies is more challenging. Examples include numerous variants of static and dynamic load balancing methods, such as optimization problems based on a parallel branch-and-bound search.

In the following, we study the application of Foster's methodology to the computation of a *stencil code* applied on a 2-dimensional array $\text{data}(i, j)$ (also called *Jacobi iteration*), where every value in $\text{data}(i, j)$ is updated by the average of its four neighbors as shown in Eq. (2.11). This update rule is applied in an iterative fashion to calculate a sequence of matrices $\text{data}_t(i, j)$, for $t = 1, \ldots, T$.

When applying the *PCAM* approach to this problem we first define a fine-grained parallel task for every matrix element and then define the communication between them as shown in Fig. 2.15(A). These fine-grained tasks are subsequently agglomerated into coarse-grained tasks. We now analyze and compare two different agglomeration schemes for this step:

$$\text{data}_{t+1}(i, j) \leftarrow \frac{\text{data}_t(i-1, j) + \text{data}_t(i+1, j) + \text{data}_t(i, j-1) + \text{data}_t(i, j+1)}{4}. \tag{2.11}$$

- **Method 1:** Fig. 2.15(B) agglomerates all tasks within the same row. The resulting larger tasks are mapped onto processors by combining several neighboring rows (tasks) in Fig. 2.15(C).

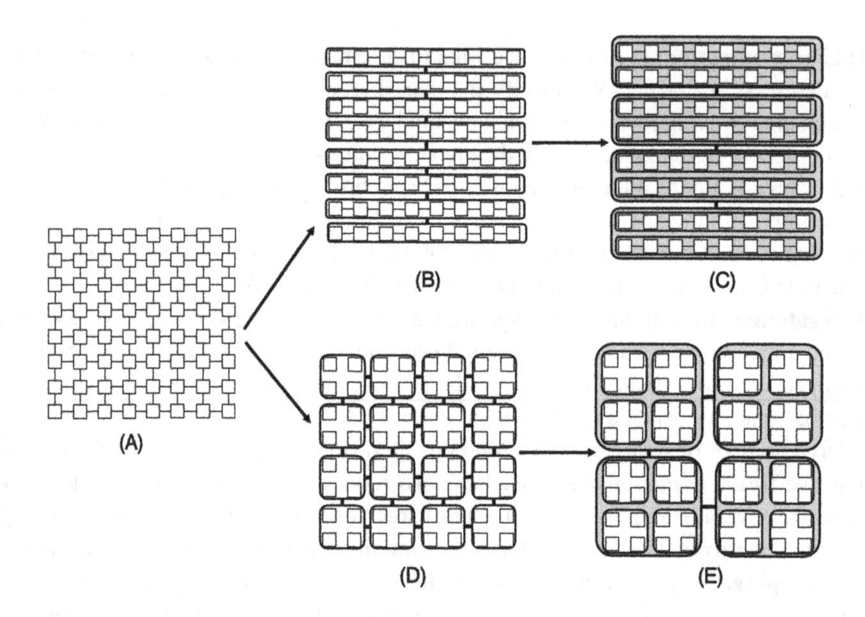

FIGURE 2.15

Two different agglomeration schemes for Jacobi iteration. We start with the same partition and communication pattern (A). Method 1 agglomerates all tasks along the same row (B) and then maps several of them to processors (C). Method 2 agglomerates a square grid of tasks (D) and then maps several of them to processors (E).

- **Method 2:** Fig. 2.15(D) agglomerates several tasks within a square grid. The resulting larger tasks are mapped onto processors by combining several neighboring squares (tasks) within a rectangle in Fig. 2.15(E).

Note that the mapping step in this example is straightforward since this example features a fixed number of equal-sized tasks and structured local communication. We now compare the communication complexity of both approaches. The time required to sent n bytes between two processors is thereby specified in Eq. (2.12), where s denotes the *latency* (or startup time) for a communication and r the inverse of the available bandwidth.

$$T_{\text{comm}}(n) = s + r \cdot n \tag{2.12}$$

Using p processors, Eq. (2.12) yields $2(s + r \cdot n)$ communication time between two processors for Method 1 while it yields $4(s + r(\frac{n}{\sqrt{p}}))$ for Method 2. Thus, for a large number of processors the agglomeration/mapping Method 2 is superior since the required communication time decreases with larger p while it remains constant for Method 1.

We have included two more examples for applying Foster's parallel algorithm methodology in the exercise section 2.5 for the matrix chain ordering problem and the computation of frequent itemsets form market-basked transactions.

K	1	1	1	1	2	2	2	3	3	4	5	5	5	5	6	7
V	v_0	v_1	v_2	v_3	v_4	v_5	v_6	v_7	v_8	v_9	v_{10}	v_{11}	v_{12}	v_{13}	v_{14}	v_{15}

FIGURE 2.16

An example for 16 entries of key–value pairs stored in K and V.

$\overline{K}$	1	2	3	4	5	6	7									
P	0	4	7	9	10	14	15									
V	v_0	v_1	v_2	v_3	v_4	v_5	v_6	v_7	v_8	v_9	v_{10}	v_{11}	v_{12}	v_{13}	v_{14}	v_{15}

FIGURE 2.17

The compressed representation of K and V using unique keys.

2.5 ADDITIONAL EXERCISES

1. **Matrix Multiplication using a CREW PRAM.** Let $A, B \in \mathbb{R}^{n\times n}$ be two square matrices and $C = A \cdot B$ their matrix product given in coordinates

$$C_{ij} := \sum_{k=0}^{n-1} A_{ik} \cdot B_{kj} \text{ for all } i, j \in \{0, \dots, n-1\}.$$

Design a CREW PRAM algorithm that uses $\mathcal{O}(n^3)$ processors and $\mathcal{O}(n^3)$ memory to compute the matrix C in logarithmic time.

(i) State the pseudo-code that is needed to compute the result for n being a power of two.
(ii) Give a simple solution if n is not a power of two. Is the asymptotic complexity affected?
(iii) Is your solution cost-optimal? If not, make it so.

2. **Compressing Arrays of Sorted Key–Value Pairs.** Let $K = (k_i)_i$ and $V = (v_i)_i$ be two arrays each of length $n = 2^m$ for some natural number $m > 0$. Further, assume that the keys in K are already sorted such that $k_0 \leq k_1 \leq \cdots \leq k_{n-1}$ but not necessarily unique, i.e., two equal keys $k_i = k_j$ for $i \neq j$ might occur. An example is given in Fig. 2.16.
The key array K can be compressed to a novel key array $\overline{K}$ containing only unique keys. In order to take account for the appearance of multiple values, one needs to compute another index array $P = (p_j)_j$ storing the first position $p_j = i$ of a value $v_i \in V$ which belongs to the key $k_i \in K$ (see Fig. 2.17).
Design an efficient parallel algorithm based on a CREW PRAM using n processors. Is your implementation cost-efficient? Discuss memory consumption. Justify your claims.

3. **All-to-All Comparison using a CREW PRAM.**
Assume two sequences of m real-valued vectors $A := (a^{(0)}, a^{(1)}, \dots, a^{(m-1)})$ and $B := (b^{(0)}, b^{(1)}, \dots, b^{(m-1)})$ where both $a^{(l)} \in \mathbb{R}^n$ and $b^{(l)} \in \mathbb{R}^n$ for all $0 \leq l < m$. The k-th coordinate of each $a^{(i)}$ and $b^{(j)}$ are denoted as $a_k^{(i)}$ and $b_k^{(j)}$, respectively. In the following, the all-pair distance matrix C_{ij}

shall be computed using the Euclidean distance measure (ED):

$$C_{ij} := \mathrm{ED}(a^{(i)}, b^{(j)}) = \sqrt{\sum_{k=0}^{n-1} (a_k^{(i)} - b_k^{(j)})^2} \quad \text{for all} \quad i, j \in \{0, \ldots, m-1\}.$$

Design a CREW PRAM algorithm that uses $\mathcal{O}(m^2 \cdot n)$ processors and constant memory per processor to compute the matrix C in $\mathcal{O}(\log n)$ time.

(i) State the pseudo-code that is needed to compute the result for n being a power of two. Can you make it work for an arbitrary dimension n?

(ii) Analyze your algorithm in terms of speedup, efficiency, and cost.

(iii) Repeat the analysis using $\mathcal{O}(m^2 \frac{n}{\log n})$ processors.

4. **PRAM Binary Search.** The well-known sequential binary search algorithm for searching an element x in a *sorted* array of n numbers $A[]$ works as follows:

- Compare x to the middle value of A.
- If equal, return the index of the identified element in A and the procedure stops.
- Otherwise, if x is larger, then the bottom half of A is discarded and the search continues recursively with the top half of A.
- Otherwise, if x is smaller, then the top half of A is discarded and search continues recursively with the bottom half of A.

(i) Design a parallel version of binary search on a CRCW PRAM using n processors. Is your solution cost-optimal?

(ii) Design a parallel version of binary search on a CRCW PRAM using $N < n$ processors. Analyze the runtime of your algorithm.

5. **Merging on a PRAM.** Consider the problem of merging two sorted arrays of numbers A and B of length m and n, respectively. Design a parallel merging algorithm on a CREW PRAM using $\frac{m+n}{\log(m+n)}$ processors. What is the time complexity of your algorithm?

6. **De Bruijn Graph.** The r-dimensional de Bruijn graph consists of 2^r nodes and 2^{r+1} directed edges. Each node corresponds to a unique r-bit binary number $u_1u_2\ldots u_r$. There is a directed edge from each node $u_1u_2\ldots u_r$ to nodes $u_2\ldots u_r0$ and $u_2\ldots u_r1$.

(i) Draw a three-dimensional de Bruijn graph.

(ii) What is the diameter of an r-dimensional de Bruijn graph? Justify your answer.

7. **Hypertree Network.** Several parallel machines have used a hypertree network topology as interconnect. A hypertree of degree k and depth d can be described as a combination of a complete top-down k-ary tree (front-view) and a bottom-up complete binary tree of depth d (side-view). Fig. 2.18 shows an example for $k = 3$ and $d = 2$. Determine the degree, diameter, and bisection-width of a hypertree of degree k and depth d.

8. **Butterfly Network.** A butterfly network of order k consists of $(k+1)2^k$ nodes arranged in $k+1$ ranks, each rank containing $n = 2^k$ nodes. We now assign each node a unique label $[i, j]$ for $i = 0, \ldots, k$ and $j = 0, \ldots, n-1$. We connect the nodes as follows:

- Every node $[i, j]$ is connected to $[i+1, j]$ for all $i \in \{0, \ldots, k-1\}$ and $j \in \{0, \ldots, n-1\}$.

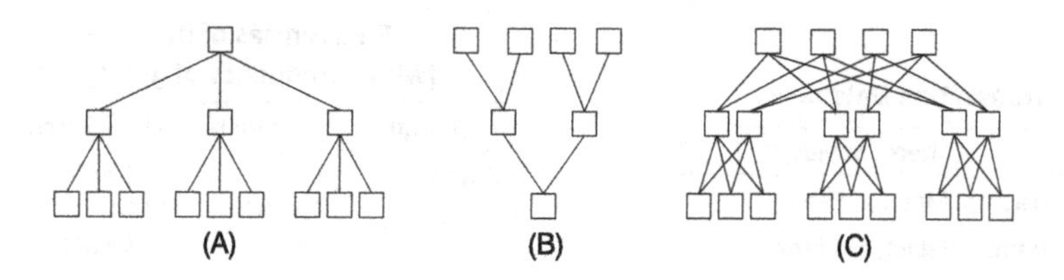

FIGURE 2.18

Front-view (A), side-view (B), and complete topology of a hypertree of degree $k = 3$ and depth $d = 2$.

- Every node $[i, j]$ is connected to $[i + 1, (j + 2^{k-i-1}) \mod 2^k]$ for all $i \in \{0, \dots, k-1\}$ and $j \in \{0, \dots, n-1\}$.

(i) Draw a butterfly network of order 3.
(ii) What is the diameter and the bisection-width of butterfly network of order k? Justify your answer.
(iii) Look up the FFT (*Fast Fourier transform*) algorithm. Explain how a butterfly network can be used for the efficient computation of the FFT of an input signal of size n.

9. **Prefix Sum on a Hypercube.** Design a parallel algorithm to compute the prefix sum of a given array A of $n = 2^d$ numbers on a d-dimensional hypercube. At the beginning of the algorithm each processor should store exactly one element of A and at the end of the algorithm each processor should store one element of the resulting prefix array. Analyze the runtime of your algorithm.
10. **Amdahl and Gustafson.** Assume you want to parallelize a given sequential program and you want achieve a speedup of at least 10 on sixteen processors. What is the maximum serial fraction of the program under consideration of
    (i) Amdahl's law.
    (ii) Gustafson's law.
11. **More general scaling.** Suppose you have implemented a parallel version of a sequential program with the following scaling properties:
    (i) The inherently sequential part of the program has a runtime of $(2500 + n)$ ms, where n is the problem size.
    (ii) The parallelizable part of the program has a runtime of n^2 ms.
    Calculate the maximum speedup achievable by this program on a problem size of $n = 10{,}000$.
12. **Iso-efficiency Analysis of Jacobi Iteration.** In an iso-efficiency analysis we are interested in by how much the problem size n has to increase in order to preserve the parallelization efficiency for increasing values of p (the number of processors). Consider the Jacobi iteration of an $n \times n$ array and the two methods of parallelization as discussed in Fig. 2.15(C) and Fig. 2.15(E).
    (i) Specify the runtime of each method as $T_{\text{comp}} + T_{\text{comm}}$ (where T_{comp} denotes the computing time and T_{comm} denotes the communication time) as a function of n and p.
    (ii) Determine the iso-efficiency function of both methods (i.e., the necessary scaling of the problem size to keep efficiency constant). Which of the two parallelization schemes scales better?

Transaction Database

ID	Items bought
1:	{Laptop, Monitor, Cable}
2:	{Printer, Tablet, Monitor}
3:	{Laptop, Tablet, Cable}
4:	{Laptop, Tablet, Monitor, Cable}
5:	{Laptop, Cable}
6:	{Laptop, Tablet, Monitor}
7:	{Printer, Tablet}
8:	{Laptop, Tablet, Monitor, Cable}
9:	{Printer, Tablet, Cable}
10:	{Laptop, Monitor, Cable}

⇨

Frequencies of itemsets (with a frequency of at least 4)

1 item		2 items		3 items	
{Laptop}	7	{Laptop, Tablet}	4	{Laptop, Monitor, Cable}	4
{Tablet}	7	{Laptop, Cable}	6		
{Monitor}	6	{Tablet, Monitor}	4		
{Cable}	7	{Tablet, Cable}	4		
		{Monitor, Cable}	4		
		{Laptop, Monitor}	5		

FIGURE 2.19

Example of itemset frequencies computed from a given database with 10 market-basket transactions.

13. **Parallel Design for the Matrix Chain Ordering Problem.** Consider a sequence of 2D matrices. The goal of the *matrix chain ordering problem (MCOP)* is to find the most efficient order to multiply these matrices.
For example, when given three matrices M_1, M_2, and M_3 of size 20×50, 50×8 matrix, and 8×80, respectively, there are two possible orderings of computing their product: (i) $(M_1 \times M_2) \times M_3$ and (ii) $M_1 \times (M_2 \times M_3)$. However, these orderings differ in the number of required arithmetic operations: (i) requires $(20 \cdot 50 \cdot 8) + (20 \cdot 8 \cdot 80) = 12{,}800$ operations; while (ii) requires $(50 \cdot 8 \cdot 80) + (20 \cdot 50 \cdot 80) = 112{,}000$ operations. Obviously, the first ordering is more efficient.
In general, given n matrices M_i of sizes $d_{i-1} \times d_i$ for $i \in \{1, \ldots n\}$, the following recurrence relation defines a dynamic programming matrix $F[i, j]$.

$$F[i,j] = \begin{cases} 0 & \text{if } i = j \\ \min_{i \leq k < j}\{F[i,k] + F[k+1,j] + d_{i-1} \cdot d_k \cdot d_j\} & \text{if } i < j \end{cases}$$

The value stored in $F[i, j]$ $(i \leq j)$ is equal to the number of operations (the cost) required in the optimal ordering for computing the matrix product $M_i \times \ldots \times M_j$. Thus, the overall minimum cost is stored in $F[1, n]$.
   - **(i)** Compute the dynamic programming matrix for four matrices of size 35×40, 40×20, 20×10, and 10×15.
   - **(ii)** Design a parallel algorithm for MCOP using Foster's methodology. What is the runtime of your algorithm?
14. **Parallel Design for Itemset Frequencies.** An important task in data mining is the computation of frequencies of itemsets from a given database of transactions (e.g. for learning association rules). An example is illustrated in Fig. 2.19 using market-basked transactions. Use Foster's methodology

to design two parallel algorithms for computing itemset frequencies from a given transaction database and compare them.

REFERENCES

[1] Gene M. Amdahl, Validity of the single processor approach to achieving large scale computing capabilities, in: Proceedings of the April 18–20, 1967, Spring Joint Computer Conference, ACM, 1967, pp. 483–485.
[2] Ian Foster, Designing and Building Parallel Programs, vol. 191, Addison Wesley Publishing Company, Reading, 1995.
[3] Ananth Grama, Anshul Gupta, Vipin Kumar, Isoefficiency Function: A Scalability Metric for Parallel Algorithms and Architectures, 1993.
[4] John L. Gustafson, Reevaluating Amdahl's law, Communications of the ACM 31 (5) (1988) 532–533.
[5] Joseph JáJá, An Introduction to Parallel Algorithms, vol. 17, Addison-Wesley, Reading, 1992.
[6] Nadathur Satish, Mark J. Harris, Michael Garland, Designing efficient sorting algorithms for manycore GPUs, in: 23rd IEEE International Symposium on Parallel and Distributed Processing, IPDPS 2009, Rome, Italy, May 23–29, 2009, 2009, pp. 1–10.
[7] Leslie G. Valiant, A bridging model for parallel computation, Communications of the ACM 33 (8) (1990) 103–111.

CHAPTER

MODERN ARCHITECTURES

3

Abstract

In the past decades, the increased single-core performance of every released processor generation has been the main contributor towards runtime improvements of applications. Nowadays, single-threaded CPU performance is stagnating and performance improvements of modern CPUs can mainly be attributed to an increasing number of cores. Nevertheless, taking full advantages of modern architectures requires in many cases not only sophisticated parallel algorithm design but also knowledge of modern architectural features. An important example is the memory system. Modern microprocessors can process data at a much faster rate than reading data from main memory – known as the *von Neumann bottleneck*. As a consequence, many programs are limited by memory accesses and not the actual computation. In this chapter, you will learn about the memory hierarchy where a fast cache hierarchy is located in-between the CPU and the main memory in an attempt to mitigate the von Neumann bottleneck. From the viewpoint of a (parallel) programmer it is therefore of high importance to be aware of the structure and functioning of cache-based memory subsystems. This, in turn, can lead to programs making effective use of the available memory system. In multi-core systems with several levels of shared/local caches the situation becomes more complex requiring the consideration of cache coherency and false sharing of data. Besides the steadily growing number of cores, modern architectures contain a number of architectural features concerning the improvement of compute performance. A prominent example is fine-grained parallelism based on the SIMD (*Single-Instruction Multiple-Data*) concept – meaning that several values can be processed at the same time using a single instruction. We will study the basics of SIMD parallelism and Flynn's taxonomy of computer architectures. Common microprocessors support SIMD parallelism by means of vector registers. You will learn about the vectorization of algorithms on standard CPUs using intrinsics. The design of efficient vectorized code often requires advanced techniques such as data layout transformations.

Keywords

CPU, Cache, von Neumann bottleneck, Memory bound, Bandwidth, Latency, Matrix multiplication, Main memory, Compute bound, Cache algorithm, Cache hit, Cache miss, Cache coherence, False sharing, Prefetching, SIMD, MIMD, Vectorization, Flynn's taxonomy, Instruction-level parallelism

CONTENTS

Parallel Programming. DOI: 10.1016/B978-0-12-849890-3.00003-4

3.1 MEMORY HIERARCHY

VON NEUMANN BOTTLENECK

In the classical *von Neumann architecture* a processor (CPU) is connected to main memory through a bus as shown in Fig. 3.1. In early computer systems timings for accessing main memory and for computation were reasonably well balanced. However, during the past few decades computation speed grew at a much faster rate compared to main memory access speed resulting in a significant performance gap. This discrepancy between CPU compute speed and main memory (DRAM) speed is commonly known as the **von Neumann bottleneck**.

Let us demonstrate that with an example. Consider a CPU with eight cores and a clock frequency of 3 GHz. Moreover, assume that 16 double-precision *floating-point operations* (Flop) can be performed per core in each clock cycle. This results in a peak compute performance of 3 GHz $\times$ 8 $\times$ 16 = 384 GFlop/s. The CPU is connected to a DRAM module with a peak memory transfer rate of 51.2 GB/s. We want to establish an upper bound on the performance of this system for calculating the dot product of two vectors u and v containing n double-precision numbers stored in main memory:

```
double dotp = 0.0;
for (int i = 0; i < n; i++)
   dotp += u[i] * v[i];
```

We further assume that the length of both vectors $n = 2^{30}$ is sufficiently large. Since two operations (one multiplication and one addition) are performed within each iteration of the loop, we have a total of $2 \cdot n = 2^{31}$ Flops to compute. Furthermore, the two vectors have to be transferred from main memory to the CPU. Thus, a total volume of $2^{31} \times 8\text{ B} = 16\text{ GB}$ of data has to be transferred. Based on our system specification, we can determine the minimum amount of time required for computation and data transfer as follows:

- **Computation:** $t_{\text{comp}} = \frac{2\text{ GFlops}}{384\text{ GFlop/s}} = 5.2\text{ ms}$
- **Data transfer:** $t_{\text{mem}} = \frac{16\text{ GB}}{51.2\text{ GB/s}} = 312.5\text{ ms}$

If we overlap computation and data transfer, a lower bound on the total execution time can be derived:

$$t_{\text{exec}} \geq \max(t_{\text{comp}}, t_{\text{mem}}) = \max(5.2\text{ ms}, 312.5\text{ ms}) = 312.5\text{ ms}. \tag{3.1}$$

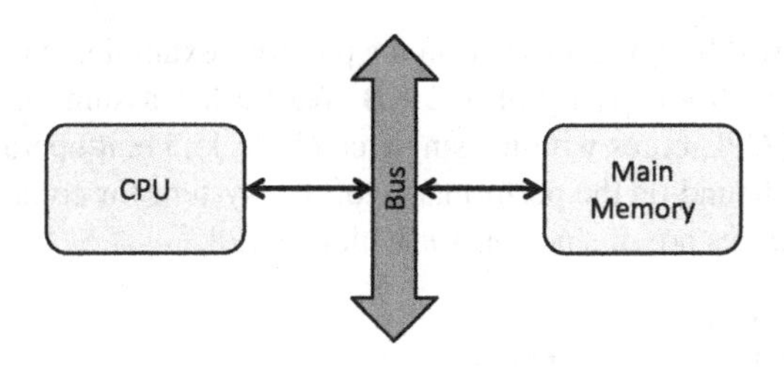

FIGURE 3.1

Basic structure of a classical von Neumann architecture.

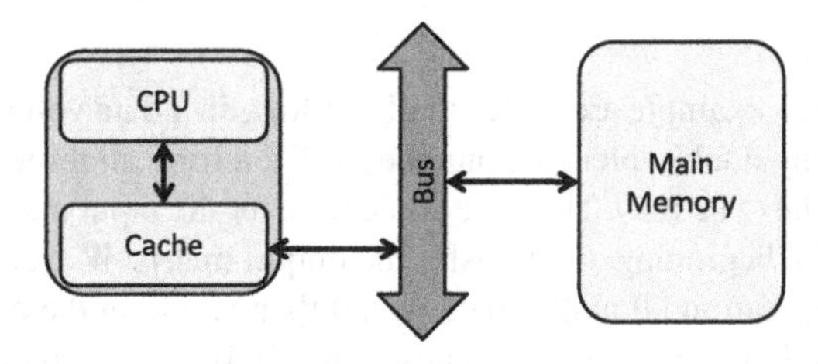

FIGURE 3.2

Basic structure of a CPU with a single cache.

The data transfer time clearly dominates. Note that each matrix element is only used once and there is no data re-usage. Thus, dot product computation is *memory bound*. The corresponding upper bound on the achievable performance can be calculated as $\frac{2^{31}\text{ Flop}}{312.5\text{ ms}} = 6.4$ GFlop/s – meaning that merely less than 2% of the available peak performance can be achieved.

In order to overcome the von Neumann bottleneck, computer architects have introduced a number of extensions to the classical architecture. One of those is the addition of a fast memory between CPU and main memory called **cache**. Modern CPUs typically contain a hierarchy of three levels of cache (L1, L2, L3) and current CUDA-enabled GPUs contain two levels. There is the usual trade-off between capacity and speed, e.g. L1-cache is small but fast and the L3-cache is big but slow. In addition caches could be private for a single core or shared between several cores. In the following, we learn more about how caches work and study a number of programs that make effective use of them.

CACHE MEMORY

Cache is fast memory added between CPU and main memory. It features significantly higher bandwidth and lower latency compared to main memory. However, its capacity is much smaller. As illustrated in Fig. 3.2, the CPU is no longer directly connected to the main memory. All loads and stores have to go through the cache. Furthermore, there is a dedicated connection between CPU and cache for fast communication. The cache is typically integrated together with the CPU cores on the same chip.

Let us again have a look at an example to demonstrate the potential benefits of the introduced cache memory. Consider the same 8-core 3 GHz CPU with a peak performance of 384 GFlop/s and a main memory peak bandwidth of 51.2 GB/s as in the previous example. Our system now additionally contains a (shared) cache with a capacity of 512 KB. We further assume that the cache is very fast and can be accessed by all CPU cores within a single clock cycle; i.e. it operates at register-speed. We want to establish an upper bound on the performance of this system for computing the matrix product $W = U \times V$ where all matrices are of shape $n \times n$ with $n = 128$:

```
for (int i = 0; i < n ; i++)
    for (int j = 0; j < n; j++) {
        double dotp = 0.0;
        for (int k = 0; k < n; k++)
            dotp += U[i][k] * V[k][j];
        W[i][j] = dotp;
    }
```

The matrix dimensions in this example are quite small resulting in a data volume of $128 \times 128 \times 8$ B = 128 KB per matrix (assuming double-precision numbers). Therefore, all three matrices (3×128 KB = 384 KB) can fit into the cache (512 KB). Thus, we could transfer the input matrices U and V from main memory to cache once at the beginning and transfer the output matrix W from cache to main memory at the end. During the computation all matrix data would then reside in the cache without the need of any main memory transfers. For each of the n^2 values in matrix W a scalar product with $2 \cdot n$ operations needs to be calculated resulting in a total of $2 \cdot n^3 = 2 \cdot 128^3 = 2^{22}$ Flops to be performed. Based on the system specification, we can determine the minimum amount of time required for computation and data transfer as follows:

- **Computation:** $t_{\text{comp}} = \frac{2^{22}\text{ Flop}}{384\text{ GFlop/s}} = 10.4\,\mu\text{s}$
- **Data transfer:** $t_{\text{mem}} = \frac{384\text{ KB}}{51.2\text{ GB/s}} = 7.5\,\mu\text{s}$

Due to the usage of the cache memory, we are now able to re-use each element of the input matrices during the computation of n different scalar products in the output matrix. As a result, the computation time is now longer than the data transfer. Matrix multiplication is therefore *compute bound*. Assuming that there is no overlap of computation and communication in this example, we establish a lower bound on the total execution time:

$$t_{\text{exec}} \geq t_{\text{comp}} + t_{\text{mem}} = 10.4\ \mu\text{s} + 7.5\ \mu\text{s} = 17.9\ \mu\text{s}\,. \tag{3.2}$$

The corresponding maximum upper bound on the achievable performance can be calculated as $\frac{2^{22}\text{ Flop}}{17.9\ \mu\text{s}} = 223$ GFlop/s. This corresponds to almost 60% of the available peak performance – over an order-of-magnitude improvement compared to our previous example. Highly optimized implementations – such as the general matrix multiplication (GEMM) of the BLAS library [1] – can even reach close to 100% of the peak performance of modern CPUs.

However, this example has a major limitation: we have assumed that all data can fit into the cache. In most situations this would not actually be the case. Thus, we need to consider what should happen if the matrices are bigger than the limited size of the cache. In such situations a set of suitable strategies are required to determine what data should be placed into cache and evicted from cache during program execution. This is discussed in the next subsection.

CACHE ALGORITHMS

The cache memory is a resource that does not need to be explicitly managed by the user. Instead, the cache is managed by a set of cache replacement policies (also called *cache algorithms*) that determine which data is stored in the cache during the execution of a program. To be both cost-effective and efficient, caches are usually several orders-of-magnitude smaller than main memory (e.g., there are typically a few KB of L1-cache and a few MB of L3-cache versus many GB or even a few TB of main memory). As a consequence, the dataset that we are currently working on (the working set) can easily exceed the cache capacity for many applications. To handle this limitation, cache algorithms are required that address the questions of

- Which data do we load from main memory and where in the cache do we store it?
- If the cache is already full, which data do we evict?

If the CPU requests a data item during program execution, it is first determined whether it is already stored in cache. If this is the case, the request can be serviced by reading from the cache without the need for a time-consuming main memory transfer. This is called a **cache hit**. Otherwise, we have a **cache miss**. Cache algorithms aim at optimizing the **hit ratio**; i.e. the percentage of data requests resulting in a cache hit. Their design is guided by two principles:

- **Spatial locality.** Many algorithms access data from contiguous memory locations with high spatial locality. Consider the following code fragment to determine the maximum value of an array `a` of size n (whereby the elements of `a[]` are stored contiguously):

```
for (int i = 0; i < n; i++)
    maximum = max(a[i], maximum);
```

  Assume the cache is initially empty. In the first iteration the value `a[0]` is requested resulting in a cache miss. Thus, it needs to be loaded from main memory. Instead of requesting only a single value, an entire **cache line** is loaded with values from neighboring addresses. Assuming a typical cache line size of 64 B and double-precision floating-point values, this would mean that the eight consecutive values `a[0]`, `a[1]`, `a[2]`, `a[3]`, `a[4]`, `a[5]`, `a[6]`, and `a[7]` are loaded into cache. The next seven iterations will then result in cache hits. The subsequent iteration requests `a[8]` resulting again in a cache miss, and so on. Overall, the hit ratio in our example is as high as 87.5% thanks to the exploitation of spatial locality.
- **Temporal locality.** The cache is organized into a number of blocks (cache lines) of fixed size (e.g. 64 B). The cache mapping strategy decides in which location in the cache a copy of a particular entry of main memory will be stored. In a *direct-mapped cache*, each block from main memory can be stored in exactly one cache line. Although this mode of operation can be easily implemented, it generally suffers from a high miss rate. In a *two-way set associative cache*, each block from main memory can be stored in one of two possible cache lines (as illustrated in Fig. 3.3). A commonly used policy in order to decide which of the two possible locations to choose is based on temporal locality and is called *least-recently used* (LRU). LRU simply evicts the least recently accessed entry. Going from a direct-mapped cache to a two-way set associative cache can improve the hit ratio significantly [2]. A generalization of the two-way set associative cache is called *fully associative*. In

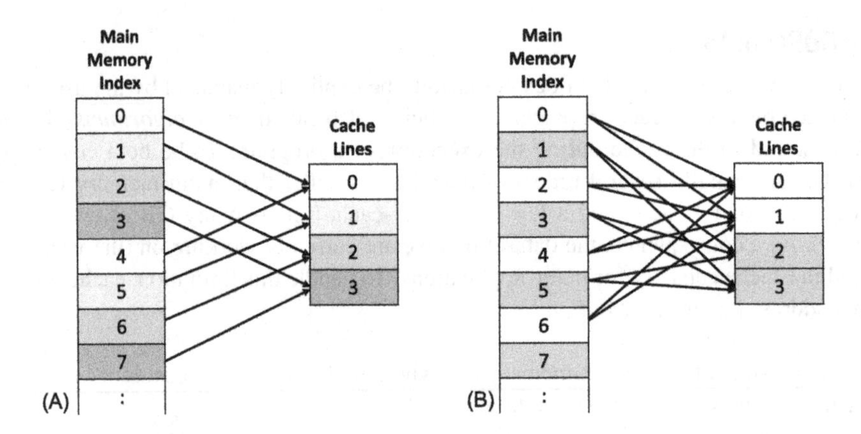

FIGURE 3.3

Illustration of (A) a direct-mapped cache and (B) a two-way associative cache.

this approach, the replacement strategy is free to choose any cache line to hold the copy from main memory. Even though the hit rate might be improved further, the costs associated with implementing a fully associative cache are often prohibitive. Therefore, n-way associative cache designs with $n = 2, 4,$ or 8 are usually preferred in practice.

OPTIMIZING CACHE ACCESSES

We have just learned about how a cache works in principle. We will now apply our knowledge to improve the performance of a given (sequential) program using matrix multiplication as a case study.

```
1  #include <iostream>
2  #include <cstdint>
3  #include <vector>
4  #include "../include/hpc_helpers.hpp"
5
6  int main () {
7
8      // matrix shapes
9      const uint64_t m = 1 << 13;
10     const uint64_t n = 1 << 13;
11     const uint64_t l = 1 << 13;
12
13     TIMERSTART(init)
14     // sum_k A_ik * B_kj = sum_k A_ik * B^t_jk = C_ij
15     std::vector<float> A (m*l, 0); // m x l
16     std::vector<float> B (l*n, 0); // l x n
17     std::vector<float> Bt(n*l, 0); // n x l
18     std::vector<float> C (m*n, 0); // m x n
19     TIMERSTOP(init)
20
```

```
21      TIMERSTART(naive_mult)
22      for (uint64_t i = 0; i < m; i++)
23          for (uint64_t j = 0; j < n; j++) {
24              float accum = 0;
25              for (uint64_t k = 0; k < l; k++)
26                  accum += A[i*l+k]*B[k*n+j];
27              C[i*n+j] = accum;
28          }
29
30      TIMERSTOP(naive_mult)
31
32      TIMERSTART(transpose_and_mult)
33      TIMERSTART(transpose)
34      for (uint64_t k = 0; k < l; k++)
35          for (uint64_t j = 0; j < n; j++)
36              Bt[j*l+k] = B[k*n+j];
37      TIMERSTOP(transpose)
38
39      TIMERSTART(transpose_mult)
40      for (uint64_t i = 0; i < m; i++)
41          for (uint64_t j = 0; j < n; j++) {
42              float accum = 0;
43              for (uint64_t k = 0; k < l; k++)
44                  accum += A[i*l+k]*Bt[j*l+k];
45              C[i*n+j] = accum;
46          }
47
48      TIMERSTOP(transpose_mult)
49      TIMERSTOP(transpose_and_mult)
50
51  }
```

Listing 3.1: Two different implementations of matrix multiplication: naive and transpose-and-multiply.

Listing 3.1 shows a C++ program for the matrix multiplication $A \cdot B = C$ using the matrix dimensions $m \times l$, $l \times n$, and $m \times n$, respectively. The three matrices are stored in linear arrays in row-major order. The first set of for-loops implements matrix multiplication in a straightforward but naive way. Studying the access patterns in the inner loop over the index k shows that the elements of A are accessed contiguously. However, the accesses of matrix B do not exhibit as much spatial locality: after accessing the value in row k and column j, the subsequent iteration accesses the value in row $k+1$ and column j. These values are actually stored by $l\times$ sizeof(float) byte-addresses apart in main memory. Thus, they will not be stored in the same cache line. The access of a value that is stored within the same cache line, such as the one stored in row k and column $j+1$ of B will only happen l iteration steps later. If l is sufficiently large (such as $l = 2^{13}$ as in our example), the corresponding cache line would have already been evicted from L1-cache. This results in a low hit rate for large values of l. This situation is illustrated on the left side of Fig. 3.4.

The subsequent part of Listing 3.1 implements matrix multiplication in a different way. First, an additional temporary array `Bt[]` is used to store the *transposed* matrix B^T. The final set of for-loops multiplies A and B^T. Due to the applied transposition, the same access pattern for both arrays `A[]` and

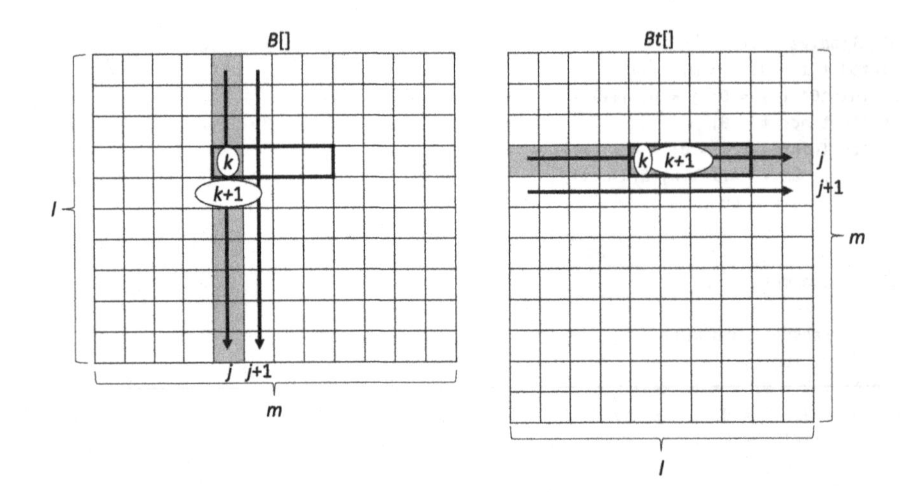

FIGURE 3.4

Access patterns of the array `B[]` in the naive implementation (left) and array `Bt[]` in the transposed multiplication implementation (right) within the inner loop iterating over the variable `k` and the outer loop over the variable `j`. Matrix cells within the same cache line (assuming a cache line can hold four float values) at the current iteration step are indicated by bold rectangular borders.

`Bt[]` can now be used. This results in a high hit rate throughout the whole computation. The access patterns for `Bt[]` is shown on the right side of Fig. 3.4.

Actual execution of our program on an Intel i7-6800K CPU using the matrix dimensions $m = n = l = 2^{13}$ produces the following output:

```
# elapsed time (init): 0.164688s
# elapsed time (naive_mult):5559.46s
# elapsed time (transpose):  0.752526s
# elapsed time (transpose_mult): 497.911s
# elapsed time (transpose_and_mult): 498.663s
```

First, note that the size of each matrix ($2^{13} \times 2^{13} \times$ sizeof(float) B = 256 MB) exceeds the available cache capacity. Thus, we expect to have a low cache hit rate for accessing the array `B[]` in our naive implementation. The cache line size of our CPU is 64 B. Thus, the system can load 16 float values during a main memory read access. This results in a cache hit rate of 93.75% for accessing `Bt[]` in the transposed multiplication. The overhead of transposing matrix B is negligible (but requires additional memory space). The achieved speedup of the transposed multiplication (including the transposition) over the naive implementation is a whopping 11.1. Using smaller values of l can actually improve the hit rate of the naive method. Executing our program for $m = n = 2^{13}$ and $l = 2^8$ outputs the following output:

```
# elapsed time (init): 0.0944191s
# elapsed time (naive_mult): 28.0118s
# elapsed time (transpose): 0.0133463s
```

```
# elapsed time (transpose_mult): 12.9108s
# elapsed time (transpose_and_mult): 12.9242s
```

We can see that the speedup of the transposed program compared to the naive program is reduced to a factor of 2.2. As we have discussed above, the access of a value that is stored within the same cache line, such as the one stored in row k and column $j+1$ of B, will happen l iteration steps later. In the case of a small dimension l this value might still be cached (since fewer cache lines have been evicted in the meantime) resulting in a higher hit rate compared to the execution with a large l-value.

In summary, we have achieved a considerable speedup by means of a relatively simple code transformation that optimizes cache accesses.

CACHE COHERENCE

So far we have only talked about reading data from main memory and cache. But what about writing? Assume we want to overwrite a value that is already cached. Only modifying the cached value would lead to an *inconsistency* between the copy stored in cache and the original value kept in main memory. Two main write policies are used in order to guarantee *coherence* between cache and main memory:

- **Write-through:** If we want to write to a main memory address whose data is cached, the associated cache line is modified as well as the original address in main memory. In a subsequent operation this cache line could be evicted without producing any inconsistencies. A disadvantage of this policy is that excessive writes slow down a program since every write operation results in a main memory access.
- **Write-back:** In this mode of operation, the main memory address is not modified immediately. Instead, only the associated cache line is modified and marked as *dirty*. Only when a dirty cache line is evicted, the stored data will be written to main memory.

The situation becomes more complex in a multi-core CPU system. Modern multi-core chips usually contain several cache levels, whereby each core has a private, i.e. not shared, (small but fast) lower-level cache such as the L1-cache but all cores share a common (larger but slower) higher-level cache such as the L2-cache as shown in Fig. 3.5. Since there is a private cache per core, it is now possible to have several copies of shared data. For example, one copy stored in the L1-cache of Core 0 and one copy stored in the L1-cache of Core 1 – in addition to the one kept in main memory. If Core 0 now writes to the associated cache line either using the write-through or the write-back policy, only the value in the L1-cache of Core 0 is updated but not the value in the L1-cache of Core 1 resulting in a cache inconsistency. This means that the caches store different values of a single address location. A simple example of a cache inconsistency is illustrated in Fig. 3.6.

Incoherent cache values can have unwanted effects when executing parallel programs. Thus, parallel architectures need to ensure cache coherence. This means that a mechanism is required to propagate changes made to one local copy of data, to the other copies in the system. One possible implementation would be to update the private caches of all cores in the system. Unfortunately, providing direct access to the caches of different cores would be extremely slow – especially for a large number of cores. Another approach could just invalidate the cache lines containing copies of the modified data in all other cores. Future references to an invalid cache line will then require the cache line to be reloaded from main memory. Various protocols have been proposed for implementing cache coherence in parallel

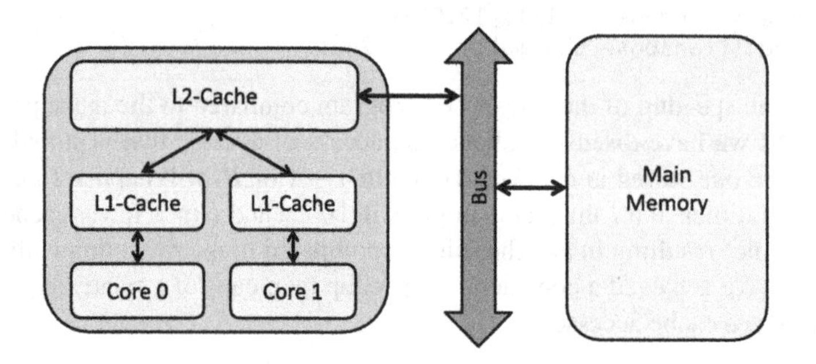

FIGURE 3.5

Illustration of a CPU-chip with two cores and two levels of cache: a private L1-cache and a shared a L2-cache.

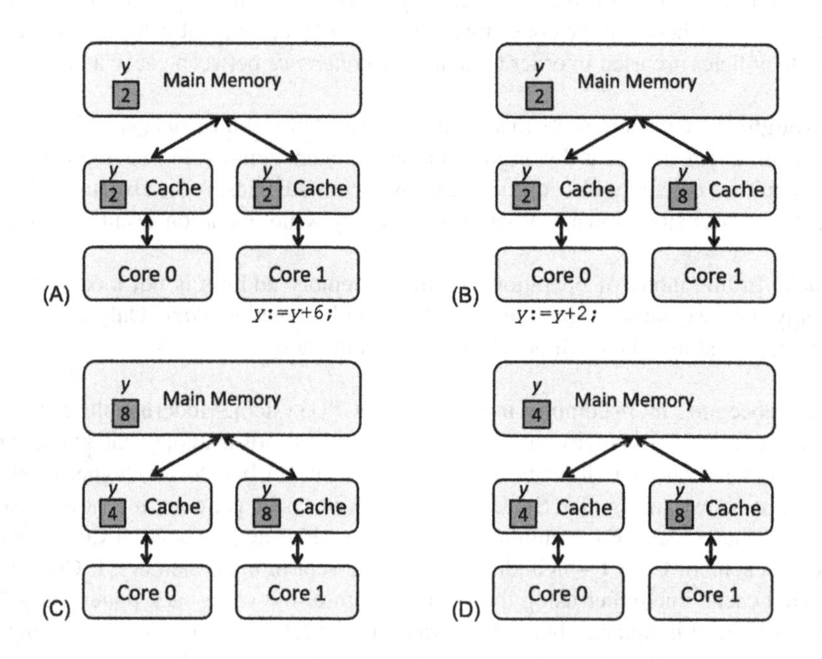

FIGURE 3.6

Example of cache incoherence: (A) The variable y is stored in (shared) main memory. It has already been cached by both Core 0 and Core 1 in their respective private caches. Core 1 now executes the instruction `y:=y+6;` thereby reading the value 2 for y from the copy in its private cache. (B) Core 1 has now completed the execution and stores the result (8) in its private cache. At the same time Core 0 executes the instruction `y:=y+2;` thereby reading the value 2 for y from the copy in its private cache. (C) Core 0 has now completed the execution and stores the result (4) in its private cache. At the same time Core 1 writes the result (8) from the previous instruction through to the main memory location of y. (D) Core 0 writes the result (4) from the previous instruction through to the main memory location of y. Since the two caches store different values for the same variable, there is a cache inconsistency.

systems such as modern multi-core CPUs. A prominent example is the *MESI* protocol which marks a cache line using four possible states: Modified(M), Exclusive(E), Shared(S), and Invalid(I). More details about such protocols are outside the scope of our book but can be found in the literature such as [3].

FALSE SHARING

We have learned that a cache is partitioned into a number of cache lines, whereby each line can hold several values (e.g. 16 float numbers for a 64 B-long cache line). A cache coherence protocol typically works on cache line level; i.e., if a core modifies a single value, all associated cache lines (in the core itself as well as in other cores) will be invalidated as a whole. Let us now imagine a situation where several cores simultaneously operate on distinct data items that are stored within the same main memory address region (fitting into a single cache line). Any write operation would now invalidate the corresponding cache line not only in the core that is writing but also in other cores. This means in turn that these other cores would need to read the required data from (slow) main memory again even though the actual data item required has not been modified by them. This is an artifact known as *false sharing* (also called *cache line ping-pong*). It can cause severe performance degradation.

We will be discussing false sharing again in more detail in Section 4.3. For now, consider a simple example of a 128-bit packed data structure that fits into a cache line holding two integers:

```
struct pack_t {
    uint64_t ying;
    uint64_t yang;

    pack_t() : ying(0), yang(0) {}
};
```

Listing 4.12 contains a sequential code that increments the two integers `ying` and `yang` as well as a parallel code that independently increments each of them using two threads. Measuring the execution time shows that the multi-threaded code runs around six times slower. This clearly demonstrates the potential performance degradation that false sharing can cause.

A helpful guideline for programmers is therefore to avoid excessive updates of entries stored in the same cache line when using multiple threads in parallel. Furthermore, it is often helpful to store intermediate results in registers instead of cache.

SIMULTANEOUS MULTI-THREADING AND PREFETCHING

If data is not cached (e.g. since there is insufficient data re-usage in your algorithm), it must still be transferred from slow main memory. On modern CPU (and GPU) architectures such an access incurs very high latency of often a few hundreds of clock cycles. *Simultaneous multi-threading* (SMT) and *hardware prefetching* are approaches utilized by modern architectures to hide such latencies.

SMT executes several threads on the same core simultaneously which need to share the available resources. If one of the threads is stalled since it needs to wait for data requested from main memory, the system can schedule another thread performing computation on already loaded data. Thus, servicing of memory requests can be overlapped with computation through multiplexing of threads. This implies that the system needs to be capable of quickly switching between threads. Current Intel CPUs implement a two-way SMT known as *hyper-threading*. Furthermore, modern CUDA-enabled GPUs execute a large number of threads per SM (streaming multiprocessor) which we will study in more detail in Section 7.2.

Prefetching can be applied in situations where it is known in advance which data will be needed by the processor. In this case the corresponding data can already be loaded from main memory before it is actually needed and then stored in the cache. This data transfer can then be overlapped by computation on already loaded data. Once the prefetched data is needed, the corresponding request could be serviced by the cache (cache hit). Thus, prefetching can in certain situations be beneficial for hiding main memory access latency. Prefetching can either be realized explicitly by the programmer (software prefetching) or automatically (hardware prefetching).

OUTLOOK

You can apply the gained knowledge about the memory system in several chapters in this book. Sections 4.3, 5.3, and 6.3 use caching on multi-core CPUs with multi-threaded C++ and OpenMP. Furthermore, Chapter 7 applies memory optimization on CUDA-enabled GPUs extensively such as in Sections 7.2, 7.3, and 7.4.

Furthermore, we will use the overlapping of data transfers and computation as an important concept on different architectures throughout this book. Examples include the interleaving communication and computation on multiple GPUs using CUDA in Section 63 and overlapping on multiple nodes of a distributed memory architecture using MPI in Section 9.6.

3.2 LEVELS OF PARALLELISM

We have discussed the memory hierarchy as an important feature of modern architectures to address the von Neumann bottleneck for transferring data between processors and memory. In addition, modern microprocessors (such as CPUs and GPUs) also incorporate several levels of parallelism in order to improve their compute performance.

FLYNN'S TAXONOMY

A frequently used and established classification of different types of parallelism is Flynn's taxonomy [4]. It identifies four classes of architectures according to their instruction and data streams:

- **SISD** (***Single Instruction, Single Data***) refers to the traditional von Neumann architecture where a single sequential processing element (PE) operates on a single stream of data.
- **SIMD** (***Single Instruction, Multiple Data***) performs the same operation on multiple data items simultaneously.
- **MIMD** (***Multiple Instruction, Multiple Data***) uses multiple PEs to execute different instructions on different data streams.
- **MISD** (***Multiple Instruction, Single Data***) employs multiple PEs to execute different instructions on a single stream of data. This type of parallelism is not so common but can be found in pipelined architectures such as systolic arrays [10].

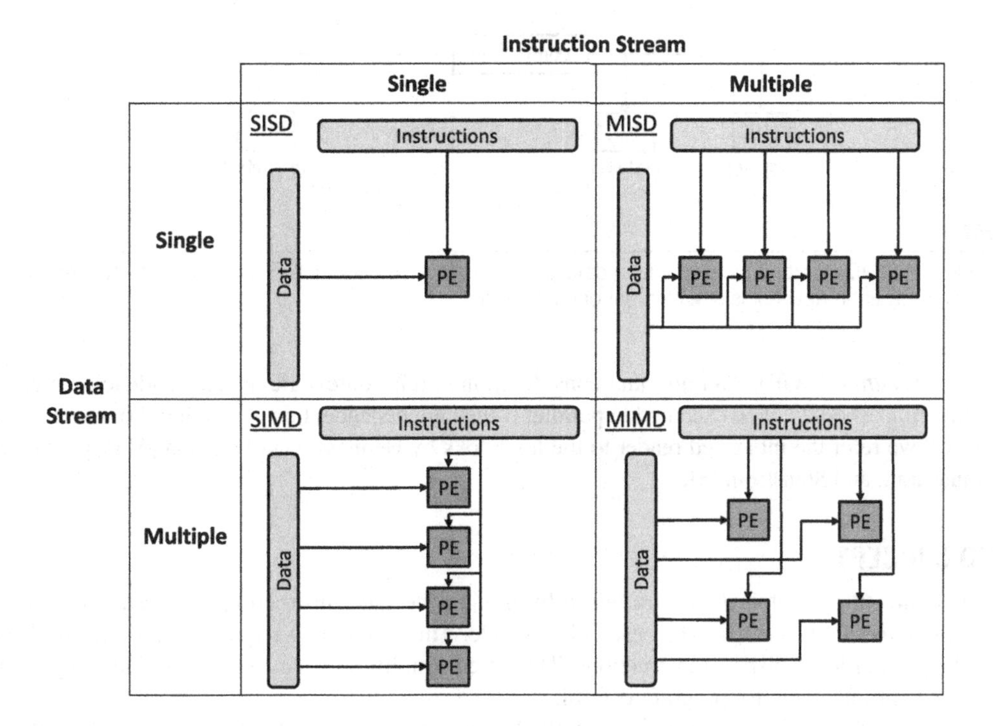

FIGURE 3.7

Flynn's taxonomy classifies computer architectures into four classes according to the number of instruction streams and data streams: SISD, MISD, SIMD, and MIMD.

We have further illustrated Flynn's taxonomy in Fig. 3.7. Modern CPUs and GPUs contain a number of features that exploit different levels of parallelism.

- **Multiple cores.** Modern microprocessors use MIMD parallelism by incorporating a number of cores (or streaming multi-processors) that can execute threads asynchronously and independently. You will learn about the programming with multiple threads of execution on CPUs and GPUs in Chapters 4, 6, and 7.
- **Vector units.** Modern architectures exploit data-level parallelism by incorporating SIMD-based vector units within each core. Vector units can execute vector instructions on a number of data items simultaneously; e.g., a 512-bit vector unit can perform an addition of 16 pairs of single-precision floating-point numbers in parallel.
- **Instruction-level parallelism.** Current processors further exploit instruction-level parallelism (ILP) by *pipelining* and *superscalar execution* of instructions. Pipelining overlaps the execution of the different stages of multiple instructions such as instruction fetch, instruction decode and register fetch, execute, memory access, and register write-back. In superscalar parallelism multiple execution units are used to execute multiple (independent) instructions simultaneously. In order to take advantage of superscalar parallelism, execution instructions might be re-ordered by the system through *out-of-*

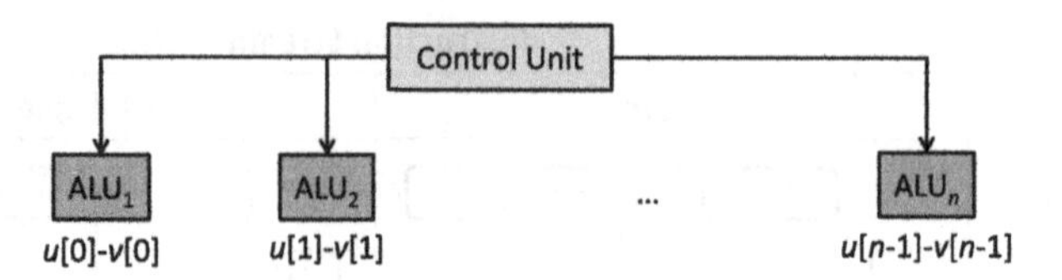

FIGURE 3.8

Example of a SIMD architecture executing a data-parallel subtraction instruction for subtracting two vectors u and v whereby each ALU holds one element of each vector.

order execution (OOE). An important consideration in this context are data dependencies, e.g. two instructions should not be executed in parallel if they are dependent on each other. For more details on ILP we refer the interested reader to the textbooks by Hennessy and Paterson [5] or by Dubois, Annavaran, and Stenström [3].

SIMD CONCEPT

SIMD architectures exploit *data parallelism* by issuing the same instruction to all available PEs or ALUs (*Arithmetic Logical Units*) in each clock cycle. Thus, they only require a single (centralized) control unit in order to dispatch instructions. This often simplifies the design of PEs since they do not need any additional logic for program control.

Let us consider a simple illustrative example. We want to map the following sequential loop for the element-wise subtraction of two vectors u and v onto a SIMD machine:

```
for (i = 0; i < n; i++)
    w[i] = u[i] - v[i];
```

Obviously, all iterations of the loop are independent and regular. It is therefore easily *SIMD-izable*. Consider n ALUs operating under a single control unit as shown in Fig. 3.8. In this case ALU_i $(0 \leq i < n)$ can compute the value $w[i]$ by first loading the values $u[i]$ and $v[i]$ in the internal registers U and V and then simply performing the subtraction instruction $U - V$ in parallel.

However, not all algorithms are SIMD-friendly. Examples include tasks containing conditional statements in the inner loop. Let us for instance introduce an if-then-else statement to our previous example:

```
for (i = 0; i < n; i++)
    if (u[i] > 0)
        w[i] = u[i] - v[i];
    else
        w[i] = u[i] + v[i];
```

In order to map a for-loop containing a conditional statement onto a SIMD architecture we now allow an ALU to *idle* (besides the execution of the instruction broadcasted by the control unit). Using this new feature, we can implement the above if-then-else statement in three stages (as illustrated in Fig. 3.9):

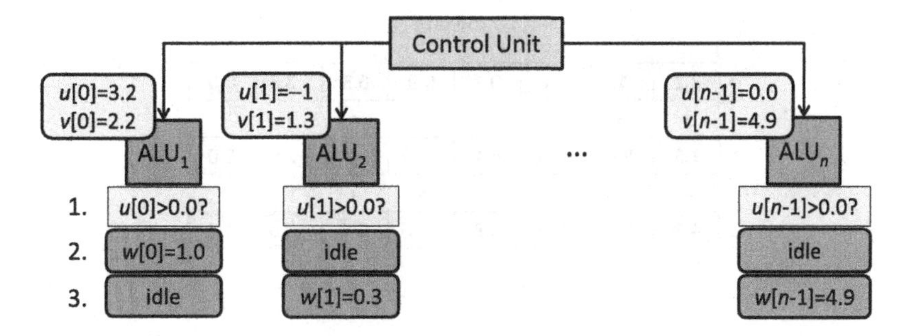

FIGURE 3.9

Example of a SIMD architecture executing a conditional statement for subtracting the corresponding values of two arrays u and v if the u-value is positive and adding them otherwise.

1. Each ALU compares its register U to 0 and sets a corresponding larger flag.
2. The instruction $U - V$ is executed by all ALUs but the result is only stored in W by ALUs with the flag set.
3. The instruction $U + V$ is executed by all ALUs but the result is only stored in W by ALUs with the flag not set.

Overall, frequent usage of conditional statements can significantly reduce the performance of SIMD systems. In our above example the SIMD efficiency is 50% since on average half of the ALUs are idle. In case of nested conditional statements the efficiency reduces further; e.g. to 25% for a doubly nested if-then-else statement.

A number of early massively parallel machines were based on the SIMD concept such as the CM-1/CM-2 of Thinking Machines and the MasPar MP-1/MP-2 [8]. Today's microprocessors usually include SIMD parallelism at a smaller scale. For example, each CPU core typically contains a *vector unit* that can operate a number of data items in parallel (which we discuss in the subsequent subsection). On CUDA-enabled GPUs all threads within a so-called warp operate in SIMD fashion which you will learn about in more detail in Section 7.2.

VECTORIZATION ON COMMON MICROPROCESSORS

The support of SIMD operations in x86-based CPUs started in 1997 with the introduction of the MMX (*Multi Media Extension* – Intel) and 3DNow! (AMD) technologies by incorporating 64-bit wide (vector) registers that could perform arithmetic on packed data types such as two 32-bit integers, four 16-bit integers, or four 8-bit integers. The size of these vector registers subsequently increased over the years. SSE (*Streaming SIMD Extensions* – launched in 1999) added 128-bit wide registers and instructions supporting operations on packed integers as well as on floating-point numbers. Vector register length further increased to 256-bit with AVX (*Advanced Vector Extensions*) in 2011 and to 512-bit with AVX-512 in 2015.

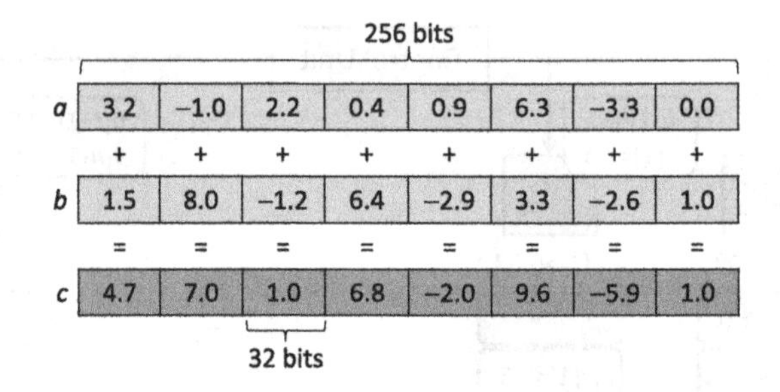

FIGURE 3.10

Parallel addition of two AVX registers holding eight single-precision (32-bit) floating-point numbers each. Each of the eight entries is also called a *vector lane*.

One way to exploit the compute power of the available vector registers is through the use of **intrinsics**. Intrinsics consist of assembly-coded functions[1] and data type definitions that can be used in C and C++ source code. For example, two 256-bit AVX registers can be added as follows.

```
__m256 a,b,c;            // declare AVX registers
...                      // initialize a and b
c = _mm256_add_ps(a,b);  // c[0:8] = a[0:8] + b[0:8]
```

A variable of the `__m256` data type represents a 256-bit long AVX register holding eight 32-bit floating-point values. The AVX intrinsic `_mm256_add_ps` performs an addition in 8-way SIMD fashion as illustrated in Fig. 3.10. Similar intrinsics also exist for other data types and arithmetic operations. The following examples are based on AVX2 technology (an expansion of the AVX instruction set introduced in Haswell microarchitectures) and single-precision floating-point numbers. The applied techniques can be easily transferred to other technologies and data formats by replacing the corresponding intrinsic functions, data types, and offsets. For example, for AVX-512 technology using the `__m512` data type and the `_mm512_add_ps` intrinsic would add 16 single-precision floating-point numbers or `_mm512_add_pd` would add 8 double-precision floating-point numbers. A complete overview of available intrinsics is provided with the Intel C++ Compiler reference [7].

In Listing 3.1 we have shown a C++ program for the matrix multiplication $A \times B = C$ using the matrix dimensions $m \times l$, $l \times n$, and $m \times n$, respectively. Our runtime comparison demonstrated a speedup of the transposed multiplication (including the transposition) over the naive implementation of up to one order-of-magnitude. We now want to further improve the runtime of transposed matrix multiplication on a standard CPU by implementing a SIMD vectorization based on 256-bit registers with AVX2.

```
1  #include <cstdint>     // uint32_t
2  #include <iostream>    // std::cout
3  #include <immintrin.h> // AVX intrinsics
```

[1]Note that not all intrinsic functions directly map to a single assembly instruction, but can be implemented using several assembly instructions. Nevertheless, intrinsics generally provide reliably high performance implementations.

```
4
5  void plain_tmm(float * A,
6                 float * B,
7                 float * C,
8                 uint64_t M,
9                 uint64_t L,
10                uint64_t N) {
11
12     for (uint64_t i = 0; i < M; i++)
13         for (uint64_t j = 0; j < N; j++) {
14             float accum = float(0);
15             for (uint64_t k = 0; k < L; k++)
16                 accum += A[i*L+k]*B[j*L+k];
17             C[i*N+j] = accum;
18         }
19 }
20
21 void avx2_tmm(float * A,
22               float * B,
23               float * C,
24               uint64_t M,
25               uint64_t L,
26               uint64_t N) {
27
28     for (uint64_t i = 0; i < M; i++)
29         for (uint64_t j = 0; j < N; j++) {
30
31             __m256 X = _mm256_setzero_ps();
32             for (uint64_t k = 0; k < L; k += 8) {
33                 const __m256 AV = _mm256_load_ps(A+i*L+k);
34                 const __m256 BV = _mm256_load_ps(B+j*L+k);
35                 X = _mm256_fmadd_ps(AV,BV,X);
36             }
37
38             C[i*N+j] = hsum_avx(X);
39         }
40 }
```

Listing 3.2: Two C++ functions for transposed matrix multiplication: `plain_tmm` (sequential non-vectorized version) and `avx2_tmm` (AVX2-vectorized).

Listing 3.2 shows two C++ functions for the computation of transposed matrix multiplication: `plain_tmm` is a straightforward sequential non-vectorized implementation and `avx2_tmm` is an AVX2-vectorized implementation. We first need to include the `immintrin.h` header file to build an application that uses intrinsics. The inner loop computes the scalar product of row i of A and row j of B. The vectorized function operates on eight values from A and B simultaneously while its non-vectorized counterpart operates only on single elements. The commands `_mm256_load_ps(A+i*L+k)` and `_mm256_load_ps(B+j*L+k)` load eight consecutive single-precision floating-point numbers from the matrices A and B^T into the 256-bit registers `AV` and `BV`. Note that this intrinsic only loads data from *aligned memory addresses*. Thus, we need to guarantee that the two matrices have actually been allocated on 32-byte boundaries. This can for example be realized with the `_mm_malloc` command:

AV	A[i*L]	A[i*L+1]	A[i*L+2]	A[i*L+3]	A[i*L+4]	A[i*L+5]	A[i*L+6]	A[i*L+7]
	*	*	*	*	*	*	*	*
BV	B[j*L]	B[j*L+1]	B[j*L+2]	B[j*L+3]	B[j*L+4]	B[j*L+5]	B[j*L+6]	B[j*L+7]
	+	+	+	+	+	+	+	+
X	X[0]	X[1]	X[2]	X[3]	X[4]	X[5]	X[6]	X[7]
	=	=	=	=	=	=	=	=
X	X[0]	X[1]	X[2]	X[3]	X[4]	X[5]	X[6]	X[7]

FIGURE 3.11

Illustration of the `_mm256_fmadd_ps(AV,BV,X)` intrinsic used in the inner loop of Listing 3.2.

```
auto A = static_cast<float*>(_mm_malloc(M*L*sizeof(float), 32));
auto B = static_cast<float*>(_mm_malloc(N*L*sizeof(float), 32));
```

The intrinsic function `_mm256_fmadd_ps(AV,BV,X)` then multiplies the eight floats of `AV` by the eight floats of `BV` and adds each of them to the values stored in the vector `X` (as illustrated in Fig. 3.11). At the end of the inner loop `X` stores eight partial scalar products. These eight values are then added horizontally with the call to a user-defined function `hsum_avx(X)` (the implementation of which is left as an exercise in Section 3.3) to produce the full scalar product.

Actual execution of our program on an Intel i7-6800K CPU using the matrix dimensions $m = 1024, l = 2048, n = 4096$ produces the following runtimes:

```
# elapsed time (plain_tmm): 12.2992
# elapsed time (avx2_tmm): 2.133s
```

Thus, our AVX2 version achieves a speedup of around 5.8 over the non-vectorized implementation clearly demonstrating the benefits of SIMD computation on a standard CPU. So far only a single core of the CPU is used. Thus, multi-threading can be employed to further reduce the runtime based on MIMD parallelism. For example, parallelizing outer loop of the `avx2_tmm` function using 12 threads on our hyper-threaded 6-core i7-6800K CPU improves the runtime by a factor of 6.7:

```
# elapsed time (avx2_tmm_multi): 0.317763s
```

Concluding, we can achieve an overall speedup of two orders-of-magnitudes over the naive baseline implementation on a consumer-grade CPU if we consider data layout transformations, vectorization, and multi-threading. Note that the implementation of the function `avx2_tmm_multi` uses OpenMP for multi-threading which you will learn about in detail in Chapter 6.

AOS AND SOA

In order to exploit the power of SIMD parallelism it is often necessary to modify the layout of the employed data structures. In this subsection we study two different ways to store a sequence record where each record consists of a (fixed) number of elements:

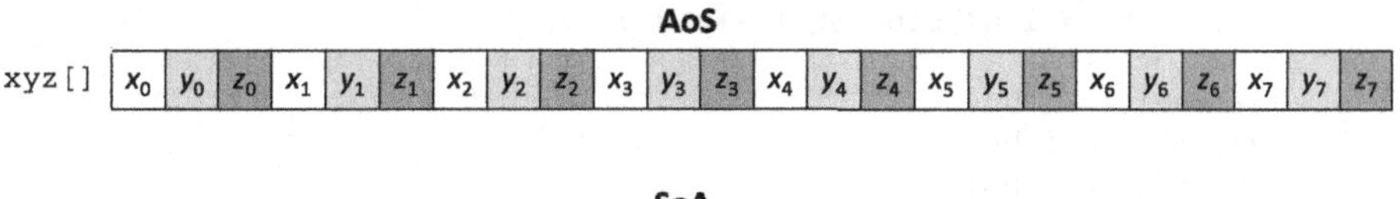

FIGURE 3.12

Comparison of the AoS and the SoA memory layout of a collection of eight 3D vectors: $(x_0, y_0, z_0), \ldots, (x_7, y_7, z_7)$.

- **AOS (Array of Structures)** simply stores the records consecutively in single array.
- **SOA (Structure of Arrays)** uses one array per dimension. Each array only stores the values of the associated element dimension.

As a case study we use a collection of n real-valued 3D vectors (i.e. each vector has x, y, and z coordinates) to compare the SIMD-friendliness of AoS and SoA. A definition of the corresponding AoS would be:

```
auto xyz = new float[3*n];
```

A definition of the corresponding SoA would look like:

```
auto x = new float[n];
auto y = new float[n];
auto z = new float[n];
```

Fig. 3.12 illustrates the memory layout of AoS and SoA for a collection of 3D vectors.

We now want to **normalize** each vector; i.e., we want to map each vector $v_i = (x_i, y_i, z_i)$ to

$$\hat{v}_i = \frac{v_i}{\|v_i\|} = \left(\frac{x_i}{\rho_i}, \frac{y_i}{\rho_i}, \frac{z_i}{\rho_i}\right) \quad \text{where} \quad \rho_i = \sqrt{x_i^2 + y_i^2 + z_i^2} \quad . \tag{3.3}$$

Normalization of vectors is a common operation in computer graphics and computational geometry. Using n 3D vectors stored in the AoS data layout in the array xyz as defined above, this can be performed sequentially in a straightforward non-vectorized way by the following function plain_aos_norm.

```
void plain_aos_norm(float * xyz, uint64_t length) {

    for (uint64_t i = 0; i < 3*length; i += 3) {
        const float x = xyz[i+0];
        const float y = xyz[i+1];
        const float z = xyz[i+2];
```

```
        float irho = 1.0f/std::sqrt(x*x+y*y+z*z);

        xyz[i+0] *= irho;
        xyz[i+1] *= irho;
        xyz[i+2] *= irho;
    }
}
```

Unfortunately, vectorization of 3D vector normalization based on the AoS format would be relatively inefficient because of the following reasons:

1. Vector registers would not be fully occupied; e.g., for a 128-bit register and single-precision floating-point numbers, a single vector would only occupy three of the four available vector lanes.
2. Summing the squares (for the computation of `irho` in `plain_aos_norm`) would require operations between neighboring (horizontal) lanes resulting in only a single value for the inverse square root calculation.
3. Scaling to longer vector registers becomes increasingly inefficient.

On the contrary, SIMD parallelization is more efficient when the 3D vectors are stored in SoA format. The following function `avx_soa_norm` stores the n vectors in the SoA data layout using the three arrays `x`, `y`, and `z` to implement normalization using AVX2 registers.

```
void avx_soa_norm(float * x, float * y, float * z,
                  uint64_t length) {

    for (uint64_t i = 0; i < length; i += 8) {

        // aligned loads
        __m256 X = _mm256_load_ps(x+i);
        __m256 Y = _mm256_load_ps(y+i);
        __m256 Z = _mm256_load_ps(z+i);

        // R <- X*X+Y*Y+Z*Z
         __m256 R = _mm256_fmadd_ps(X, X,
                    _mm256_fmadd_ps(Y, Y,
                    _mm256_mul_ps  (Z, Z)));

        // R <- 1/sqrt(R)
        R = _mm256_rsqrt_ps(R);

        // aligned stores
        _mm256_store_ps(x+i, _mm256_mul_ps(X, R));
        _mm256_store_ps(y+i, _mm256_mul_ps(Y, R));
        _mm256_store_ps(z+i, _mm256_mul_ps(Z, R));
    }
}
```

During each loop iteration, eight vectors are normalized simultaneously. This is made possible by the SoA layout where each lane of the arrays `x[]`, `y[]`, and `z[]` store the corresponding coordinate of a vector as illustrated in Fig. 3.12 leading to an efficient SIMD implementation.

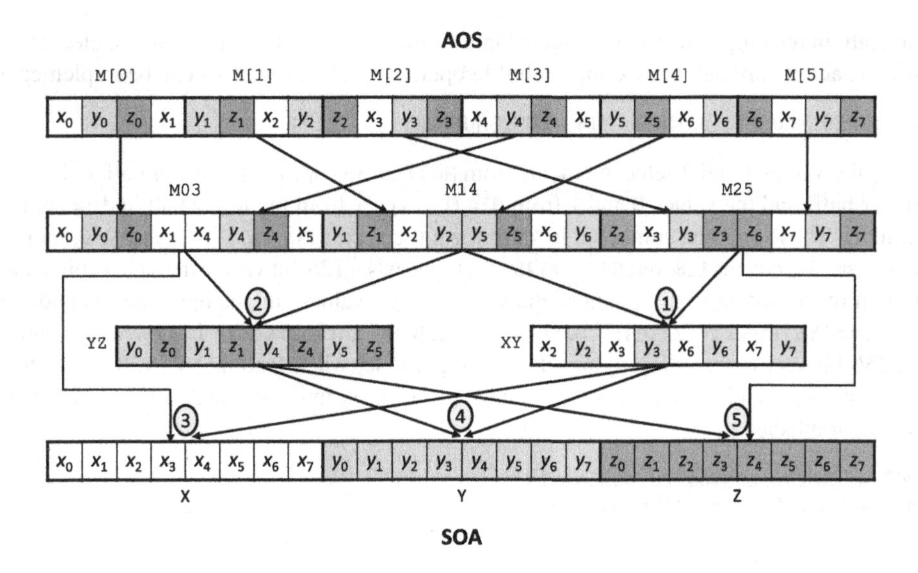

FIGURE 3.13

Transposition of eight 3D vectors: $(x_0, y_0, z_0), \ldots, (x_7, y_7, z_7)$ stored in AoS format into SoA format using 256-bit registers. Indicated register names correspond to the variables used in Listing 3.3. The upper part loads vector elements into the registers `M03`, `M14`, and `M25`. The lower part performs five shuffle operations to store the vectors into the 256-bit registers `X`, `Y`, and `Z`. The indicated number of each shuffle operation corresponds to the order of shuffle intrinsics used in Listing 3.3.

However, certain applications still prefer to arrange their geometric data in the compact AoS format since other operations might benefit from more densely packed vectors. Nevertheless, we still would like to take advantage of the efficient SoA-based SIMD code. In this case, a possible solutions using 256-bit registers could work as follows:

1. Transpose eight consecutive 3D vectors stored in AoS format into SoA format using three 256-bit registers.
2. Perform the vectorized SIMD computation using the SoA format.
3. Transpose the result from SoA back to AoS format.

The transposition of data between AoS and SoA requires a permutation of values. A possible implementation is shown in Fig. 3.13. In order to implement the illustrated data rearrangement from AoS to SoA using AVX2, we will take advantage the following three intrinsics:

- `__m256 _mm256_shuffle_ps(__m256 m1, __m256 m2, const int sel)`: selects elements from `m1` and `m2` according to four 2-bit values (i.e. four numbers between 0 and 3) stored in `sel` to be placed in the output vector. The first two elements of the output vectors are selected from the first four elements of `m1` according to the first two bit-pairs of `sel`. The third and fourth elements are selected from the first four elements of `m2` according to the third and fourth bit-pair of `sel`. The elements five to eight in the output vector are selected in a similar way but choosing from the elements five to eight in `m1` and `m2` instead. For example, shuffle operation "2" in Fig. 3.13 can be implemented by:

```
YZ = _mm256_shuffle_ps(M03, M14, _MM_SHUFFLE(1,0,2,1));
```

whereby the values 1 and 2 select elements from `M03` (i.e. y_0, z_0 from the lower half and y_4, z_4 from the upper half) and the values 0 and 1 from `M14` (i.e. y_1, z_1 from the lower half and y_5, z_5 from the upper half). Those are then combined in `YZ` to form the vector $(y_0, z_0, y_1, z_1, y_4, z_4, y_5, z_5)$.

- `__m256 _mm256_castps128_ps256(__m128 a)`: typecasts a 128-bit vector into a 256-bit vector. The lower half of the output vector contains the source vector values and the upper half is undefined.
- `__m256 _mm256_insertf128_ps(__m256 a, __m128 b, int offset)`: inserts a 128-bit vector into a 256-bit vector according to an offset. For example, when loading two 128-bit vectors `M[0]` and `M[3]` storing the elements x_0, y_0, z_0, x_1 and x_4, y_4, z_4, x_5 into a single 256-bit AVX register, `M03` can be accomplished by:

```
M03 = _mm256_castps128_ps256(M[0]);
M03 = _mm256_insertf128_ps(M03 ,M[3],1);
```

```
1  #include <random>       // prng
2  #include <cstdint>      // uint32_t
3  #include <iostream>     // std::cout
4  #include <immintrin.h> // AVX intrinsics
5
6  // timers distributed with this book
7  #include "../include/hpc_helpers.hpp"
8
9  void aos_init(float * xyz, uint64_t length) {
10
11     std::mt19937 engine(42);
12     std::uniform_real_distribution<float> density(-1, 1);
13
14     for (uint64_t i = 0; i < 3*length; i++)
15         xyz[i] = density(engine);
16 }
17
18 void avx_aos_norm(float * xyz, uint64_t length) {
19
20     for (uint64_t i = 0; i < 3*length; i += 3*8) {
21
22 /////////////////////////////////////////////////////////////////
23 // AOS2SOA: XYZXYZXYZXYZXYZXYZXYZXYZ --> XXXXXXX YYYYYYY ZZZZZZZZ
24 /////////////////////////////////////////////////////////////////
25
26         // registers: NOTE: M is an SSE pointer (length 4)
27         __m128 *M = (__m128*) (xyz+i);
28         __m256 M03, M14, M25;
29
30         // load lower halves
31         M03 = _mm256_castps128_ps256(M[0]);
```

```
32        M14 = _mm256_castps128_ps256(M[1]);
33        M25 = _mm256_castps128_ps256(M[2]);
34
35        // load upper halves
36        M03 = _mm256_insertf128_ps(M03 ,M[3],1);
37        M14 = _mm256_insertf128_ps(M14 ,M[4],1);
38        M25 = _mm256_insertf128_ps(M25 ,M[5],1);
39
40        // everyday I'm shuffling...
41        __m256 XY = _mm256_shuffle_ps(M14, M25,
42                        _MM_SHUFFLE( 2,1,3,2));
43        __m256 YZ = _mm256_shuffle_ps(M03, M14,
44                        _MM_SHUFFLE( 1,0,2,1));
45        __m256 X  = _mm256_shuffle_ps(M03, XY ,
46                        _MM_SHUFFLE( 2,0,3,0));
47        __m256 Y  = _mm256_shuffle_ps(YZ , XY ,
48                        _MM_SHUFFLE( 3,1,2,0));
49        __m256 Z  = _mm256_shuffle_ps(YZ , M25,
50                        _MM_SHUFFLE( 3,0,3,1));
51
52 ////////////////////////////////////////////////////////////////
53 // SOA computation
54 ///////////////////////////////////////////////////////////////
55
56        // R <- X*X+Y*Y+Z*Z
57        __m256 R = _mm256_fmadd_ps(X, X,
58                   _mm256_fmadd_ps(Y, Y,
59                   _mm256_mul_ps  (Z, Z)));
60
61        // R <- 1/sqrt(R)
62        R = _mm256_rsqrt_ps(R);
63
64        // normalize vectors
65        X = _mm256_mul_ps(X, R);
66        Y = _mm256_mul_ps(Y, R);
67        Z = _mm256_mul_ps(Z, R);
68
69 /////////////////////////////////////////////////////////////////
70 // SOA2AOS: XXXXXXX YYYYYYY ZZZZZZZZ -> XYZXYZXYZXYZXYZXYZXYZXYZ
71 /////////////////////////////////////////////////////////////////
72
73        // everyday I'm shuffling...
74        __m256 RXY = _mm256_shuffle_ps(X,Y,
75                      _MM_SHUFFLE(2,0,2,0));
76        __m256 RYZ = _mm256_shuffle_ps(Y,Z,
77                      _MM_SHUFFLE(3,1,3,1));
78        __m256 RZX = _mm256_shuffle_ps(Z,X,
79                      _MM_SHUFFLE(3,1,2,0));
80        __m256 R03 = _mm256_shuffle_ps(RXY, RZX,
81                      _MM_SHUFFLE(2,0,2,0));
82        __m256 R14 = _mm256_shuffle_ps(RYZ, RXY,
83                      _MM_SHUFFLE(3,1,2,0));
```

```
84            __m256 R25 = _mm256_shuffle_ps(RZX, RYZ,
85                                 _MM_SHUFFLE(3,1,3,1));
86
87            // store in AOS (6*4=24)
88            M[0] = _mm256_castps256_ps128(R03);
89            M[1] = _mm256_castps256_ps128(R14);
90            M[2] = _mm256_castps256_ps128(R25);
91            M[3] = _mm256_extractf128_ps(R03, 1);
92            M[4] = _mm256_extractf128_ps(R14, 1);
93            M[5] = _mm256_extractf128_ps(R25, 1);
94        }
95    }
96
97    int main () {
98        const uint64_t num_vectors = 1UL << 28;
99        const uint64_t num_bytes = 3*num_vectors*sizeof(float);
100
101       auto xyz = static_cast<float*>(_mm_malloc(num_bytes , 32));
102
103       aos_init(xyz, num_vectors);
104
105       TIMERSTART(avx_aos_normalize)
106       avx_aos_norm(xyz, num_vectors);
107       TIMERSTOP(avx_aos_normalize)
108
109       _mm_free(xyz);
110   }
```

Listing 3.3: Vectorized normalization of an array of 3D vectors stored in AoS format by on-the-fly transposition into SoA format and inverse transposition of the results.

The code for vectorized normalization of an array of 3D vectors stored in AoS format using transposition into SoA format is shown in Listing 3.3. Our solution transposes eight subsequent 3D vectors at a time. The corresponding function `avx_aos_norm` consists of three stages: AoS2SoA, SoA computation, and SoA2AoS. The AoS2SoA stage starts by loading pairs of four subsequent vector elements into three 256-bit registers using intrinsics `_mm256_castps128_ps256` and `_mm256_insertf128_ps` as described above and as illustrated in the upper part of Fig. 3.13. Subsequently, we apply five `_mm256_shuffle_ps` operations to implement the necessary shuffling of vector elements as illustrated in the lower part of Fig. 3.13. The efficiently vectorized SoA computation can then proceed in the way we have studied earlier. Since the result is stored in SoA, it needs to be transposed back to AoS format. This is implemented by the six corresponding shuffling operations in the SoA2AoS part of the function.

Actual execution of our AVX program and the corresponding plain AoS program on an Intel i7-6800K CPU using $n = 2^{28}$ vectors produces the following runtimes:

```
# elapsed time (plain_aos_normalize): 0.718698s
# elapsed time (avx_aos_normalize): 0.327667s
```

We can see that despite the transposition overhead, the vectorized implementation can still achieve a speedup of around 2.2.

OUTLOOK

In this subsection we have learned that data layout and associated transformations can be crucial to enable the power of SIMD processing. Some further examples are stencil codes [6] and molecular dynamics simulations [9]. We have also included data layout transformation for a 1D Jacobi stencil as an exercise in Section 3.3.

To achieve even higher performance vector intrinsics can be mixed with multi-threaded code. We will study how AVX can be combined with OpenMP to implement a highly efficient softmax regression classifier in Section 6.5.

Besides the usage of intrinsics, more higher-level language constructs are becoming increasingly popular for SIMD programming. One example is the SIMD pragma of OpenMP which we will learn about in Section 6.7. Furthermore, warp intrinsics in CUDA studied in Section 8.1 are based on the SIMD concept.

3.3 ADDITIONAL EXERCISES

1. **Performance of Convolution.** Consider a CPU with a single core operating at 2.5 GHz that can perform eight single-precision floating-point operations per clock cycle. The CPU is connected to a DRAM module with a peak memory transfer rate of 25.6 GB/s. It has a cache of size 256 KB operating at the same speed as a register. We want to compute the convolution of a 2D image I of size $n \times n$ stored in main memory with a mask M of size $(2k + 1) \times (2k + 1)$; i.e., we want to compute an image C of size $n \times n$ with

$$C[x][y] = \sum_{i=-k}^{k} \sum_{j=-k}^{k} I[x+i][y+j] \times M[k-i][k-j] \text{ for all } 0 \leq x, y \leq n-1 .$$

In the above equation, we set $I[x+i][y+j] = 0$ if $x+i < 0$ or $x+i \geq n$ or $y+j < 0$ or $y+j \geq n$.

   (i) Establish an upper bound on the achievable performance of this system to compute a convolution of a real-valued image I of size 256×256 with a mask M of size 1×1.

   (ii) Do the same for a mask M of size 5×5.

   (iii) What would be a good caching strategy if I is many times bigger than the cache?

2. **Cache Lines and Vectorization.** Consider a CPU with a cache line of length `L = sizeof(float)*16` and two square matrices $A, B \in$ `float`$^{N \times N}$, each of length $N = 1024$. In the following we want to calculate $C = A + B$ by the index-wise processing of the entries $C_{ij} = A_{ij} + B_{ij}$ for all $i, j \in \{0, \ldots, N-1\}$.

   (i) Calculate the number of total cache misses in the function `row_wise_add`:

```
// row-major-order addition
void row_wise_add(float * A, float * B,
                  float * C, size_t  N) {

    for (int i = 0; i < N; i++)
        for (int j = 0; j < N; j++)
            C[i*N+j] = A[i*N+j] + B[i*N+j];

}
```

(ii) Do the same for the function `col_wise_add` and compare the result to (i). Which function do you believe to perform faster and why?

```
// col-major-order addition
void col_wise_add(float * A, float * B,
                  float * C, size_t  N) {

    for (int j = 0; j < N; j++)
        for (int i = 0; i < N; i++)
            C[i*N+j] = A[i*N+j] + B[i*N+j];
}
```

(iii) Using the intrinsic `_mm256_load_ps` you can load eight `float` values at once from an aligned array[2] and store them in a local variable `__m256 tmp`. The equivalent function `_mm256_store_ps` is used to write `tmp` back to the original array. Find the appropriate command for the addition of two variables of type `__m256`. Afterwards, implement an AVX-vectorized variant of `row_wise_add`.

(iv) Compare the runtimes obtained in (i–iii). Do they comply with your expectations?

3. **Tiled Matrix Multiplication.** We are considering again the matrix multiplication $A \cdot B = C$ using the matrix dimensions $m \times l$, $l \times n$, and $m \times n$, respectively. We have learned in this chapter that the straightforward implementation suffers from poor performance due to cache misses:

```
void mat_mult_naive(float * A, float * B, float * C,
                    size_t m, size_t l, size_t n) {
    for (uint64_t i = 0; i < m; i++)
        for (uint64_t j = 0; j < n; j++) {
            float accum = 0;
            for (uint64_t k = 0; k < l; k++)
                accum += A[i*l+k]*B[k*n+j];
            C[i*n+j] = accum;
        }
}
```

An approach to reduce cache misses is to partition the matrices A and B into small rectangular tiles. Multiplication of two of those tiles can be performed in the cache thereby exploiting data reuse and spatial locality. Modify the `mat_mult_naive` function to a cache-friendly implementation of tiled matrix multiplication.

4. **False Sharing, Speedup and Efficiency.** Consider again the addition of two square matrices $C = A + B$. We would like to utilize P processors in parallel. Get an overview of the very basics of the C++ threading model introduced in Section 4.1.

(i) The function `false_sharing_add` in `false_sharing.cpp`[3] spawns P threads and for each of them calls the method `add_interleaved` to calculate the sum. Explain in detail why false sharing occurs. The associated excerpt is given as follows:

---

[2]Make sure that you know what aligned memory means.

[3]Provided at the supplementary website of this book (https://parallelprogrammingbook.org).

```
// adds entries (threadID, threadID+P, ...)
void add_interleaved(float * A, float * B, float * C,
                     size_t  N, size_t  P, size_t ID) {

    for (int i = 0; i < N; i++)
        for (int j = ID; j < N; j += P)
            C[i*N+j] = A[i*N+j] + B[i*N+j];

}

// spawns P threads and calls add_interleaved
// for each threadID
void false_sharing_add(float * A, float * B, float * C,
                       size_t  N, size_t  P) {

    std::thread threads[P];

    for (int i = 0; i < P; i++)
        threads[i] = std::thread(add_interleaved,
                                 A, B, C, N, P, i);

    for (int i = 0; i < P; i++)
        threads[i].join();
}
```

(ii) Write a new function `coalesced_mem_add` which calculates the sum $C = A + B$ in batches of size $\frac{N}{P}$ using only coalesced memory accesses.

(iii) Measure the execution times for different values of $p \in \{1, ..., P\}$. Calculate the gained speedups and efficiencies in dependency on the number of used processors p. How many GFlop/s do you gain if you assume the $+$ to be a single floating-point operation? Discuss the results.

(iv) This task is optional: Use vectorization to improve the gained runtimes.

5. Loop Unrolling for Max Reduction. The following (sequential) function computes the maximum of a given array of floating-point numbers:

```
float plain_max(float * data,
                uint64_t length) {

    float max = -INFINITY;

    for (uint64_t i = 0; i < length; i++)
        max = std::max(max, data[i]);

    return max;
}
```

The functions `plain_max_unroll_2` and `plain_max_unroll_4` perform the same computation but unroll the loop by a factor of 2 and 4, respectively:

```
float plain_max_unroll_2(float * data,
                         uint64_t length) {
```

```
    float max_0 = -INFINITY;
    float max_1 = -INFINITY;

    for (uint64_t i = 0; i < length; i += 2) {
        max_0 = std::max(max_0, data[i+0]);
        max_1 = std::max(max_1, data[i+1]);
    }

    return std::max(max_0, max_1);
}

float plain_max_unroll_4(float * data,
                         uint64_t length) {

    float max_0 = -INFINITY;
    float max_1 = -INFINITY;
    float max_2 = -INFINITY;
    float max_3 = -INFINITY;

    for (uint64_t i = 0; i < length; i += 4) {
        max_0 = std::max(max_0, data[i+0]);
        max_1 = std::max(max_1, data[i+1]);
        max_2 = std::max(max_2, data[i+2]);
        max_3 = std::max(max_3, data[i+3]);
    }

    return std::max(max_0, std::max(max_1,
                    std::max(max_2, max_3)));
}
```

Executing these functions on a Intel i7-6800K CPU using a vector of size 2^{28} produces the following runtimes:

```
# elapsed time (plain_max): 0.230299s
# elapsed time (plain_max_unroll_2): 0.116835s
# elapsed time (plain_max_unroll_4): 0.0856038s
```

Explain why the unrolled functions are significantly faster.

6. **Horizontal Summation in SIMD.** Our transposed matrix multiplication implementation in Listing 3.2 contains the function call `hsum_avx(X)` to sum up all eight vector elements stored in the 256-bit register `X`. Provide an implementation of `hsum_avx(X)` using (a combination of AVX and SSE) intrinsics.
7. **Naming Convention of Intrinsic Functions.** The naming of intrinsic functions (such as `_mm256_set_pd` or `_mm512_sll_epi32`) can be a bit confusing at first. Describe the naming convention of intrinsic functions in a way that makes it easy to (approximately) judge what an intrinsic does by just looking at its name.
8. **Vectorized Jacobi 1D Stencil.** Jacobi iteration is an important method with a variety of applications. A simple 1D Jacobi 3-point stencil can be implemented by the following code fragment:

```
for (uint64_t t = 0; t < T; t++) {
    for (uint64_t i = 1; i < N-1; i++) {
        B[i] = 0.33*(A[i-1] + A[i] + A[i+1]);
    for (uint64_t i = 1; i < N-1; i++)
        A[i] = B[i];
}
```

The first inner loop performs the actual stencil computation while the second inner loop simply copies the output array into the input array for the subsequent iteration.

(i) Analyze the data dependencies in the sequential code.
(ii) Propose a *data layout transformation* that allows for an efficient vectorized implementation.
(iii) Which intrinsics are required for your vectorized implementation?

REFERENCES

[1] L. Susan Blackford, et al., An updated set of basic linear algebra subprograms (BLAS), ACM Transactions on Mathematical Software 28 (2) (2002) 135–151.

[2] Ulrich Drepper, What every programmer should know about memory, http://people.redhat.com/drepper/cpumemory.pdf, 2007.

[3] Michel Dubois, Murali Annavaram, Per Stenström, Parallel Computer Organization and Design, Cambridge University Press, 2012.

[4] Michael J. Flynn, Some computer organizations and their effectiveness, IEEE Transactions on Computers 100 (9) (1972) 948–960.

[5] John L. Hennessy, David A. Patterson, Computer Architecture: A Quantitative Approach, Elsevier, 2011.

[6] Tom Henretty, et al., A stencil compiler for short-vector SIMD architectures, in: Proceedings of the 27th International ACM Conference on International Conference on Supercomputing, ACM, 2013, pp. 13–24.

[7] Intel, Intel C++ compiler 17.0 developer guide and reference, https://software.intel.com/en-us/intel-cplusplus-compiler-17.0-user-and-reference-guide (visited on 10/24/2016).

[8] Neil B. MacDonald, An overview of SIMD parallel systems–AMT DAP, Thinking Machines CM-200, & MasPar MP-1, in: The Workshop on Parallel Computing, Quaid-i-Azam University, Islamabad, Pakistan, 26th–30th April 1992, Citeseer, 1992.

[9] Simon J. Pennycook, et al., Exploring SIMD for molecular dynamics, using Intel® Xeon® processors and Intel® Xeon Phi coprocessors, in: Parallel & Distributed Processing (IPDPS), 2013 IEEE 27th International Symposium on, IEEE, 2013, pp. 1085–1097.

[10] Patrice Quinton, Yves Robert, Systolic Algorithms & Architectures, Prentice Hall, 1991.

CHAPTER

C++11 MULTITHREADING

4

Abstract

During the last decade, single-core performance of modern CPUs has been stagnating due to hard architectural limitations of silicon-based semiconductors. On the one hand, the historic shrinking of the manufacturing process of three orders-of-magnitude from a few micrometers to a few nanometers cannot proceed at arbitrary pace in the future. This can be explained by emerging delocalization effects of quantum mechanics which governs physics at the nanometer scale. On the other hand, the power dissipation of integrated circuits exhibits a quadratic dependency on the voltage and a linear dependency on the frequency. Thus, significant up-scaling of the CPU's frequency is prohibitive in terms of energy consumption. As a result, we cannot expect single-threaded programs to run faster on future hardware without considering parallelization over multiple processing units. The free lunch is over, so to speak. Future improvements in runtime have to be accomplished by using several processing units.
Historically, there have been several C- and C++-based libraries that support multithreading over multiple CPU cores. POSIX Threads, or short PThreads, have been the predominant implementation over several decades in the Linux/UNIX world. Some Windows versions featured a POSIX-compatible layer which has been deprecated in newer Windows versions in favor of Microsoft's homebrew threading API. Intel's Threading Building Blocks (TBB) is another popular implementation. This heterogeneous software landscape made it difficult to write platform-portable code in C or C++. With the release of C++11 and its novel threading API it is finally possible to write platform-independent code in C++ that is supported by compilers from both the Linux/UNIX world and the Windows ecosystem without the need for third party libraries like Intel TBB. Thus, our approach to multithreading in this chapter is based on the modern C++11 and C++14 dialects of the C++ language.

Keywords

Multithreading, C++, C++11, Thread spawning, Race condition, Promise, Future, Deadlock, Task parallelism, Asynchronous task, Static scheduling, Matrix vector multiplication, Thread distribution, Closure, False sharing, Cache line, Load balancing, Dynamic scheduling, Mutex, Condition variable

CONTENTS

Parallel Programming. DOI: 10.1016/B978-0-12-849890-3.00004-6

4.1 INTRODUCTION TO MULTITHREADING (HELLO WORLD)

In this chapter, you will learn how to write multithreaded programs in the modern C++11 and C++14 dialects of the C++ programming language. This includes basic applications based on trivially parallel computation patterns, asynchronous task parallelism but also examples using advanced synchronization and inter-thread signaling techniques.

DISTINCTION BETWEEN MULTITHREADING AND MULTIPROCESSING

Before we start coding let us briefly summarize the fundamental concepts of *multithreading* and discuss differences from a related paradigm called *multiprocessing*. The historical distinction between the multithreading and multiprocessing paradigms on a hardware level can be summarized as follows:

- **Multiprocessing** parallelizes a program over multiple compute units (e.g. CPU cores) in order to exploit redundant resources such as registers and arithmetic logic units (ALUs) of different CPU cores to speed up computation.
- **Multithreading** shares the hardware resources such as caches and RAM of a single core or multiple cores in order to avoid idling of unused resources.

The two above stated definitions are not mutually exclusive which is often a major cause of confusion. A multithreaded program can but does not necessarily need to utilize distinct CPU cores and thus, depending on the situation, may also obey the definition of multiprocessing. Moreover, multiprocessing does not explicitly exclude the use of shared resources and consequently could be implemented within a multithreading scenario but could also employ heavy-weight processes communication over sockets or message passing interfaces. To complicate matters, some authors exclusively use the term *(hardware) thread* in case a CPU core exhibits redundant processing pipelines for each thread, and *hyperthread*, as coined by Intel, when a core has the capability to process independent tasks in the same processing pipeline based on sophisticated scheduling strategies.

The good news from a programming point of view is that we do not have to care about these technical details since modern operation systems (OSs) treat hardware threads and hyperthreads executed on the same CPU core, as well as threads on different CPU cores, equally. As an example, a Xeon E5-2683 v4 CPU provides 16 physical cores with hyperthreading capability which are treated by the OS as 32 independent cores. That does not imply that we can expect a 32× speedup but we do not need to artificially distinguish between hardware threads and hyperthreads during programming.

SPAWNING AND JOINING THREADS

An arbitrary number of software threads can be spawned by the master thread of a system process. It is even possible to recursively spawn threads from within already spawned ones. The actual number of concurrently running threads should be adjusted to roughly match the amount of physical cores of your system since the OS might serialize their execution using expensive context switches if their number exceeds the amount of available cores. This behavior is called *oversubscription* and should be avoided to prevent performance degradation.

All threads share the resources of the parent system process, i.e., they can access the same memory space. This is advantageous since threads can be spawned with low latency and benefit from lightweight inter-thread communication using shared registers and arrays. A drawback of threads is that a thread could easily spy on the data of another thread. As an example, consider we wanted to program a web browser with support for distinct tabs. When using threads for different tabs, a malicious website or plugin could access sensitive information of another tab or even crash the whole application. Consequently, one should use distinct system processes with independent memory spaces to implement security critical applications as realized by the popular Google Chrome browser. Concluding, threads are a lightweight shared memory mechanism for the concurrent execution within the memory space of a single system process in contrast to independent system processes acting on distributed memory and communicating over heavy-weight channels such as sockets. Process-based parallelization approaches are discussed in detail in Chapter 9 presenting the Message Passing Interface (MPI) and Chapter 10 dealing with Unified Parallel C++.

After having spawned a number of threads, you might ask how to terminate them in an appropriate manner. The instruction flow of the master thread continues independently of the work accomplished in the spawned threads until we reach the end of the main function. In order to ensure that all spawned threads have finished their work, we have to wait for them in the master thread. This is accomplished with a call to join. Alternatively, one can detach threads, i.e., the master process kills them during termination without waiting for them. The latter has to be taken with a grain of salt since we do not get any guarantee that a detached thread has finished its work. In the worst case, you might end up with partially written files or incomplete messages sent over the network. Altogether, we have to care about four things:

1. Each thread can only be joined or detached once.
2. A detached thread cannot be joined, and vice versa.

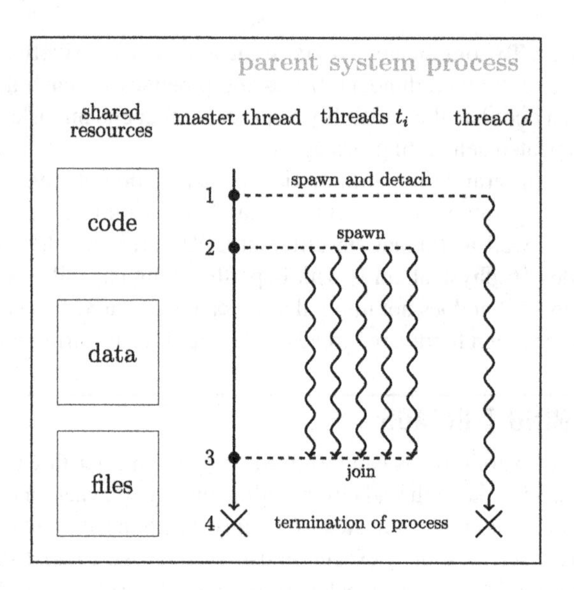

FIGURE 4.1

Exemplary workflow of a multithreaded program. First, we spawn a thread d and detach it immediately. This thread executes some work until finished or being terminated at the end of our program (4). Second, five threads t_i are spawned concurrently processing some data. Third, the master thread waits for them to finish by joining them. Fourth, we reach the end of the program resulting in the termination of the master thread and all detached threads. Code, data, and file handles are shared among all threads within their respective scopes.

3. Joined or detached threads cannot be reused.
4. All threads have to be joined or detached within the scope of their declaration.

Ignoring the aforementioned rules can result in premature termination of your program. Concerning the last rule, you have to be aware that there is no implicit joining or detaching of threads when leaving their scope of declaration. Fig. 4.1 shows a typical workflow performing the spawning, joining, and detaching of threads.

Let us make a final remark on detached threads. In practice, there is a lack of good use cases for detached threads without non-trivial synchronization. Some authors argue [4] that threads should always be detached since it might be difficult to track them along all possible execution paths in order to finally join them. Another argument for detached threads is that non-detached threads are kept alive by the OS until someone joins them which may complicate performance analysis. However, the examples in this book exclusively spawn a fixed number of threads which are easy to track and join. An actual use case for a detached thread scenario without any synchronization is a thread that monitors our application during runtime and occasionally writes some output to a log file. Nevertheless, we would have to deal with potentially incomplete log entries that have to be handled later by an error-aware parser. Putting it short, the naive use of detached threads effectively delegates your problems to someone else – in the worst case: the future you. In both cases, joinable and detached threads, the programmer has to either carefully maintain a list of spawned threads to be joined or explicitly implement synchronization mechanisms to avoid incomplete transactions to attached media.

OUR FIRST MULTITHREADED PROGRAM

Let us start coding. Our Hello World example in Listing 4.1 spawns four threads each printing a greeting message to the command line.

```
1  #include <cstdint>    // uint64_t
2  #include <vector>     // std::vector
3  #include <thread>     // std::thread
4
5  // this function will be called by the threads (should be void)
6  void say_hello(uint64_t id) {
7      std::cout << "Hello from thread: "  << id << std::endl;
8  }
9
10 // this runs in the master thread
11 int main(int argc, char * argv[]) {
12
13     const uint64_t num_threads = 4;
14     std::vector<std::thread> threads;
15
16     // for all threads
17     for (uint64_t id = 0; id < num_threads; id++)
18         // emplace the thread object in vector threads
19         // using argument forwarding, this avoids unnecessary
20         // move operations to the vector after thread creation
21         threads.emplace_back(
22             // call say_hello with argument id
23             say_hello, id
24         );
25
26     // join each thread at the end
27     for (auto& thread: threads)
28         thread.join();
29 }
```

Listing 4.1: Multithreaded Hello World example using four threads.

The above stated source code is straightforward. Initially, we reserve some memory for the thread handles. This is accomplished in our example with an empty `std::vector` from the standard library which maintains `std::thread` objects (see Line 14). Afterwards, we spawn `num_threads` many threads each executing the method `say_hello` with the argument `id` (Lines 21–24) and subsequently store them in the vector `threads`. Alternatively, we could have moved the thread objects implicitly using the member function `push_back` of the vector `threads`:

```
threads.push_back(std::thread(say_hello, id));
```

We need to store the thread handles explicitly in order to be able to access them a second time during the join phase in Line 28. The program can be compiled on the command line with a C++11 compliant compiler (here GCC 5.4.0):

```
g++ -O2 -std=c++11 -pthread hello_world.cpp -o hello_world
```

The compiler flag `-O2` activates standard code optimization, `-std=c++11` enables support for C++11, and `-pthread` adds support for multithreading based on the PThreads library. Remarkably, the C++11 multithreading API in the Linux/Unix ecosystem is just a wrapper of the traditional PThreads library. As a result, we can write modern C++ compliant code without sacrificing performance. The output of our program is four greeting messages printed in potentially random order:

```
Hello from thread: 3
Hello from thread: 1
Hello from thread: 0
Hello from thread: 2
```

Finally, let us briefly discuss the used data structures and methods. The thread handles could have been stored alternatively in a traditional dynamic array

```
std::thread * threads = new std::thread[num_threads];
```

which has to be manually freed using `delete [] threads;` at the end of the program. An advantage of `std::vector` is that its destructor is called automatically when leaving the associated scope, thus relieving us from the burden of the manual freeing of memory in order to avoid memory leaks. The constructor of the std::thread class accepts an arbitrary number of arguments. The first argument corresponds to the called function which should return `void` since `std::thread` provides no mechanism to directly access return values. The remaining ones enumerate the arguments of the called function. Note that we have to explicitly specify potential template parameters of function templates since they cannot be deduced automatically from the argument list at compile time. Assume we implement a function template `say_hello<T>` which accepts arbitrary integer types for the thread identifier:

```
template <typename index_t>
void say_hello(index_t id) {
    std::cout << "Hello from thread: "  << id << std::endl;
}
```

The template argument of `say_hello` has now to be specified explicitly during the thread launch:

```
std::thread my_thread_handle(say_hello<uint64_t>, id);
```

In the final loop in Line 28 we join all spawned threads. The correct data type of the thread objects is determined automatically. This is achieved with the `auto` keyword which infers the corresponding element type (std::thread) from the container object `std::vector<std::thread> threads` on the right-hand side. Since objects of type `std::thread` are move-only (i.e. not copyable) we have to use the reference type `auto&`. Nevertheless, threads can be moved using `std::move` as follows:

```
std::thread yang(some_function);
auto ying = std::move(yang);
ying.join();
```

Note that after having accomplished the move from `yang` to `ying` we cannot safely access `yang` anymore. The naive copy `auto ying = yang;` results in a compile time failure. However, declaring a reference `auto& ying = yang;` is an allowed operation. In the latter case, `ying` and `yang` refer to the same object. A constant reference `const auto& ying = yang;` is not advisable since `join` is not a `const` member function of `std::thread`.

4.2 HANDLING RETURN VALUES (FIBONACCI SEQUENCE)

Threads can execute functions with arbitrary arguments and return values. However, thread objects do not offer a straightforward way to access the return value. This might be an acceptable behavior in our Hello World example but limits our code to fire-and-forget scenarios where we launch a certain amount of threads that execute some work without providing any feedback. This section discusses different approaches that are applicable to functions with return values. For the sake of simplicity we chose a basic scalar function that iteratively computes the n-th Fibonacci number. The sequence of Fibonacci numbers is recursively defined by the implicit equation

$$a_n = a_{n-1} + a_{n-2} \quad \text{with initial conditions} \quad a_0 = 0, a_1 = 1 \quad . \tag{4.1}$$

The sequence is evaluated stepwise starting from $n = 0$ producing the well-known series $(0, 1, 1, 2, 3, 5, 8, 13, \ldots)$. Individual entries of the sequence a_n could be computed more efficiently based on a square-and-multiply approach in $\mathcal{O}(\log(n))$ time or even $\mathcal{O}(1)$ when using floating-point arithmetic. Fibonacci numbers and the closely related *golden ratio* $\Phi = \lim_{n\to\infty} \frac{a_n}{a_{n-1}} = \frac{1+\sqrt{5}}{2}$ are ubiquitous in informatics and popular science. They can be used for the construction of hash functions, the analysis of AVL tree balancing but also influenced architecture, art, and music. A corresponding linear time implementation written in the C++ language is shown in Listing 4.2.

```
1  template <
2      typename value_t,
3      typename index_t>
4  value_t fibo(
5      value_t n) {
6
7      // initial conditions
8      value_t a_0 = 0;
9      value_t a_1 = 1;
10
11     // iteratively compute the sequence
12     for (index_t index = 0; index < n; index++) {
13         const value_t tmp = a_0; a_0 = a_1; a_1 += tmp;
14     }
15
16     return a_0;
17 }
```

Listing 4.2: A basic function template iteratively computing the n-th Fibonacci number.

In the following, we concurrently compute Fibonacci numbers by spawning a thread for each number and subsequently communicate the result back to the master thread.

THE TRADITIONAL WAY

The traditional error handling model in the C programming language reserves the return value of a function for the error code. An example is the `main` function which returns an integer indicating whether it terminated successfully or unsuccessfully. Hence, other computed quantities are usually passed via pointers in the argument list which are subsequently manipulated inside the function body. A similar approach is applicable to functions called by threads: we simply pass a pointer to the result value and write the computed value to the associated address. The corresponding implementation is shown in Listing 4.3.

```
#include <iostream>       // std::cout
#include <cstdint>        // uint64_t
#include <vector>         // std::vector
#include <thread>         // std::thread

template <
    typename value_t,
    typename index_t>
void fibo(
    value_t n,
    value_t * result) {   // <- here we pass the address

    value_t a_0 = 0;
    value_t a_1 = 1;

    for (index_t index = 0; index < n; index++) {
        const value_t tmp = a_0; a_0 = a_1; a_1 += tmp;
    }

    *result = a_0;        // <- here we write the result
}

// this runs in the master thread
int main(int argc, char * argv[]) {

    const uint64_t num_threads = 32;
    std::vector<std::thread> threads;

    // allocate num_threads many result values
    std::vector<uint64_t> results(num_threads, 0);

    for (uint64_t id = 0; id < num_threads; id++)
        threads.emplace_back(
            // specify template parameters and arguments
            fibo<uint64_t, uint64_t>, id, &(results[id])
        );

    // join the threads
    for (auto& thread: threads)
        thread.join();

    // print the result
    for (const auto& result: results)
        std::cout << result << std::endl;
}
```

Listing 4.3: Passing return values the traditional way.

Let us discuss the source code. First, we reserve storage for `num_threads` elements by declaring a `std::vector` in Line 30. Second, the address of each element in the vector is passed during the spawning of threads (see Line 35). Finally, we adjust the value stored at the address of the pointer (Line 11) with the computed result (Line 20) in the `fibo` function.

Note that the communication over pointers is possible since all threads share the same memory space in a multithreading scenario. Nevertheless, you have to be aware of a potential pitfall: the memory passed via a pointer in the first for-loop (Line 38) has to be persistent during the execution of threads. In contrast, a variable or object defined within the body of the first for-loop (Line 33) would be destroyed immediately after each iteration since we leave its scope. Hence, the thread would operate on potentially freed memory resulting in a segmentation fault. Moreover, we have to ensure that we do not alter the result values from within the master thread during thread execution to avoid potential race conditions. Concluding, you have to guarantee that the objects manipulated in threads remain existent during execution and that there are no data races on shared resources.

THE MODERN WAY USING PROMISES AND FUTURES

C++11 provides a mechanism for return value passing specifically designed to fit the characteristics of asynchronous execution. The programmer may define so-called *promises* that are *fulfilled* in the *future*. This is achieved with a pair of tied objects $s = (p, f)$ where p is a writable view of the state s, the promise, which can be set to a specific value. This signaling step can only be accomplished once and thus is called *fulfilling the promise*. The object f, the future, is a readable view of the state s that can be accessed after being signaled by the promise. Hence, we establish a causal dependency between the promise p and the future f that can be used as synchronization mechanism between a spawned thread and the calling master thread. The overall workflow can be described as follows:

1. We create the state $s = (p, f)$ by initially declaring a promise p for a specific data type `T` via `std::promise<T> p;` and subsequently assign the associated future with `std::future<T> f = p.get_future();`.
2. The promise p is passed as rvalue reference in the signature of the called function via `std::promise<T> && p`. Hence, p has to be moved using `std::move()` from the master thread to the spawned thread.
3. The promise p is fulfilled within the body of the spawned thread by setting the corresponding value with `p.set_value(some_value);`.
4. Finally, we can read the future f in the master thread using `f.get()`. The master thread blocks its execution until f is being signaled by p.

Keep in mind that the state $s = (p, f)$ establishes a non-trivial relationship between the spawned thread and the master thread since both share different views p and f to the same object s. Consequently, you could deadlock your program when trying to read the future f without ever fulfilling the promise p (see Fig. 4.2). The corresponding implementation of our Fibonacci example using promises and futures is shown in Listing 4.4. Note that promises and futures are defined in the `future` header.

```
1  #include <iostream>                  // std::cout
2  #include <cstdint>                   // uint64_t
3  #include <vector>                    // std::vector
```

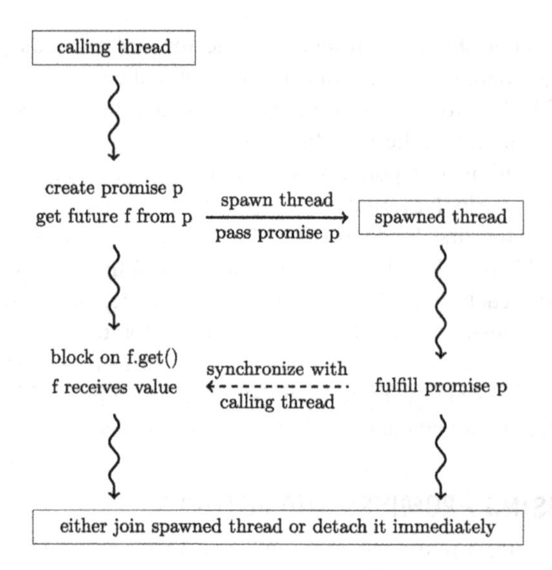

FIGURE 4.2

Synchronization of two threads using a promise and its associated future.

```
 4  #include <thread>                         // std::thread
 5  #include <future>                         // std::promise/future
 6
 7  template <
 8      typename value_t,
 9      typename index_t>
10  void fibo(
11      value_t n,
12      std::promise<value_t> && result) {  // <- pass promise
13
14      value_t a_0 = 0;
15      value_t a_1 = 1;
16
17      for (index_t index = 0; index < n; index++) {
18          const value_t tmp = a_0; a_0 = a_1; a_1 += tmp;
19      }
20
21      result.set_value(a_0);                // <- fulfill promise
22  }
23
24  int main(int argc, char * argv[]) {
25
26      const uint64_t num_threads = 32;
27      std::vector<std::thread> threads;
28
29      // storage for futures
30      std::vector<std::future<uint64_t>> results;
```

```
31
32      // for each thread
33      for (uint64_t id = 0; id < num_threads; id++) {
34
35          // define a promise and store the associated future
36          std::promise<uint64_t> promise;
37          results.emplace_back(promise.get_future());
38
39          // emplace the spawned thread
40          threads.emplace_back(
41              // move the promise to the spawned thread
42              // note that promise is now moved elsewhere
43              // and cannot be accessed safely anymore
44              fibo<uint64_t, uint64_t>, id, std::move(promise)
45          );
46      }
47
48      // read the futures resulting in synchronization of threads
49      // up to the point where promises are fulfilled
50      for (auto& result: results)
51          std::cout << result.get() << std::endl;
52
53      // this is mandatory since threads have to be either
54      // joined or detached at the end of our program
55      for (auto& thread: threads)
56          thread.join();
57  }
```

Listing 4.4: Passing return values using promises and futures.

The code is similar to the traditional passing of pointers in Listing 4.3. First, we reserve storage for the `num_threads` many futures in Line 30. Second, the state $s = (p, f)$ is created within the body of the first for-loop for each thread (see Lines 36–37). The associated promise is subsequently moved to the spawned thread as rvalue reference. Note that `promise` cannot be safely accessed anymore after Line 44 since it is now no longer in a valid state. Third, the promise `result` is fulfilled within the body of the `fibo` function (Line 21) by setting it to the computed value `a_0`. Fourth, we read the futures in the master thread (Line 51) with a call to `get()`. The master thread waits for all futures to be signaled and thus enforces synchronization up to the point where promises are fulfilled. More precisely, the threads are synchronized in the exact same order the futures are read. Finally, all threads have to be either joined or detached at the end of our program. In the first case, a second barrier is established which waits for all threads to terminate. In contrast, detaching the threads at the end of our program does not guarantee that all threads have finished potential work after having fulfilled their promise. However, detaching is a viable option here since the function `fibo` does not execute any work after fulfilling the promise (see Line 21).

This mechanism can be applied to functions that communicate one or more values to the master thread. In the multivariate case you can simply pass multiple promises in the argument list. Nevertheless, the described approach seems to be unhandy when returning only a single value. Fortunately, C++11 offers a mechanism that transforms functions to tasks with a corresponding future object handling the return value. The `future` header provides the function `std::packaged_task` that allows for the convenient construction of task objects. Assume we want to create a task that maps a Boolean function comp which decides whether a certain floating-point value `float value` is smaller than some given integer threshold `int64_t threshold`:

```
bool comp(float value, int64_t threshold) {
    return value < threshold;
}
```

The task can now be created and used in a straightforward manner:

```
// create the task and assign future
std::packaged_task<bool(float, int64_t)> task(comp);
auto future = task.get_future();

// call the task with arguments
task(value, threshold); // WARNING: this is sequential!

// access future object (either true or false)
std::cout << future.get() << std::endl;
```

A disadvantage of the described approach is that you have to hard-code the signature of the called function in the template parameter of `std::packaged_task` which, moreover, affects the types of the arguments. This might become a problem if you want to store several tasks wrapping distinct functions with different arguments in the same container class. It would be more convenient if all tasks exhibit the same signature `void task(void)` independent of the assigned functions. This can be achieved using a homebrew task factory function template as shown in Listing 4.5.

```
1  #include <iostream>
2  #include <future>
3
4  template <
5      typename Func,       // <-- type of function func
6      typename ... Args,   // <-- type of arguments arg0,arg1,...
7      typename Rtrn=typename std::result_of<Func(Args...)>::type>
8  auto make_task(          // ^-- type of return value func(args)
9      Func &&    func,
10     Args && ...args) -> std::packaged_task<Rtrn(void)> {
11
12     // basically build an auxilliary function aux(void)
13     // without arguments returning func(arg0,arg1,...)
14     auto aux = std::bind(std::forward<Func>(func),
15                          std::forward<Args>(args)...);
16
17     // create a task wrapping the auxilliary function:
18     // task() executes aux(void) := func(arg0,arg1,...)
19     auto task = std::packaged_task<Rtrn(void)>(aux);
20
21     // the return value of aux(void) is assigned to a
22     // future object accessible via task.get_future()
23     return task;
24 }
```

Listing 4.5: Homebrew task factory function template.

The first template parameter `Func` of the function template `make_task` specifies the type of the function `func`. The second parameter `...Args` is variadic, i.e., we might pass none or more arguments `args`. The three dots `...` left of `args` (see Line 10) denote that the arguments are *packed* (interpreted as a single entity). In contrast, three dots right of `args` (see Line 15) indicate that the arguments are unpacked one by another. Do not feel confused by the notation, variadic templates are just a fancy way to pass an arbitrary amount of arguments to functions.[1] As an example, the expression `func(args...)` refers to `comp(value, threshold)` in the aforementioned example. Next, we create an auxiliary function `aux` in Line 14 which evaluates `func(args...)` using `std::bind`. Note that `aux` accepts no arguments as intended by design. Afterwards, we create the task in Line 19 and finally return it in Line 23. The task factory can now be used as follows:

```
// create the task and assign future
auto task = make_task(comp, value, threshold);
auto future = task.get_future();

// call the task with NO arguments
task(); // WARNING: this is sequential!

// alternatively spawn a thread and detach it
// std::thread thread(std::move(task)); thread.detach();

// access future object (either true or false)
std::cout << future.get() << std::endl;
```

Admittedly, the readability of the source code has enhanced significantly. We will later use our custom `make_task` factory for the implementation of a thread pool that maintains tasks in a queue. Nevertheless, it is also useful in our Fibonacci example which can be now written in a clear and simple fashion (see Listing 4.6).

```
1  #include <iostream> // std::cout
2  #include <cstdint>  // uint64_t
3  #include <vector>   // std::vector
4  #include <thread>   // std::thread
5  #include <future>   // std::packaged_task
6
7  // fill in custom make_task factory here
8
9  // traditional signature of fibo without syntactic noise
10 uint64_t fibo(uint64_t n) {
11
12     uint64_t a_0 = 0;
13     uint64_t a_1 = 1;
14
15     for (uint64_t index = 0; index < n; index++) {
16         const uint64_t tmp = a_0; a_0 = a_1; a_1 += tmp;
17     }
18
```

[1] An interested reader might refer to [6] for more details on variadic templates.

```
19      return a_0;
20  }
21
22  int main(int argc, char * argv[]) {
23
24      const uint64_t num_threads = 32;
25
26      // storage for threads and futures
27      std::vector<std::thread> threads;
28      std::vector<std::future<uint64_t>> results;
29
30      // create tasks, store futures and spawn threads
31      for (uint64_t id = 0; id < num_threads; id++) {
32          auto task = make_task(fibo, id);
33          results.emplace_back(task.get_future());
34          threads.emplace_back(std::move(task));
35      }
36
37      for (auto& result: results)
38          std::cout << result.get() << std::endl;
39
40      for (auto& thread: threads)
41          thread.detach();
42  }
```

Listing 4.6: Passing return values using packaged tasks.

THE ASYNCHRONOUS WAY

C++11 features another out-of-the-box mechanism, namely `std::async`, which is similar to our `make_task` factory for the convenient creation of task objects. The command `std::async` executes a task asynchronously using either dedicated threads or the calling master thread. Task creation and assignment of the corresponding future is as simple as

```
auto future = std::async(fibo, id);
```

Unfortunately, the command `std::async` has to be taken with a grain of salt since its behavior can be vastly unintuitive despite its simple syntax. Let us briefly enumerate severe pitfalls related to its use:

1. A naive call to `std::async` does not necessarily imply that a novel thread is spawned. The execution might be performed by the calling thread.
2. The execution of the task could be delayed forever if we do not access the corresponding future via `future.get()`.
3. The execution of distinct tasks could be serialized if we do not pay attention to the scopes of the corresponding futures.

This sounds pretty disastrous on first sight. However, each of the aforementioned issues can be resolved in a straightforward manner. First, if not specified otherwise, `std::async` uses a default launch policy whose behavior depends on the implementation. Hence, the called function (in our case `fibo`) could be executed on the master thread, as well as on a spawned thread, at any time. We only get the guarantee that the task is executed asynchronously. This does not imply that it is executed at all. Mathematically speaking, we can only rely on the fact that two or more issued tasks are commutative in their execution order (if executed at all). Fortunately, we can change this behavior by explicitly specifying one of the two possible launch policies:

- `std::launch::async` spawns a thread and immediately executes the task.
- `std::launch::deferred` executes the task in a lazy evaluation fashion on the calling (same) thread at the first time the future is accessed with `get()`.

The latter might become handy if we want to delegate some tasks for computation without prior knowledge which result is used first. This behavior is futile in a High Performance Computing environment where we want to concurrently exploit the full computational capabilities of all available CPU cores. Obviously, we should use the `std::launch:async` policy in this case. The policy can be passed as first argument in an overloaded variant of `std::async`:

```
auto future = std::async(std::launch::async, fibo, id);
```

In contrast, the following task could be delayed forever if we do not access the future via `future.get()`; since lazy evaluation is performed on request:

```
auto future = std::async(std::launch::deferred, fibo, id);
```

Even when using the `std::launch::async` policy which enforces the creation of new threads, we can still encounter serialization caused by implicit synchronization. In particular, the task is synchronized if the destructor of the future is called. This happens if we leave the scope of its declaration. Hence, the following code serializes the issued tasks:

```
for (uint64_t id = 0; id < num_threads; id++) {
    auto future = std::async(std::launch::async, fibo, id);
} // <- here, the destructor of future is called
```

Note that the same is true for the following code fragment:

```
for (uint64_t id = 0; id < num_threads; id++)
    std::async(std::launch::async, fibo, id);
```

The only way to prevent this behavior is to store the futures outside of the for-loop body. As a result, the use of `std::async` is not advisable for methods without return value unless you intend to waste memory on storing future objects which exclusively yield `void`. Our Fibonacci example can be rewritten using the described techniques as follows:

```
1  #include <iostream> // std::cout
2  #include <cstdint>  // uint64_t
3  #include <vector>   // std::vector
4  #include <future>   // std::async
5
6  // traditional signature of fibo without syntactic noise
7  uint64_t fibo(uint64_t n) {
8
```

```
9       uint64_t a_0 = 0;
10      uint64_t a_1 = 1;
11
12      for (uint64_t index = 0; index < n; index++) {
13          const uint64_t tmp = a_0; a_0 = a_1; a_1 += tmp;
14      }
15
16      return a_0;
17  }
18
19  int main(int argc, char * argv[]) {
20
21      const uint64_t num_threads = 32;
22      std::vector<std::future<uint64_t>> results;
23
24      // for each thread
25      for (uint64_t id = 0; id < num_threads; id++)
26          // directly emplace the future
27          results.emplace_back(
28              std::async(
29                  std::launch::async, fibo, id
30              )
31          );
32
33      // synchronization of spawned threads
34      for (auto& result: results)
35          std::cout << result.get() << std::endl;
36  }
```

Listing 4.7: Passing return values using `std::async`.

You might have noticed that there is no need to provide storage for `std::thread` objects since they are completely hidden inside the implementation of `std::async`.

4.3 SCHEDULING BASED ON STATIC DISTRIBUTIONS (MATRIX VECTOR MULTIPLICATION)

In this section, you will learn how to statically schedule for-loops in cases where the amount of tasks is significantly higher than the number of available CPU cores. In particular, we investigate static block distributions, cyclic distributions, and block-cyclic distributions. Moreover, we discuss how to pass many arguments to a thread in an elegant way. This can be achieved by exploiting the capture mechanism of anonymous functions, so-called *lambdas*. The techniques are demonstrated with the help of dense matrix vector products (DMV), a common operation in linear algebra with manifold applications in informatics and natural science. Notably, DMV is a basic building block in neural networks, softmax regression, and discrete Markov chains. Examples from natural science include linear transformations of coordinate systems, propagation of finite dimensional approximation of quantum mechanical states, and modeling of linear systems in general.

Let $A \in \mathbb{R}^{m \times n}$ be a real-valued matrix of shape $m \times n$ and $x \in \mathbb{R}^n$ be a vector in n dimensions; then A linearly maps x from an n-dimensional vector space to an m-dimensional vector space (i.e. parallel lines are mapped onto parallel lines). The entries of A will be denoted as A_{ij} where the index i enumerates the rows and j refers to its columns. The product $b := A \cdot x$ can be written in coordinate form as follows:

$$b_i = \sum_{j=0}^{n-1} A_{ij} \cdot x_j \quad \text{for all} \quad i \in \{0, \ldots, m-1\} \quad . \tag{4.2}$$

The sum over j can be evaluated independently for each fixed index i resulting in overall $m \cdot n$ additions. Hence, it is advisable to parallelize the matrix vector product over the $\mathcal{O}(m)$ outer indices enumerated by i. Another approach would parallelize the inner sum over j using a parallel reduction over $\mathcal{O}(\log(n))$ steps. The latter, however, involves non-trivial synchronization after each reduction step which would unnecessarily complicate this example.

THE SEQUENTIAL PROGRAM

Let us first write down the sequential program. The matrix A and the vectors x and b are stored in linear memory. Here we choose a `std::vector` container for the storage of the values. Listing 4.8 shows the corresponding sequential source code.

```
1   #include <iostream>                    // std::cout
2   #include <cstdint>                     // uint64_t
3   #include <vector>                      // std::vector
4   #include <thread>                      // std::thread (not used yet)
5
6   #include "../include/hpc_helpers.hpp" // custom timers
7
8   // initialize A as lower triangular matrix
9   // simulating prefix summation and vector x
10  // with consecutive values (0, 1, 2, 3, ...)
11  template <
12      typename value_t,
13      typename index_t>
14  void init(
15      std::vector<value_t>& A,
16      std::vector<value_t>& x,
17      index_t m,
18      index_t n) {
19
20      for (index_t row = 0; row < m; row++)
21          for (index_t col = 0; col < n; col++)
22              A[row*n+col] = row >= col ? 1 : 0;
23
24      for (index_t col = 0; col < n; col++)
25          x[col] = col;
26  }
27
```

```
28  // the sequential matrix vector product
29  template <
30      typename value_t,
31      typename index_t>
32  void sequential_mult(
33      std::vector<value_t>& A,
34      std::vector<value_t>& x,
35      std::vector<value_t>& b,
36      index_t m,
37      index_t n) {
38
39      for (index_t row = 0; row < m; row++) {
40          value_t accum = value_t(0);
41          for (index_t col = 0; col < n; col++)
42              accum += A[row*n+col]*x[col];
43          b[row] = accum;
44      }
45  }
46
47  int main(int argc, char* argv[]) {
48
49      const uint64_t n = 1UL << 15;
50      const uint64_t m = 1UL << 15;
51
52      TIMERSTART(overall)
53
54      TIMERSTART(alloc)
55      std::vector<uint64_t> A(m*n);
56      std::vector<uint64_t> x(n);
57      std::vector<uint64_t> b(m);
58      TIMERSTOP(alloc)
59
60      TIMERSTART(init)
61      init(A, x, m, n);
62      TIMERSTOP(init)
63
64      TIMERSTART(mult)
65      sequential_mult(A, x, b, m, n);
66      TIMERSTOP(mult)
67
68      TIMERSTOP(overall)
69
70      // check if summation is correct
71      for (uint64_t index = 0; index < m; index++)
72          if (b[index] != index*(index+1)/2)
73              std::cout << "error at position "
74                        << index << std::endl;
75
76  }
```

Listing 4.8: Sequential matrix vector multiplication.

Let us discuss the source code. First, we include necessary headers from the standard library and a custom `hpc_helpers.hpp` header file for the convenient time measurement which is distributed with this book. The function template `init` in Line 11 fills the matrix A with ones below the diagonal and with zeros otherwise simulating prefix summation. The vector $x = (0, 1, 2, \ldots)$ is initialized with ascending integers. Consequently, we expect the entries b_i of $b = A \cdot x$ to be the partial sums from zero to i. Second, the function template `sequential_mult` in Line 29 processes the actual matrix vector product by consecutively computing scalar products of the i-th row of A with the vector x. Third, we allocate the storage for the matrix A and the vectors x and b within the `main` function (see Line 55). Afterwards, we initialize them using `init` in Line 61 and finally execute the matrix vector product in Line 65. The program can be compiled on the command line with a call to

```
g++ -O2 -std=c++11 matrix_vector.cpp -o matrix_vector
```

and produces the following output when executed on a Xeon E5-2683 v4 CPU:

```
# elapsed time (alloc): 2.74034s
# elapsed time (init): 1.31006s
# elapsed time (mult): 1.2569s
# elapsed time (overall): 5.30746s
```

Both, the initialization phase and multiplication step, run equally long since they exhibit the same theoretical time complexity of $\mathcal{O}(m \cdot n)$. The initial allocation (see Lines 55–57) seems to be quite expensive though. This can be explained by the default behavior of `std::vector` whose `std::allocator` initializes all entries by either calling their constructor or in case of plain old data types setting their default value. However, this is completely unnecessary in our case since we initialize the memory in `init` anyway. Unfortunately, there is no straightforward way to fix that issue unless you implement a custom allocator class which is way beyond the scope of this book. Alternatively, you could write a class template `no_init_t<T>` for plain old data types that wraps all operators but leaves the constructor empty as shown in Line 13 of Listing 4.9. Note that the `no_init_t` wrapper is part of the `hpc_helpers.hpp` header file and has to be compiled with enabled C++14 support.

```
1  #include <type_traits> // std::is_fundamental, ...
2
3  template<class T>
4  class no_init_t {
5  public:
6
7      // check whether it is a fundamental numeric type
8      static_assert(std::is_fundamental<T>::value &&
9                    std::is_arithmetic<T>::value,
10                   "must be a fundamental, numeric type");
11
12     //do nothing
13     constexpr no_init_t() noexcept { /* HERE WE DO NOTHING! */ }
14
15     //convertible from a T
16     constexpr no_init_t(T value) noexcept: v_(value) {}
17
18     //act as a T in all conversion contexts
19     constexpr operator T () const noexcept { return v_; }
```

```
20
21     // negation on a value-level and bit-level
22     constexpr no_init_t& operator - () noexcept {
23         v_ = -v_; return *this;
24     }
25     constexpr no_init_t& operator ~ () noexcept {
26         v_ = ~v_; return *this;
27     }
28
29     // increment/decrement operators
30     constexpr no_init_t& operator ++ ()    noexcept {
31         v_++; return *this;
32     }
33     constexpr no_init_t& operator ++ (int) noexcept {
34         v_++; return *this;
35     }
36     constexpr no_init_t& operator -- ()    noexcept {
37         v_--; return *this;
38     }
39     constexpr no_init_t& operator -- (int) noexcept {
40         v_--; return *this;
41     }
42
43     // assignment operators
44     constexpr no_init_t& operator  += (T v) noexcept {
45         v_  += v; return *this;
46     }
47     constexpr no_init_t& operator  -= (T v) noexcept {
48         v_  -= v; return *this;
49     }
50     constexpr no_init_t& operator  *= (T v) noexcept {
51         v_  *= v; return *this;
52     }
53     constexpr no_init_t& operator  /= (T v) noexcept {
54         v_  /= v; return *this;
55     }
56
57     // bitwise assignment operators
58     constexpr no_init_t& operator  &= (T v) noexcept {
59         v_  &= v; return *this;
60     }
61     constexpr no_init_t& operator  |= (T v) noexcept {
62         v_  |= v; return *this;
63     }
64     constexpr no_init_t& operator  ^= (T v) noexcept {
65         v_  ^= v; return *this;
66     }
67     constexpr no_init_t& operator >>= (T v) noexcept {
68         v_ >>= v; return *this;
69     }
70     constexpr no_init_t& operator <<= (T v) noexcept {
71         v_ <<= v; return *this;
```

```
72     }
73
74 private:
75     T v_;
76 };
```

Listing 4.9: Wrapper class template for plain old data types.

Afterwards, you have to wrap the data type `uint64_t` with `no_init<uint64_t>`

```
std::vector<no_init_t<uint64_t>> A(n*n);
std::vector<no_init_t<uint64_t>> x(n);
std::vector<no_init_t<uint64_t>> b(n);
```

and recompile the program with C++14 support as follows:

```
g++ -O2 -std=c++14 matrix_vector.cpp -o matrix vector
```

The corresponding runtime is now significantly reduced since we do not spend double the time on the initialization. The allocation is now basically free and initialization is performed exclusively in `init`:

```
# elapsed time (alloc): 1.8089e-05s
# elapsed time (init): 2.88586s
# elapsed time (mult): 1.29033s
# elapsed time (overall): 4.17636s
```

Another viable option would be the use of a dynamic array instead of a vector

```
uint64_t * A = new uint64_t[m*n];
```

which could be wrapped with a `std::unique_ptr` in order to avoid manual freeing of memory. Concluding, you have to be aware of considerable initialization overhead during memory allocation of container objects from the standard library.

BLOCK DISTRIBUTION OF THREADS

After having implemented a sequential version of DMV we can now proceed with its parallelization. As mentioned before, it is advisable to parallelize the individual scalar products $b_i = \langle a_i | x \rangle$ of the rows a_i of the matrix A with the vector x. In contrast to the Fibonacci example from the previous section where we concurrently processed a few tasks, we now have to handle an increased level of parallelism since the number of rows $m = 2^{15} = 32{,}768$ is significantly higher than the amount of available CPU cores. The naive spawning of 2^{15} threads would result in excessive oversubscription which forces the OS to time-slice the execution of threads using expensive context switches. This could be avoided if we limit ourselves to process the rows in small batches by iteratively spawning a few number of threads. Nevertheless, the described computation pattern would consecutively spawn $m/p = 4096$ groups of $p = 8$ threads on a machine with $p = 8$ CPU cores which involves a non-negligible overhead of thread creation.

A better approach would spawn p threads once where each thread processes m/p rows. If m is not an exact multiple of p (i.e. $m\%p \neq 0$), each thread should at least compute $\lfloor \frac{m}{p} \rfloor$ many tasks. We define

thread 0				thread 1				···	thread $p-1$			
0	1	2	3	4	5	6	7	···	$m-4$	$m-3$	$m-2$	$m-1$

FIGURE 4.3

An example of a static block distribution assigning $c=4$ consecutive tasks to each of the p threads in order to concurrently process $p \cdot c = m$ overall tasks.

the chunk size as $c = \lceil \frac{m}{p} \rceil$. The latter can be written using exclusively integer arithmetic in terms of the following closed expression for safe integer division:

$$\texttt{SDIV}(x, y) = \left\lfloor \frac{x+y-1}{y} \right\rfloor \geq \frac{x}{y} \quad \text{for all } x, y \in \mathbb{R}^+ \quad . \tag{4.3}$$

The `SDIV` macro is defined in the `hpc_helper.hpp` header file. At this point we can choose different assignment strategies to distribute c rows of the matrix A to a thread. An obvious choice is to compute c consecutive rows within a single thread. This pattern is called *static block distribution* of threads with a chunk size of c. Fig. 4.3 shows an example of a static block distribution using a chunk size of $c = \frac{m}{p} = 4$.

Before we start coding let us briefly discuss the arguments that have to be passed to the threads. Obviously, we should pass the matrix A, and the vectors x and b with their corresponding shape. Furthermore, we have to pass the thread identifier `id`. The chunk size c, as well as the first and last task identifier of a corresponding chunk, can be computed on-the-fly as a function of the thread identifier `id`, the overall thread count p, and the shape of the matrix $m \times n$ within each thread. Nevertheless, the complete list of arguments A, x, b, m, n, p and `id` is agonizingly long, resulting in barely readable code. Notably, the thread identifier `id` is the only argument that has to be passed explicitly by value since it is different for each thread – the remaining arguments could be passed as references or accessed within a shared scope. The latter can be accomplished by declaring A, x, b, m, n, p globally which is unfavorable from a software engineering point of view. Alternatively, we can exploit the capturing mechanism of anonymous functions, so-called *closures* or *lambdas*, to pass references in an elegant way. As an example, let us declare a lambda expression `add_one` that increments a given value `uint64_t v`:

```
auto add_one = [] (const uint64_t& v) { return v+1; };
```

The anonymous function `add_one` can be called within its scope of declaration like any other traditional function, e.g. `auto two = add_one(1)`. It is worth mentioning that we cannot directly access variables or objects declared in the outer scope within the body of the lambda (in curly brackets), i.e., the following code fragment results in a compile-time error:

```
uint64_t w = 1;
// ERROR: w is not declared within the body of add_w!
auto add_w = [] (const uint64_t& v) { return v+w; };
```

This issue can be resolved using the capturing mechanism of lambdas. We can specify a list of variables or objects within the leading angle brackets that are passed either by value (=) or by reference (&) to the closure.

```
uint64_t w = 1;  // w will be accessed inside lambdas

// capture w by value
auto add_w_0 = [w]  (const uint64_t& v) { return v+w; };

// capture w by reference
auto add_w_1 = [&w] (const uint64_t& v) { return v+w; };

// capture everything accessed in add_w_2 by reference
auto add_w_2 = [&]  (const uint64_t& v) { return v+w; };

// capture everything accessed in add_w_3 by value
auto add_w_3 = [=]  (const uint64_t& v) { return v+w; };
```

In the case of DMV, the automatic capturing of the scope by reference (as realized in add_w_2) is a viable option: we pass the thread identifier id as const reference in the arguments and seamlessly transfer the remaining variables by capturing the whole scope on a reference level. Note that capturing the whole scope by value (as realized in add_w_3) is not a tractable option since it would perform redundant copies of the huge matrix A and further would write the result to a copy of b that cannot be accessed in the scope of the master thread. The final implementation for concurrent DMV using a static block distribution is pleasantly simple:

```
1   template <
2       typename value_t,
3       typename index_t>
4   void block_parallel_mult(
5       std::vector<value_t>& A,
6       std::vector<value_t>& x,
7       std::vector<value_t>& b,
8       index_t m,
9       index_t n,
10      index_t num_threads=8) {
11
12      // this function is called by the threads
13      auto block = [&] (const index_t& id) -> void {
14          //         ^-- capture whole scope by reference
15
16          // compute chunk size, lower and upper task id
17          const index_t chunk = SDIV(m, num_threads);
18          const index_t lower = id*chunk;
19          const index_t upper = std::min(lower+chunk, m);
20
21          // only computes rows between lower and upper
22          for (index_t row = lower; row < upper; row++) {
23              value_t accum = value_t(0);
24              for (index_t col = 0; col < n; col++)
25                  accum += A[row*n+col]*x[col];
26              b[row] = accum;
27          }
28      };
29
30      // business as usual
```

```
31     std::vector<std::thread> threads;
32
33     for (index_t id = 0; id < num_threads; id++)
34         threads.emplace_back(block, id);
35
36     for (auto& thread : threads)
37         thread.join();
38 }
```

Listing 4.10: DMV template using a static block distribution.

Let us discuss the code. The closure `block` in Line 13 performs DMV on a chunk of consecutive rows. Since `block` captures the whole scope by reference (via `[&]`) we can access variables and vectors without passing them explicitly. An exception is the thread identifier `id` which has to be passed later as argument because it is not yet declared in Line 13. Within the closure we first calculate the chunk size using safe integer division (Line 17). Second, we proceed with the computation of the initial row `lower` (Line 18) and the corresponding `upper` row (exclusive). Note that we have to clip `upper` if it exceeds the overall number of rows in the last block (Line 19) since `chunk` is an overestimation of n/p. The partial matrix vector product is subsequently performed in the for-loops (Lines 22–27). The remaining code fragment is straightforward: we spawn `num_threads` many threads calling the closure `block` and join them afterwards. Note that detaching is not a viable option here due to lacking synchronization. The block parallel multiplication runs in approximately 230 ms when executed on a Xeon E5-2683 v4 CPU using eight threads. This corresponds to a speedup of roughly 5.6 and a parallelization efficiency of about 70%.

CYCLIC DISTRIBUTION OF THREADS

Alternatively, we could have chosen a different distribution of threads where c tasks are assigned to each of the p threads using a stride of p. As an example, thread 0 would consecutively process task 0, task p, task $2p$, and so forth. In general, thread `id` processes the list (task `id`, task `id` $+ p$, task `id` $+ 2p$, $\ldots$) up to the point where we exhaust the list of available tasks. According to this round-robin fashion the distribution pattern is called *static cyclic distribution* of threads. Fig. 4.4 visualizes the described assignment strategy.

The corresponding implementation of DMV with closures using a cyclic distribution is shown in Listing 4.11. Notably, the code gets even simpler since we do not have to compute the chunk size explicitly and the associated variables `lower` and `upper`. The cyclic assignment of rows is exclusively handled by the first for-loop in the closure and works for an arbitrary number of threads and rows. The provided parallelization performs at the same speed as the block cyclic variant ($\approx$ 230 ms using eight threads).

```
1 template <
2     typename value_t,
3     typename index_t>
4 void cyclic_parallel_mult(
5     std::vector<value_t>& A,
6     std::vector<value_t>& x,
7     std::vector<value_t>& b,
8     index_t m,
```

threads	0	1	2	...	$p-1$	0	1	2	...	$p-1$	...
tasks	0	1	2	...	$p-1$	p	$p+1$	$p+2$	...	$2p-1$	...

FIGURE 4.4

An example of a static cyclic distribution assigning c tasks to each thread in a round-robin fashion using a stride of p.

```
 9        index_t n,
10        index_t num_threads=8) {
11
12        // this function is called by the threads
13        auto cyclic = [&] (const index_t& id) -> void {
14
15            // indices are incremented with a stride of p
16            for (index_t row = id; row < m; row += num_threads) {
17                value_t accum = value_t(0);
18                for (index_t col = 0; col < n; col++)
19                    accum += A[row*n+col]*x[col];
20                b[row] = accum;
21            }
22        };
23
24        // business as usual
25        std::vector<std::thread> threads;
26
27        for (index_t id = 0; id < num_threads; id++)
28            threads.emplace_back(cyclic, id);
29
30        for (auto& thread : threads)
31            thread.join();
32    }
```

Listing 4.11: DMV template using a static cyclic distribution.

FALSE SHARING

The cyclic indexing scheme is easier to implement than the block distribution and performs at the same speed. Nevertheless, we have to be aware of a common pitfall. Let us assume we would have implemented the closure `cyclic` as follows:

```
// WARNING: This code fragment is suboptimal
auto cyclic = [&] (const index_t& id) -> void {

    for (index_t row = id; row < n; row += num_threads) {

        // initialize result vector to zero
        b[row] = 0;
```

```
            // directly accumulate in b[row]
            for (index_t col = 0; col < n; col++)
                b[row] += A[row*n+col]*x[col];
        }
    };
```

On first sight, the code looks equivalent and indeed it computes the correct result. The only difference from the original `cyclic` closure is that we accumulate the contributions of the scalar product directly in the result vector b and do not declare a dedicated register `accum` in order to cache the intermediate results. You might argue that this should not affect performance too much since modern CPUs provide sophisticated caching strategies that aim to minimize the actual accesses to the attached RAM. In the particular case of a cyclic distribution, however, this assumption might be wrong.

Assume we perform cyclic DMV using two or more threads executed on distinct cores of the same CPU which each exhibit their own cache hierarchy realized with low-latency on-chip memory being orders-of-magnitude faster (and more expensive [2]) than attached off-chip RAM. For the sake of simplicity, we further assume that each CPU core features only a single cache level (omitting potential L2 and L3 caches). The caches of all cores are coherent over the whole CPU, i.e. the cores can mutually tag cache entries as dirty using a suitable coherence protocol. The mutual propagation of changed cache entries to peer caches is crucial in order to ensure that a core A does not reuse an outdated value which has been modified earlier by another core B. As a result, a dirty cache entry has to be reloaded by core A from slow off-chip RAM after having been updated by core B. Putting it short, if at least one core modifies a value residing in its associated cache which is subsequently read from a peer cache, then the updated value has to be communicated over slow RAM.

In theory, this is never the case using cyclic DMV since all write and read operations to `b[row]` are issued without data races, i.e., each thread operates exclusively on its own list of rows. Nevertheless, as a result of performance considerations, cache entries are not updated individually in practice but in blocks of several bytes. These *cache lines* have a length of 64 bytes on modern Intel CPUs. Consequently, whole memory regions are marked as dirty which might invalidate neighboring values despite never having been modified in a peer cache.

The result vector b in our DMV implementation is of type `uint64_t` which occupies eight bytes of memory. Consequently, $64/8 = 8$ consecutive entries of b reside in the same cache line despite being frequently accessed by eight distinct threads. This excessive invalidation of shared cache lines is called *false sharing* in the literature. False sharing may drastically influence the runtime of your program. As an example, cyclic DMV for $m = 8$ rows and $n = 2^{27}$ columns performs at roughly 240 ms without false sharing and 290 ms with false sharing. This is a performance degradation of more than 20%.

Let us summarize that observation with a short code fragment demonstrating false sharing in a minimal example: We independently increment two integers `ying` and `yang` that are stored together in a struct (see Listing 4.12). Note that the use of the `volatile` keyword (Line 15 and Line 25) avoids the optimization of the for-loops in Line 17, Line 28, and Line 33. Without the use of `volatile` the compiler would simply substitute the for-loops by `ying += 1 << 30;` and `yang += 1 << 30;`. Be aware that the `volatile` keyword in C++ has a completely different purpose and functionality than in the Java programming language. In particular, it does not guarantee atomic or race condition-free access to memory.

```
1  #include "../include/hpc_helpers.hpp" // timers
2  #include <thread>                      // std::thread
3
4  // this is a 128-bit packed data structure that fits
5  // into a cache line of 64 bytes holding two integers
6  struct pack_t {
7      uint64_t ying;
8      uint64_t yang;
9
10     pack_t() : ying(0), yang(0) {}
11 };
12
13 // sequentially increment the integers
14 void sequential_increment(
15     volatile pack_t& pack) {
16
17     for (uint64_t index = 0; index < 1UL << 30; index++) {
18         pack.ying++;
19         pack.yang++;
20     }
21 }
22
23 // use one thread for each member of the packed data type
24 void false_sharing_increment(
25     volatile pack_t& pack) {
26
27     auto eval_ying = [&pack] () -> void {
28         for (uint64_t index = 0; index < 1UL << 30; index++)
29             pack.ying++;
30     };
31
32     auto eval_yang = [&pack] () -> void {
33         for (uint64_t index = 0; index < 1UL << 30; index++)
34             pack.yang++;
35     };
36
37     std::thread ying_thread(eval_ying);
38     std::thread yang_thread(eval_yang);
39     ying_thread.join();
40     yang_thread.join();
41 }
42
43 int main(int argc, char* argv[]) {
44
45     pack_t seq_pack;
46
47     TIMERSTART(sequential_increment)
48     sequential_increment(seq_pack);
49     TIMERSTOP(sequential_increment)
50
51     std::cout << seq_pack.ying << " "
```

```
52              << seq_pack.yang << std::endl;
53
54      pack_t par_pack;
55
56      TIMERSTART(false_sharing_increment_increment)
57      false_sharing_increment(par_pack);
58      TIMERSTOP(false_sharing_increment_increment)
59
60      std::cout << par_pack.ying << " "
61              << par_pack.yang << std::endl;
62  }
```

Listing 4.12: Minimal example for false sharing.

The output of the corresponding program reveals a significant increase of the runtime when using false sharing:

```
# elapsed time (sequential_increment): 0.542641s
1073741824 1073741824
# elapsed time (parallel_increment): 3.26746s
1073741824 1073741824
```

Concluding, make sure that you avoid excessive updates of entries stored in the same cache line when using more than one thread in parallel. Moreover, try to cache intermediate results in registers in order to reduce the update frequency to cached entities as demonstrated with the auxiliary variable `accum` in the `cyclic` and `block` closure.

BLOCK-CYCLIC DISTRIBUTION OF THREADS

Block distributions and cyclic distributions can be combined to a so-called *block-cyclic distribution*. This approach assigns a fixed block of c tasks to each of the p threads which are processed sequentially by each thread. Doing so, we initially assign $s := p \cdot c$ tasks to be processed in parallel. If we cannot exhaust the overall number of m tasks in the first round, i.e. $m > s$, we simply repeat the described procedure with a stride of s. Loosely speaking, we employ a cyclic distribution on the blocks of fixed length c (see Fig. 4.5). Pure block and cyclic distributions can be seen as special cases of the block-cyclic variant. If we choose one element per block then the block-cyclic distribution with $c = 1$ corresponds to a pure cyclic distribution. In contrast, the choice $c = \lceil m/p \rceil$ realizes a pure block distribution.

The parameter c can be used to adjust the desired properties of your distribution. A small value for c is beneficial in the case of many heterogeneous tasks that have to be implicitly load-balanced as demonstrated in the next section. However, if we choose c too small then the resulting memory access pattern might exhibit suboptimal caching behavior or even false sharing. In contrast, high values of c correspond to cache-friendly access patterns but could lead to load-imbalance for heterogeneous tasks with heavily varying execution times. A reasonable trade-off is the following: we choose c to cover a complete cache line, i.e. $c = 16$ for 32 bit data types and $c = 8$ for 64 bit data types. As a result, we can guarantee cache-friendly access patterns while providing a high granularity of parallelism in order to avoid potential load-imbalance. The corresponding implementation of block-cyclic DMV is shown in Listing 4.13.

thread 0		thread 1		...	thread $p-1$		thread 0		thread 1		...
0	1	2	3	...	$s-2$	$s-1$	s	$s+1$	$s+2$	$s+3$	...

FIGURE 4.5

An example of a static block-cyclic distribution assigning $c = 2$ tasks to each thread. The blocks are executed in a round-robin fashion using a stride of $s = p \cdot c$.

```
1  template <
2      typename value_t,
3      typename index_t>
4  void block_cyclic_parallel_mult(
5      std::vector<value_t>& A,
6      std::vector<value_t>& x,
7      std::vector<value_t>& b,
8      index_t m,
9      index_t n,
10     index_t num_threads=8,
11     index_t chunk_size=64/sizeof(value_t)) {
12
13     // this function is called by the threads
14     auto block_cyclic = [&] (const index_t& id) -> void {
15
16         // precompute offset and stride
17         const index_t offset = id*chunk_size;
18         const index_t stride = num_threads*chunk_size;
19
20         // for each block of size chunk_size in cyclic order
21         for (index_t lower = offset; lower < m; lower += stride) {
22
23             // compute the upper border of the block (exclusive)
24             const index_t upper = std::min(lower+chunk_size, m);
25
26             // for each row in the block
27             for (index_t row = lower; row < upper; row++) {
28
29                 // accumulate the contributions
30                 value_t accum = value_t(0);
31                 for (index_t col = 0; col < n; col++)
32                     accum += A[row*n+col]*x[col];
33                 b[row] = accum;
34             }
35         }
36     };
37
38     // business as usual
39     std::vector<std::thread> threads;
40
```

```
41        for (index_t id = 0; id < num_threads; id++)
42            threads.emplace_back(block_cyclic, id);
43
44        for (auto& thread : threads)
45            thread.join();
46    }
```

Listing 4.13: DMV template using a static block-cyclic distribution.

The enumeration of the blocks is accomplished in the outer for-loop (Line 21) using a cyclic indexing scheme with a stride of $s = p \cdot c$. The access to the entries of a block is realized in the second for-loop (Line 27) with a step size of one. The final for-loop in Line 31 computes the contributions of the scalar product for a fixed value of `row`. Concluding, the block-cyclic distribution is the most versatile but also the longest solution in terms of source code. One could joke that besides the programming of parallel hardware High Performance Computing is actually the art of correct indexing.

4.4 HANDLING LOAD IMBALANCE (ALL-PAIRS DISTANCE MATRIX)

The static distributions in the previous section are called *static* because the assignment pattern of tasks to threads is predetermined at program start. This implies that we have carefully analyzed the problem beforehand and have subsequently chosen a suitable distribution of threads. This might become difficult if the time taken to process a certain task varies heavily. The case where a few threads still process their corresponding chunk of tasks while others have already finished their computation is called *load imbalance*. This section shows you how to balance heavily skewed work distributions using static and dynamic distributions of threads. In our example, we compute the all-pairs distance matrix for the MNIST dataset [3]. MNIST consists of 65,000 handwritten digits stored as gray-scale images of shape 28×28 with their corresponding labels ranging from zero to nine. Fig. 4.6 depicts 20 typical samples. For our purpose we interpret each of the $m = 65{,}000$ images as plain vector with $n = 784$ intensity values. Assume we store the images in a data matrix $D_{ij} = x_j^{(i)}$ of shape $m \times n$ where i denotes the m image indices and the index j enumerates the n pixels of each image.

Common tasks in data mining and machine learning are supervised classification and unsupervised clustering of data. Numerous classification algorithms, among others *k-nearest neighbor classifiers (kNN)* and *support vector machines (SVM)*, but also clustering algorithms such as *DBSCAN* and *spectral clustering*, rely on all-pair distance information. In detail, we are interested in the distances/similarities between all pairwise combinations of instances $\Delta_{ii'} = d\left(x^{(i)}, x^{(i')}\right)$ for all $i, i' \in \{0, \ldots, m-1\}$. The used distance/similarity measure $d : \mathbb{R}^n \times \mathbb{R}^n \to \mathbb{R}$ might be a traditional metric induced by the L_p-norm family such as Euclidean distance or Manhattan distance but could also be realized by any symmetric binary function that assigns a notion of similarity to pair of instances. An example for the latter are scalar products, correlation measures such as Pearson or Spearman correlation coefficients, and mutual information. Furthermore, the matrix Δ of shape $m \times m$ can be used to construct kernels for SVMs. Popular examples are linear kernels $K_{ii'} = \langle x^{(i)}, x^{(i')} \rangle$ or kernels based on radial basis functions $K_{ii'} = \exp\left(-\frac{1}{2\sigma^2} \|x^{(i)} - x^{(i')}\|^2\right)$. Putting it short, there are manifold applications that rely on all-pair distance information.

From a theoretical point of view we have to calculate m^2 distance/similarity scores between vectors in $\mathbb{R}^n$. Assuming that the computation of a single score values takes $\mathcal{O}(n)$ time, we have to spend

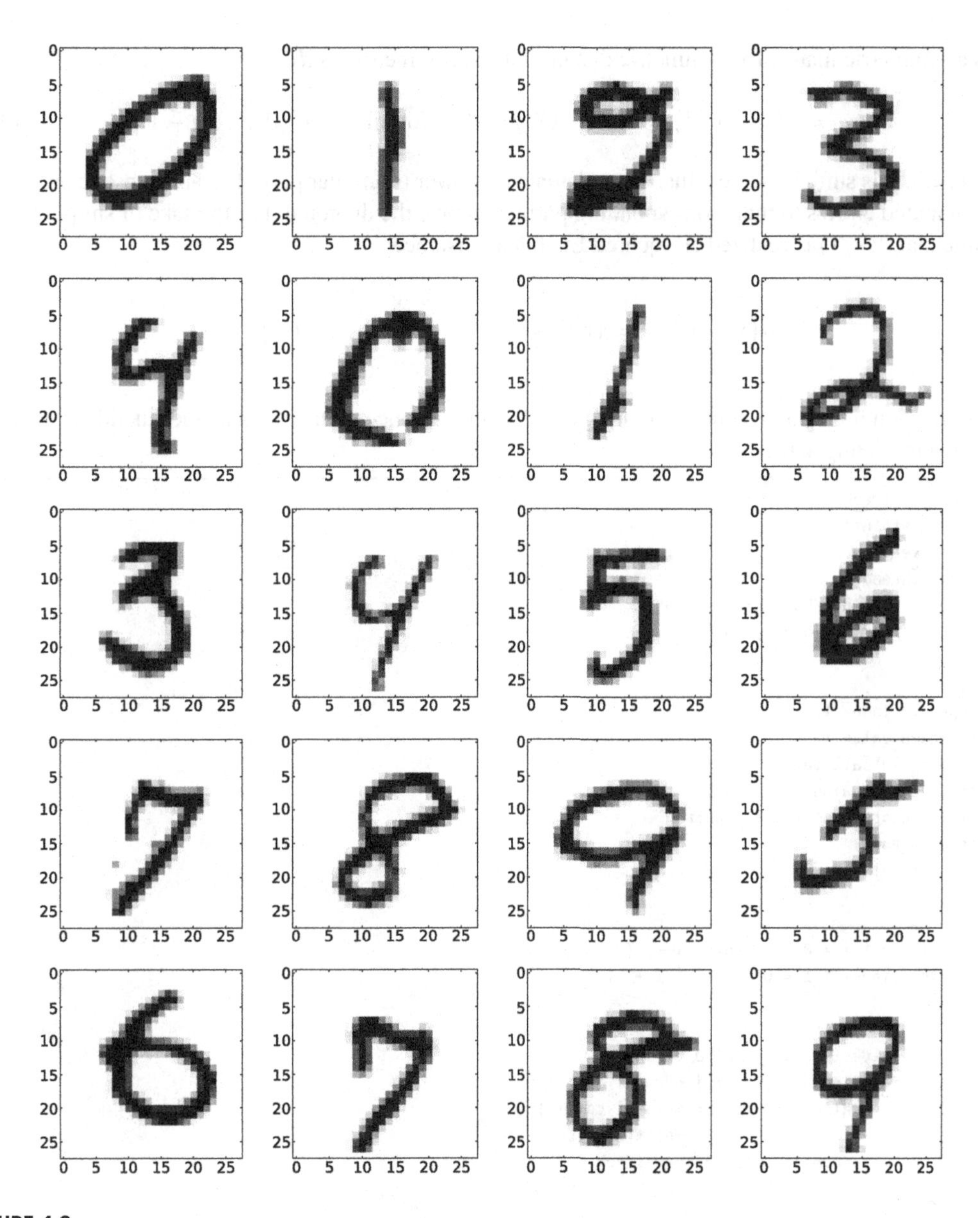

FIGURE 4.6

Twenty exemplary samples from the MNIST dataset consisting of 65,000 handwritten digits. The images of shape 28×28 are embedded as plain vectors in $\mathbb{R}^{784}$.

asymptotically $\mathcal{O}(m^2 \cdot n)$ many operations to compute the whole all-pairs distance matrix Δ. Note that all entries above its diagonal are determined by their corresponding partners below the diagonal since

Δ is a symmetric matrix for symmetric distance/similarity measures $d(\cdot, \cdot)$:

$$\Delta_{ii'} = d\big(x^{(i)}, x^{(i')}\big) = d\big(x^{(i')}, x^{(i)}\big) = \Delta_{i'i} \quad \text{for all } i, i' \in \{0, \ldots, m-1\} \ . \tag{4.4}$$

As a result, it is sufficient to exclusively calculate the lower triangular part of Δ and subsequently copy the computed entries to their corresponding partners above the diagonal. For the sake of simplicity, we assume that $d(\cdot, \cdot)$ is realized as squared Euclidean distance:

$$d\big(x^{(i)}, x^{(i')}\big) := \|x^{(i)} - x^{(i')}\|^2 = \sum_{j=0}^{n-1} \big(D_{ij} - D_{i'j}\big)^2 \quad . \tag{4.5}$$

The corresponding source code computing the all-pairs distance matrix for squared Euclidean distance is shown in Listing 4.14.

```
1  #include <iostream>                   // std::cout
2  #include <cstdint>                    // uint64_t
3  #include <vector>                     // std::vector
4  #include <thread>                     // std::thread (not used yet)
5  #include "../include/hpc_helpers.hpp" // timers, no_init_t
6  #include "../include/binary_IO.hpp"   // load_binary
7
8  template <
9      typename index_t,
10     typename value_t>
11 void sequential_all_pairs(
12     std::vector<value_t>& mnist,
13     std::vector<value_t>& all_pair,
14     index_t rows,
15     index_t cols) {
16
17     // for all entries below the diagonal (i'=I)
18     for (index_t i = 0; i < rows; i++) {
19         for (index_t I = 0; I <= i; I++) {
20
21             // compute squared Euclidean distance
22             value_t accum = value_t(0);
23             for (index_t j = 0; j < cols; j++) {
24                 value_t residue = mnist[i*cols+j]
25                                 - mnist[I*cols+j];
26                 accum += residue * residue;
27             }
28
29             // write Delta[i,i'] = Delta[i',i] = dist(i, i')
30             all_pair[i*rows+I] = all_pair[I*rows+i] = accum;
31         }
32     }
33 }
34
35 int main() {
36
```

```
37        // used data types
38        typedef no_init_t<float> value_t;
39        typedef uint64_t          index_t;
40
41        // number of images and pixels
42        const index_t rows = 65000;
43        const index_t cols = 28*28;
44
45        // load MNIST data from binary file
46        TIMERSTART(load_data_from_disk)
47        std::vector<value_t> mnist(rows*cols);
48        load_binary(mnist.data(), rows*cols,
49                    "./data/mnist_65000_28_28_32.bin");
50        TIMERSTOP(load_data_from_disk)
51
52        // compute all-pairs distance matrix
53        TIMERSTART(compute_distances)
54        std::vector<value_t> all_pair(rows*rows);
55        sequential_all_pairs(mnist, all_pair, rows, cols);
56        TIMERSTOP(compute_distances)
57    }
```

Listing 4.14: Sequential computation of all-pairs distance matrix.

Let us talk about the source code. First, we declare the shape of the data matrix (Lines 42–43) in the `main` function and subsequently load the MNIST dataset from disk to a vector `mnist` (Lines 47–49). Second, we reserve memory for the all-pairs distance matrix and call the sequential function for the computation of all pairwise distances (Lines 54–55). Third, we compute the squared Euclidean distance (Lines 22–27) for all index combinations $i' \leq i$ (Lines 17–18) and finally write the contributions to the all-pairs distance matrix (Line 30). The code can be compiled with a C++14 compliant compiler:

```
g++ -O2 -std=c++14 -pthread all_pair.cpp -o all_pair
```

The $65{,}000^2 \approx 4 \cdot 10^9$ entries of Δ are computed in roughly 30 minutes on a single core of a Xeon E5-2683 v4 CPU. This is quite a lot of time. A straightforward parallelization using a static distribution over the rows of Δ is conceivable. However, due to the symmetry of Δ we only have to compute $i + 1$ entries in row i instead of a whole row consisting of m entries. This inhomogeneous work distribution over distinct rows might result in a suboptimal schedule. We discuss different parallelization approaches using static and dynamic distributions of threads in the following.

STATIC SCHEDULES

A static distribution such as a pure block distribution or a block-cyclic distribution with a big chunk size is not suitable for this task. Let $T(i) = \alpha \cdot (i + 1)$ be the time taken to compute row i for some positive constant $\alpha > 0$, and further $0 \leq i_0 < m - c$ be the first row to be processed in a chunk by a thread and $c > 0$ the corresponding chunk size, then the overall time to compute the whole chunk has

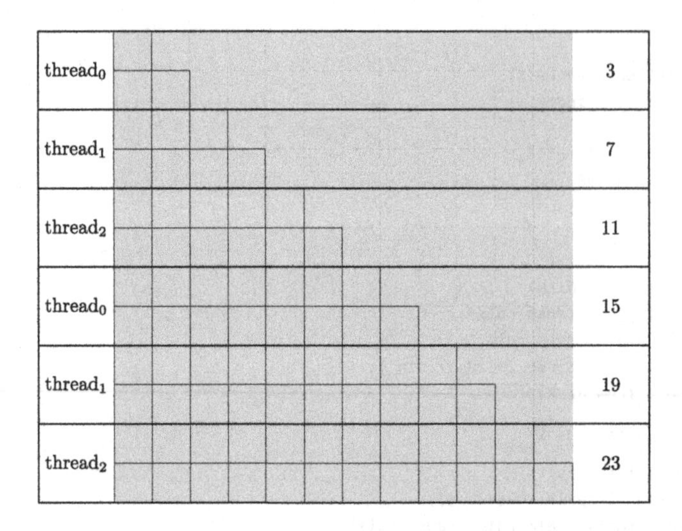

FIGURE 4.7

A block-cyclic distribution of three threads using a chunk size of 2 operating on a 12×12 all-pairs distance matrix. The numbers on the right correspond to the amount of distance measure computations in a chunk.

a quadratic dependency on c, a linear dependency on c, and the product $i_0 \cdot c$ since

$$F(i_0, c) := \sum_{l=0}^{c-1} T(i_0 + l) = \sum_{l=0}^{i_0+c-1} T(l) - \sum_{l=0}^{i_0-1} T(l)$$
$$= \alpha \left(\sum_{l=0}^{i_0+c-1} l - \sum_{l=0}^{i_0-1} l + c \right) = \frac{\alpha \left(c + 2i_0 \cdot c + c^2 \right)}{2}. \tag{4.6}$$

Here we used the well-known equality $\sum_{l=0}^{k} l = \frac{k\cdot(k+1)}{2}$ to produce the final expression for $F(i_0, c)$. If we set the chunk size c to be a small constant, as it is the case in a cyclic distribution ($c = 1$), then we obtain an overall workload $F(i_0, c)$ that scales linearly with the chunk identifier (which itself is a linear function of i_0). The impact of the mixing term $i_0 \cdot c$ increases with the chunk size, i.e. we worsen the load imbalance for increasing values of c. The worst case is a pure block distribution where the chunk size is maximal ($c = \lceil m/p \rceil$). Consequently, we should choose c as small as possible in order to approximately equalize the execution times for each chunk (see Fig. 4.7). Concluding, it is advisable to employ static schedules with a small chunk size when fighting load imbalance. Listing 4.15 shows the parallel implementation of all-pairs matrix computation using a block-cyclic distribution.

```
1  template <
2      typename index_t,
3      typename value_t>
4  void parallel_all_pairs(
5      std::vector<value_t>& mnist,
```

```
6       std::vector<value_t>& all_pair,
7       index_t rows,
8       index_t cols,
9       index_t num_threads=64,
10      index_t chunk_size=64/sizeof(value_t)) {
11
12      auto block_cyclic = [&] (const index_t& id) -> void {
13
14          // precompute offset and stride
15          const index_t off = id*chunk_size;
16          const index_t str = num_threads*chunk_size;
17
18          // for each block of size chunk_size in cyclic order
19          for (index_t lower = off; lower < rows; lower += str) {
20
21              // compute the upper border of the block (exclusive)
22              const index_t upper = std::min(lower+chunk_size,rows);
23
24              // for all entries below the diagonal (i'=I)
25              for (index_t i = lower; i < upper; i++) {
26                  for (index_t I = 0; I <= i; I++) {
27
28                      // compute squared Euclidean distance
29                      value_t accum = value_t(0);
30                      for (index_t j = 0; j < cols; j++) {
31                          value_t residue = mnist[i*cols+j]
32                                          - mnist[I*cols+j];
33                          accum += residue * residue;
34                      }
35
36                      // write Delta[i,i'] = Delta[i',i]
37                      all_pair[i*rows+I] =
38                      all_pair[I*rows+i] = accum;
39                  }
40              }
41          }
42      };
43
44      // business as usual
45      std::vector<std::thread> threads;
46
47      for (index_t id = 0; id < num_threads; id++)
48          threads.emplace_back(block_cyclic, id);
49
50      for (auto& thread : threads)
51          thread.join();
52  }
```

Listing 4.15: All-pairs matrix template using a static block-cyclic distribution.

When executed on a dual socket Xeon E5-2683 v4 CPU (2×16 physical cores + hyperthreading) using 64 software threads, we obtain the following execution times and speedups depending on the chunk size c (averaged over 10 runs). The sequential baseline implementation from the previous subsection took 30 minutes to compute the result.

Chunk size c	1	4	16	64	256	1024
Time in s	44.6	45.0	45.6	49.9	57.0	78.5
Speedup	40.5	40.0	39.5	36.1	31.6	22.9

The experiment confirms our theoretical considerations: an increased chunk size causes a higher level of load imbalance resulting in longer overall execution times. The worst-case $c = 1024$ approximates a pure block distribution where $c = \lceil \frac{65{,}000}{64} \rceil = 1016$.

DYNAMIC BLOCK-CYCLIC DISTRIBUTIONS

As demonstrated above, static schedules with a small chunk size are useful to approximately balance skewed work distributions. However, this approach is not applicable to any situation. We basically cheated in the aforementioned example since we knew the explicit time dependency $T(i)$ of each task beforehand and further could exploit the monotonicity of the work performed in a chunk $F(i_0, c)$ in the position of the chunk i_0. In detail, we relied on the fact that two adjacent chunks take approximately the same time to process. In general, $T(i)$ and thus $F(i_0, c)$ could be any function with potentially exponential dependency in the index enumerating the tasks. The situation gets even worse if we cannot robustly estimate the runtime of a task at program start.

Branch-and-bound algorithms such as the Traveling Salesman Problem (TSP), the Knapsack Problem, and other exhaustive constrained graph search algorithms using backtracking approaches can be decomposed into independent tasks with highly varying execution times. Other examples include transactions to attached media such as databases or web services which might exhibit heavily varying response times (some queries are exponentially harder than others). Let us demonstrate that with a simple toy model. Assume we have to process four tasks a, b, A, B using two processors. The lower case tasks are ten times smaller than the upper case ones, i.e. $T(A) = T(B) = 10 \cdot T(a) = 10 \cdot T(b) = 10\,\text{s}$. An optimal schedule would assign the tasks $\{A, a\}$ to thread 0 and $\{B, b\}$ to thread 1 resulting in an overall parallel runtime of 11 seconds. Nevertheless, we could also obtain a worst-case schedule such as $\{a, b\}$ and $\{A, B\}$ which takes 20 seconds to compute. Unfortunately, we cannot exclude this schedule with a static distribution since we have no a priori information about the tasks, i.e. they are indistinguishable for us at program start. We can resolve this issue with a greedy on-demand assignment strategy. Assume thread 0 starts to process task A and thread 1 begins with b. After having processed b in 1 second, thread 1 runs out of work and greedily picks the next task: let us say it is a. While thread 0 is still computing A, thread 1 subsequently picks the remaining task B. The resulting worst-case on-demand schedule $\{A\}$ and $\{b, a, B\}$ finishes after merely 12 seconds in contrast to the 20 seconds of the static worst-case schedule.

Task assignment at runtime is called *dynamic scheduling* and can be realized on a per-task basis or as demonstrated before on a chunk-level. The first can be seen as a special case of the latter for the choice $c = 1$. The same observation about the chunk size in static distributions is applicable to dynamic distributions. Smaller chunk sizes are favorable for heavily skewed work distributions. Note that determining an optimal schedule is an inherent hard problem even if we know the execution times of the tasks beforehand. Thus, dynamic assignment strategies are only heuristics to improve static schedules.

In the following, we will refine the static block-cyclic approach for the computation of the all-pairs matrix Δ to dynamically select chunks of rows until we exhaust the list of overall rows. This is achieved with a globally accessible variable `global_lower` which denotes the first row of the currently processed chunk. Whenever a thread runs out of work, it reads the value of `global_lower`, subsequently increments it by the chunk size c, and finally processes the corresponding rows of that chunk. All threads terminate if `global_lower` is greater than or equal to the number of to be processed rows m. However, we have to be aware that the global variable is shared among all threads which might result in a race condition during its incrementation by c. Hence, we have to ensure that accesses to `global_lower` are mutually exclusive to guarantee correct results.

C++11 provides a mechanism to restrict the execution of critical sections to a certain thread in a mutually exclusive manner: the so-called **mutex**. A mutex can be locked by a specific thread, i.e. the subsequent code fragment can only be executed by the issuing thread until the mutex is released. While being locked a mutex cannot be locked or unlocked by other threads which have to wait for its release causing an implicit synchronization of threads. Loosely speaking, mutexes are used to serialize certain portions of code in a parallel context to ensure the safe manipulation of shared information. Mutexes are defined in the `mutex` header. Basically, all you have to do is to define a globally accessible mutex and use it in a thread:

```
#include <mutex>
std::mutex mutex;

// to be called by threads
void some_function(...) {

    mutex.lock()
    // this region is only processed by one thread at a time
    mutex.unlock();

    // this region is processed in parallel
}
```

Be aware that you should always unlock a locked mutex to avoid deadlocks. C++11 features a convenient wrapper `std::lock_guard` which locks a specific scope and automatically handles the release on leave:

```
#include <mutex>
std::mutex mutex;

// to be called by threads
void some_function(...) {

    {
        // here we acquire the lock
        std::lock_guard<std::mutex> lock_guard(mutex);
        // this region is locked by the mutex
    } // <- here we release the lock

    // this region is processed in parallel
}
```

Listing 4.16 shows how to implement a dynamic assignment strategy for the blocks. We start with the declaration of the mutex and the shared global counter `global_lower` in Lines 15–16. It is crucial to define both here in order to be able to capture them by the closure `dynamic_block_cyclic` in Line 18. Inside the lambda, we iteratively probe the value `lower` encoding the first index of the to be processed chunk in a while-loop (Line 24) until there are no chunks left. The actual value of `lower` is incrementally read from the global counter `global_lower` (Line 29) and subsequently incremented by c (Line 30) within a locked scope (Lines 27–32). The remaining parts of the closure are almost identical to the static variant from the previous subsection. The only exception is the missing thread identifier `id` in the arguments of the closure `dynamic_block_cyclic` which is not needed anymore (Line 18 and Line 60).

```
1   #include <mutex> // std::mutex, std::lock_guard
2
3   template <
4       typename index_t,
5       typename value_t>
6   void dynamic_all_pairs(
7       std::vector<value_t>& mnist,
8       std::vector<value_t>& all_pair,
9       index_t rows,
10      index_t cols,each thrsds
11      index_t num_threads=64,
12      index_t chunk_size=64/sizeof(value_t)) {
13
14      // declare mutex and current lower index
15      std::mutex mutex;
16      index_t global_lower = 0;
17
18      auto dynamic_block_cyclic = [&] ( ) -> void {
19
20          // assume we have not done anything
21          index_t lower = 0;
22
23          // while there are still rows to compute
24          while (lower < rows) {
25
26              // update lower row with global lower row
27              {
28                  std::lock_guard<std::mutex> lock_guard(mutex);
29                          lower  = global_lower;
30                  global_lower += chunk_size;
31              } // here we release the lock
32
33              // compute the upper border of the block (exclusive)
34              const index_t upper = std::min(lower+chunk_size,rows);
35
36              // for all entries below the diagonal (i'=I)
37              for (index_t i = lower; i < upper; i++) {
38                  for (index_t I = 0; I <= i; I++) {
39
40                      // compute squared Euclidean distance
```

```
41                  value_t accum = value_t(0);
42                  for (index_t j = 0; j < cols; j++) {
43                      value_t residue = mnist[i*cols+j]
44                                      - mnist[I*cols+j];
45                      accum += residue * residue;
46                  }
47
48                  // write Delta[i,i'] = Delta[i',i]
49                  all_pair[i*rows+I] =
50                  all_pair[I*rows+i] = accum;
51              }
52          }
53      }
54  };
55
56  // business as usual
57  std::vector<std::thread> threads;
58
59  for (index_t id = 0; id < num_threads; id++)
60      threads.emplace_back(dynamic_block_cyclic);
61
62  for (auto& thread : threads)
63      thread.join();
64  }
```

Listing 4.16: All-pairs matrix template using a dynamic block-cyclic distribution.

When executed on a dual socket Xeon E5-2683 v4 CPU (2×16 physical cores + hyperthreading) using 64 software threads we obtain the following execution times and speedups depending on the chunk size c (averaged over 10 runs). The sequential baseline implementation from the previous subsection took 30 minutes to compute the result. The results for the static block-cyclic distribution are taken from the previous subsection.

Mode	Chunk size c	1	4	16	64	256	1024
Static	Time in s	44.6	45.0	45.6	49.9	57.0	78.5
	Speedup	40.5	40.0	39.5	36.1	31.6	22.9
Dynamic	Time in s	43.6	43.6	43.9	46.3	53.8	77.6
	Speedup	41.3	41.3	41.0	38.9	33.5	23.2

Obviously, the dynamic assignment of chunks to threads is beneficial for all chunk size configurations. Moreover, we observe again that small chunk sizes are favorable when processing tasks with heavily skewed load distributions.

4.5 SIGNALING THREADS WITH CONDITION VARIABLES (PING PONG)

Up to this point, we pursued a *race-to-sleep* strategy which fully utilizes all spawned threads until completion of their corresponding tasks. Race-to-sleep approaches are typically considered the most efficient computation patterns in terms of energy consumption since terminated threads do not consume any energy in theory. Nevertheless, sometimes it is necessary to put already spawned threads to sleep due to inter-thread dependencies. As an example, consider a chain of interdependent tasks that cannot proceed their computation until a certain task has finished its work. The dynamic schedule from the previous section is a typical use case where threads have to wait for the increment of a global counter variable. However, the presented dynamic schedule keeps all threads busy during the waiting for the mutex. In practice, it would be undesirable to waste valuable computational resources on a thread that accomplishes nothing but waiting for other threads. Hence, we are interested in a mechanism that allows for an easy way to put threads to sleep and subsequently signal them to wake up. C++11 features a simple mechanism to achieve that: so-called **condition variables**.

MODELING A SLEEPING STUDENT

Before we proceed with the actual implementation of condition variables, let us make an initial remark. The name *condition variable* is somehow misleading since it implies that a condition variable is always in a well-defined state. Actually, we will see that condition variables alone cannot be used to communicate the state of our program due to spurious wake-ups of threads that may occur accidentally. Let us demonstrate this behavior with a real-world example. Assume you have a really annoying alarm clock set to 7 a.m. in the morning. You can be sure that if it rings, you will be awake and thus you can proceed with your breakfast. However, it is possible that you accidentally woke up too early at 6 a.m. Hence, being awake is not a guarantee that it is past 7 a.m. The only way to distinguish if you are awake right on time or too early is to have a look at the display of the clock which tells you the correct time. So to speak, you need a distinct piece of information encoded in another state to exclude spurious wake-ups. In case of a spurious wake-up you would go back to sleep and repeat the above described procedure until you are sure it is time for breakfast. This is exactly how condition variables work: they can be seen as a remote signaling mechanism that guarantees that a thread is ready for work if being signaled. The check whether it is really time to work has to be accomplished with the help of a distinct state (usually another shared variable). You might wonder, why spurious wake-ups are allowed at all and have not been excluded from the programming language. In practice, some architectures implement spurious behavior due to performance reasons while others do not. Hence, we have to take care of this possibility in order to write correct code that is portable across platforms.

The typical workflow for the *signaling thread* looks as follows:

1. The signaling thread has to acquire a mutex using either `mutex.lock()` or by means of a scoped wrapper such as `std::lock_guard` or `std::unique_lock`.
2. While holding the lock, the shared state is modified and further inherently sequential work is performed (such as printing to the command line).
3. The lock is released either explicitly with `mutex.unlock()`, or implicitly by leaving the scope of `std::lock_guard` or `std::unique_lock`.
4. The actual signaling by means of the condition variable `cv` is performed using `cv.notify_one()` for one thread, or `cv.notify_all()` for all threads.

The workflow for the *waiting threads* that intend to be signaled reads as follows:

1. A waiting thread has to acquire a `std::unique_lock` using the same mutex as in the signaling phase. Note that `std::lock_guard` cannot be used here.
2. While being locked call either `cv.wait()`, `cv.wait_for()`, or `wait_until()` using the aforementioned condition variable `cv`. The lock is released automatically in order to ensure that other threads can acquire the mutex again.
3. In case (i) the condition variable `cv` is notified, (ii) the timeout of `cv.wait()` or `cv.wait_for()` expires, or (iii) a spurious wake-up occurs, the thread is awaken, and the lock is reacquired. At this point, we have to check whether the globally shared state indicates to proceed or to wait (sleep) again.

An exemplary implementation of a sleeping thread that is signaled by the master thread is shown in Listing 4.17. Initially, we define the mutex, the condition variable, and the globally shared state in Lines 12–14. Subsequently, we spawn a thread that iteratively probes whether the global state `time_for_breakfast` is true and afterwards is put to sleep. Hence, a spurious wake-up would result in another round of wait due to the incorrect value of the shared state. Note that the loop can be avoided if we specify the predicate with the help of a closure as demonstrated in Lines 29–30. Finally, we alter the shared state in the master thread (Line 42) and notify the waiting thread.

```
1  #include <iostream>             // std::cout
2  #include <thread>               // std::thread
3  #include <mutex>                // std::mutex
4  #include <chrono>               // std::this_thread::sleep_for
5  #include <condition_variable>   // std::condition_variable
6
7  // convenient time formats (C++14 required)
8  using namespace std::chrono_literals;
9
10 int main() {
11
12     std::mutex mutex;
13     std::condition_variable cv;
14     bool time_for_breakfast = false; // globally shared state
15
16     // to be called by thread
17     auto student = [&] ( ) -> void {
18
19         { // this is the scope of the lock
20             std::unique_lock<std::mutex> unique_lock(mutex);
21
22             // check the globally shared state
23             while (!time_for_breakfast)
24                 // lock is released during wait
25                 cv.wait(unique_lock);
26
27             // alternatively, you can specify the
28             // predicate directly using a closure
29             // cv.wait(unique_lock,
30             //         [&](){ return time_for_break_fast; });
31         } // lock is finally released
32
```

```
33            std::cout << "Time to make some coffee!" << std::endl;
34        };
35
36        // create the waiting thread and wait for 2 s
37        std::thread my_thread(student);
38        std::this_thread::sleep_for(2s);
39
40        { // prepare the alarm clock
41            std::lock_guard<std::mutex> lock_guard(mutex);
42            time_for_breakfast = true;
43        } // here the lock is released
44
45        // ring the alarm clock
46        cv.notify_one();
47
48        // wait until breakfast is finished
49        my_thread.join();
50    }
```

Listing 4.17: Signaling of a sleeping student in the morning.

PLAYING PING PONG WITH CONDITION VARIABLES

In the following, we write a simple program consisting of two threads that mutually write "ping" and "pong" to the command line. While thread 0 is writing "ping" to `stdout`, thread 1 is still sleeping and waiting to be signaled by thread 0 to perform the "pong." After thread 0 has written its "ping" it alters a shared binary variable indicating that the program is in the "pong" state, subsequently signals thread 1 using a condition variable, and finally goes to sleep. The described procedure is repeated with interchanged roles until we reach a fixed number of iterations or encounter another user-defined condition for termination. For the sake of simplicity we neglect the termination condition and let the program run forever.

The corresponding code is shown in Listing 4.18. Its structure is quite simple: we define two closures, `ping` (Line 16) and `pong` (Line 33), which are executed by two independent threads (Lines 50–53). The global binary state of the program indicating if we are in the ping or pong phase is encoded in the variable `is_ping` in Line 14. Both threads wait for the other one to perform their print statement in a wait-loop using `cv.wait`, together with a closure checking the predicate (Line 21 and Line 38). Note that in this example the notification is moved to the locked scope to avoid unnecessary clutter in the source code.

```
 1  #include <iostream>              // std::cout
 2  #include <thread>                // std::thread
 3  #include <mutex>                 // std::mutex
 4  #include <chrono>                // std::this_thread::sleep_for
 5  #include <condition_variable>    // std::condition_variable
 6
 7  // convenient time formats (C++14 required)
 8  using namespace std::chrono_literals;
 9
10  int main() {
```

```
11
12      std::mutex mutex;
13      std::condition_variable cv;
14      bool is_ping = true; // globally shared state
15
16      auto ping = [&] ( ) -> void {
17          while (true) {
18
19              // wait to be signaled
20              std::unique_lock<std::mutex> unique_lock(mutex);
21              cv.wait(unique_lock,[&](){return is_ping;});
22
23              // print "ping" to the command line
24              std::this_thread::sleep_for(1s);
25              std::cout << "ping" << std::endl;
26
27              // alter state and notify other thread
28              is_ping = !is_ping;
29              cv.notify_one();
30          }
31      };
32
33      auto pong = [&] ( ) -> void {
34          while (true) {
35
36              // wait to be signaled
37              std::unique_lock<std::mutex> unique_lock(mutex);
38              cv.wait(unique_lock,[&](){return !is_ping;});
39
40              // print "pong" to the command line
41              std::this_thread::sleep_for(1s);
42              std::cout << "pong" << std::endl;
43
44              // alter state and notify other thread
45              is_ping = !is_ping;
46              cv.notify_one();
47          }
48      };
49
50      std::thread ping_thread(ping);
51      std::thread pong_thread(pong);
52      ping_thread.join();
53      pong_thread.join();
54  }
```

Listing 4.18: Playing ping pong with condition variables.

ONE-SHOT SYNCHRONIZATION USING FUTURES AND PROMISES

Condition variables are useful in case many threads perform non-trivial, or repetitive synchronization patterns. An example for the latter is our ping pong application where two threads reuse one condition variable for potentially infinite signaling steps. The overhead of defining a condition variable, another distinct global state, the notification of the involved threads, and the subsequent probing using `cv.wait` in a locked scope seems to be a bit out of scale if we aim to signal one or more threads only once as performed in the introductory alarm clock example. The amount of source code can be drastically reduced for so-called *one-shot synchronization* scenarios by using futures and promises. This technique initially proposed by Scott Meyers [4] exploits the synchronization property of a future f and its corresponding promise p as explained in Section 4.2.

In order to demonstrate the advantages of this approach we reimplement the alarm clock example from the beginning of this section. The workflow of the communication pattern is straightforward:

1. We pass a future f of any type (assume `std::future<void>` for simplicity) to the function modeling the sleeping student.
2. The future f is immediately accessed at the beginning of the function body with a call to `f.get()`.
3. The corresponding promise p is fulfilled in the master thread by writing any value to p via `p.set_value()`. The actual value is irrelevant.

Synchronization between the issuing master thread and the sleeping student is achieved since the access to a future's values via `f.get()` blocks until the fulfilling of the corresponding promise with `p.set_value()`. The implementation of the described approach is straightforward and needs no further explanation:

```
1   #include <iostream>            // std::cout
2   #include <thread>              // std::thread
3   #include <future>              // std::future
4   #include <chrono>              // std::this_thread::sleep_for
5
6   // convenient time formats (C++14 required)
7   using namespace std::chrono_literals;
8
9   int main() {
10
11      // create pair (future, promise)
12      std::promise<void> promise;
13      auto future = promise.get_future();
14
15      // to be called by thread
16      auto student = [&] ( ) -> void {
17
18          future.get(); // blocks until fulfilling promise
19          std::cout << "Time to make coffee!" << std::endl;
20      };
21
22      // create the waiting thread and wait for 2s
23      std::thread my_thread(student);
24      std::this_thread::sleep_for(2s);
25
26      // ring the alarm clock
27      promise.set_value();
28
```

```
29     // wait until breakfast is finished
30     my_thread.join();
31 }
```

Listing 4.19: Signaling of a sleeping student in the morning.

The described approach is only applicable in situations where one thread signals another thread since a future can only be read once. Fortunately, we can apply a workaround to allow for the signaling of many threads at once. So-called *shared futures* can be used to broadcast a certain value to more than one thread. Conceptually, a shared future is treated like a traditional future on first access. Further accesses simply return the original value stemming from the first access, i.e. from this point the shared future acts like a constant. In practice, all participating threads try to concurrently access the shared future and whoever succeeds first after fulfilling the promise triggers the synchronization. Waking up several students is as easy as:

```
1  #include <iostream>             // std::cout
2  #include <thread>               // std::thread
3  #include <future>               // std::future
4  #include <chrono>               // std::this_thread::sleep_for
5
6  // convenient time formats (C++14 required)
7  using namespace std::chrono_literals;
8
9  int main() {
10
11     // create pair (future, promise)
12     std::promise<void> promise;
13     auto shared_future = promise.get_future().share();
14
15     // to be called by one or more threads
16     auto students = [&] ( ) -> void {
17
18         // blocks until fulfilling promise
19         shared_future.get();
20         std::cout << "Time to make coffee!" << std::endl;
21     };
22
23     // create the waiting thread and wait for 2s
24     std::thread my_thread0(students);
25     std::thread my_thread1(students);
26     std::this_thread::sleep_for(2s);
27
28     // get up lazy folks!
29     promise.set_value();
30
31     // wait until breakfast is finished
32     my_thread0.join();
33     my_thread1.join();
34 }
```

Listing 4.20: Signaling of several sleeping students in the morning.

4.6 PARALLELIZING OVER IMPLICITLY ENUMERABLE SETS (THREAD POOL)

We have already discussed how to fight load-imbalance in Section 4.4. The corresponding setting assumed that to be processed tasks are stored in a linear data structure. We further showed that dynamic schedules are often preferable over static ones, and that monolithic pure block approaches can result in heavily skewed load distributions. A limitation of the presented implementations so far is the assumption that the number of tasks to be processed is known beforehand.

IMPLICITLY ENUMERABLE SETS

Some algorithms traverse non-linear topologies with a potentially unknown number of tasks. As an example, consider an a priori unknown directed acyclic graph (DAG) which models dependencies between tasks. If the whole graph fits into memory, we could rewrite the set of vertices (tasks) as a list of iteratively processable jobs by determining a suitable topological sorting of the DAG. Note, this can be achieved by breadth-first search-based algorithms in linear time ($\mathcal{O}(|V| + |E|)$) of both the amount of vertices $|V|$ and edges $|E|$ as shown in [1]. Nevertheless, this preprocessing step is not always applicable because of two major limitations:

1. The DAG does not fit into the RAM of our workstation and it is intractable to load it incrementally from disk in batches.
2. We do not know any closed mathematical expression which explicitly enumerates the vertices associated to the tasks.

The latter limitation sounds rather theoretical on first sight but it is actually quite easy to find representative examples in real life. Let us play a round of the popular game *Boggle*. The task is to find all English words within a 4×4 grid of letters A to Z which lie on a connected path. A path may start in any of the 16 grid cells and can be extended in horizontal, vertical, and diagonal direction without visiting a cell twice. Fig. 4.8 shows an exemplary Boggle configuration together with one valid solution "THREAD." Other solutions include "JOB," "READ" and "HAT." Assume that someone provided us a binary criterion which decides whether a certain string is an English word or not, and further would ask us to determine all possible solutions. One could argue that this task is quite easy: all you have to do is mapping all realizable paths together with the associated strings by the aforementioned decision function onto the binary set $\{0, 1\}$.

The hard part of this problem is to determine the set of all realizable paths compliant with the rules of Boggle. A tractable solution for this problem is to implicitly enumerate the state space by recursively generating solutions candidates. Assume we begin our search in the third cell of the first row ("T") and generate the five valid extensions ("TB," "TA," "TO," "TZ," and "TH") out of the eight potential extension in the general case. Note that three extensions to the top are impossible since the cell "T" lies at the border. We can subsequently extend the five candidate solutions recursively until we find no further extensions. Each valid path has to be checked if it corresponds to a valid English word. The overall number of valid paths can only be determined after having traversed all possibilities. Hence, an a priori schedule is hard to determine.

Another real-world example is a web server. An arbitrary amount of web pages could be requested at any time. Moreover, the time to answer a query could heavily vary depending on the requested action.

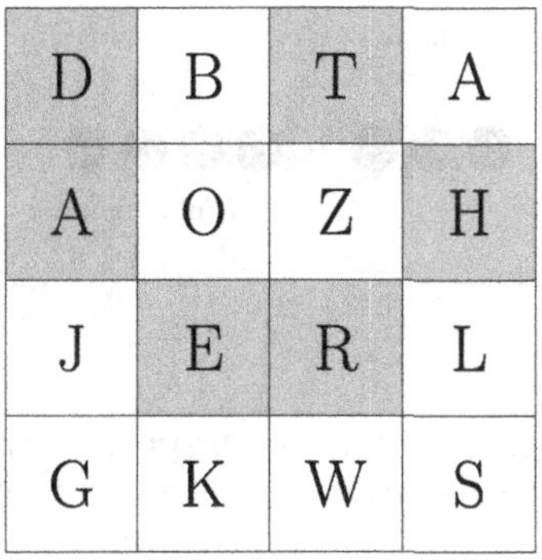

D	B	T	A
A	O	Z	H
J	E	R	L
G	K	W	S

FIGURE 4.8

An exemplary Boggle grid with one valid solution.

Again, it is impossible to provide a reasonable a priori estimation of to be processed tasks at program start. Addressing this issue, a naive approach would spawn a new thread for every incoming request resulting in an excessive oversubscription of threads and thus would significantly hurt performance. In the worst case, a user intentionally stalls our workstation with a Denial of Service (DOS) attack by submitting tons of useless requests. A practicable workaround is a thread pool which maintains a fixed number of threads used to incrementally process the list of tasks. Note, DOS attacks could be avoided by augmenting the tasks with time stamps: requests that have not been processed for a predefined time span are simply discarded.

USE CASES FOR A THREAD POOL

In this section, we implement a thread pool that maintains a list of tasks with arbitrary return values and arguments for two reasons. First, neither C++11 nor C++14 features an out-of-the-box thread pool and thus you have to reimplement it anyway in aforementioned examples. Second, a correctly working thread pool exhibits non-trivial synchronization and signaling mechanism that are ideally suited to demonstrate the benefits of condition variables, as well as futures and promises in an educational context. The proposed implementation is heavily inspired by Jakob Progsch's and Václav Zeman's thread pool library [5] and has been rewritten with a focus on readability and educational clarity.

A typical use case is the incremental submission of tasks enumerated in a data structure with linear topology (see Listing 4.21). Here we traverse a linear data structure using a for-loop. In each iteration a task is submitted to the thread pool `TP` via `TP.enqueue(square, task)` (see Line 19). The corresponding future is immediately stored in a globally accessible array `futures` (see Line 15) which is subsequently accessed in the final for-loop in Lines 24–25. Fig. 4.9 visualizes the described workflow. Note that linear topologies could also be tackled with the proposed dynamic scheduling techniques discussed in the previous sections.

```
1  #include <iostream>
2  #include "threadpool.hpp" // the to be written pool
3
4  ThreadPool TP(8);         // 8 threads in a pool
```

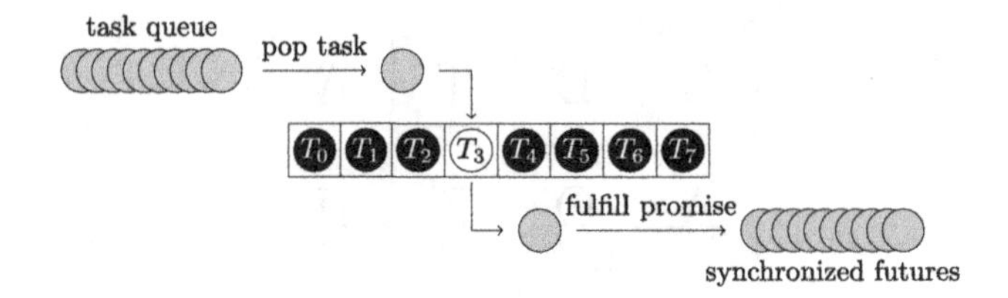

FIGURE 4.9

An exemplary thread pool executing eight threads where thread T_3 is idling. Subsequently, T_3 removes a new task from the task queue, and executes it. The associated future is synchronized by fulfilling the corresponding promise after termination of the task.

```
 5
 6 int main () {
 7
 8     // function to be processed by threads
 9     auto square = [](const uint64_t x) {
10         return x*x;
11     };
12
13     // more tasks than threads in the pool
14     const uint64_t num_tasks = 32;
15     std::vector<std::future<uint64_t>> futures;
16
17     // enqueue the tasks in a linear fashion
18     for (uint64_t task = 0; task < num_tasks; task++) {
19         auto future = TP.enqueue(square, task);
20         futures.emplace_back(std::move(future));
21     }
22
23     // wait for the results
24     for (auto& future : futures)
25         std::cout << future.get() << std::endl;
26 }
```

Listing 4.21: Thread pool: exemplary usage operating on a linear topology.

A more complex example using a tree topology is shown in Listing 4.22. Here we recursively traverse a complete binary tree using preorder enumeration (see Fig. 4.10). The closure `traverse` in Line 18 calls itself recursively for the left and right child until we meet a certain criteria for termination (see Line 19). A job is submitted to the thread pool for each implicitly enumerated node in Line 22. Finally, the futures are accessed at the end of our `main` function in Lines 35–36. Note that this kind of recursive traversal could be used to determine all solutions of a Boggle game. The only difference is a more complex traversal closure which generates possible extensions of a path.

```
1 #include <iostream>
2 #include "threadpool.hpp" // the to be written pool
3
4 ThreadPool TP(8);         // 8 threads in a pool
```

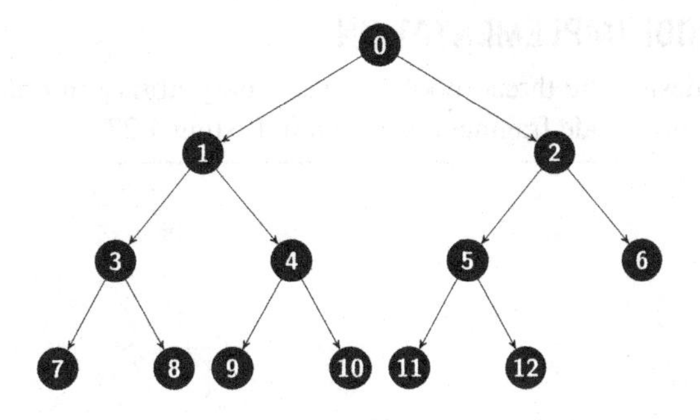

FIGURE 4.10

An example of a complete binary tree consisting of $n = 13$ nodes. A node with index k is defined as a leaf if neither the left child $(2k+1)$ nor the right child $(2k+2)$ exists. Hence, the recursion terminates if $k \geq n$.

```
5
6  int main () {
7
8      // function to be processed by threads
9      auto square = [](const uint64_t x) {
10         return x*x;
11     };
12
13     const uint64_t num_nodes = 32;
14     std::vector<std::future<uint64_t>> futures;
15
16     // preorder binary tree traversal
17     typedef std::function<void(uint64_t)> traverse_t;
18     traverse_t traverse = [&] (uint64_t node){
19         if (node < num_nodes) {
20
21             // submit the job
22             auto future = TP.enqueue(square, node);
23             futures.emplace_back(std::move(future));
24
25             // traverse a complete binary tree
26             traverse(2*node+1); // left child
27             traverse(2*node+2); // right child
28         }
29     };
30
31     // start at the root node
32     traverse(0);
33
34     // get the results
35     for (auto& future : futures)
36         std::cout << future.get() << std::endl;
37 }
```

Listing 4.22: Thread pool: exemplary usage using a tree topology.

A SIMPLE THREAD POOL IMPLEMENTATION

Let us start coding. We design the thread pool as header-only library in order to avoid inconvenient linking of libraries. The initial code fragment is shown in Listing 4.23.

```
1   #ifndef THREADPOOL_HPP
2   #define THREADPOOL_HPP
3
4   #include <cstdint>
5   #include <future>
6   #include <vector>
7   #include <queue>
8   #include <thread>
9   #include <mutex>
10  #include <condition_variable>
11
12  class ThreadPool {
13
14  private:
15
16      // storage for threads and tasks
17      std::vector<std::thread> threads;
18      std::queue<std::function<void(void)>> tasks;
19
20      // primitives for signaling
21      std::mutex mutex;
22      std::condition_variable cv;
23
24      // the state of the thread pool
25      bool stop_pool;
26      uint32_t active_threads;
27      const uint32_t capacity;
```

Listing 4.23: Thread pool: initial source code fragment.

First, we include all necessary header files providing explicit integer types, futures, container classes, threads, mutexes, and condition variables in Lines 4–10. Note that the included guard in Lines 1–2 ensures that our thread pool header can only be included once. Second, we declare the fundamental primitives of the class `ThreadPool` as private member variables. The vector `threads` maintains the threads, and the queue `tasks` stores the to be processed tasks. Moreover, the mutex `mutex` and the condition variables `cv` will be used for signaling idling threads. Finally, we declare a Boolean variable `stop_pool` indicating whether the pool is still running, an unsigned 32-bit integer for the number of active threads, and another one for the maximum number of executable threads. Now we can proceed with the definition of three private auxiliary function (templates) in Listing 4.24.

```
28      // custom task factory
29      template <
30          typename     Func,
31          typename ... Args,
32          typename Rtrn=typename std::result_of<Func(Args...)>::type>
33      auto make_task(
34          Func &&    func,
```

```
35          Args && ...args) -> std::packaged_task<Rtrn(void)> {
36
37          auto aux = std::bind(std::forward<Func>(func),
38                               std::forward<Args>(args)...);
39
40          return std::packaged_task<Rtrn(void)>(aux);
41      }
42
43      // will be executed before execution of a task
44      void before_task_hook() {
45          active_threads++;
46      }
47
48      // will be executed after execution of a task
49      void after_task_hook() {
50          active_threads--;
51      }
```

Listing 4.24: Thread pool: private helper functions.

The task factory template make_task in Lines 29–41 is already known from Listing 4.5 in Section 4.2. It creates a packaged task from a function with arbitrary signature. The return value of the corresponding function can be accessed via an associated future object. The two auxiliary functions, before_task_hook in Lines 44–46 and after_task_hook in Lines 49–51, specify the actions to be performed before and after execution of a submitted task. In our case, we simply increment and decrement the active_threads member variable. We will later use the hooks to implement a non-blocking mechanism for work-sharing among threads. The public constructor of the class ThreadPool in Listing 4.25 contains the majority of the scheduling logic.

```
52  public:
53      ThreadPool(
54          uint64_t capacity_) :
55          stop_pool(false),     // pool is running
56          active_threads(0),    // no work to be done
57          capacity(capacity_) { // remember size
58
59          // this function is executed by the threads
60          auto wait_loop = [this] ( ) -> void {
61
62              // wait forever
63              while (true) {
64
65                  // this is a placeholder task
66                  std::function<void(void)> task;
67
68                  { // lock this section for waiting
69                      std::unique_lock<std::mutex>
70                          unique_lock(mutex);
71
72                      // actions must be performed on
73                      // wake-up if (i) the thread pool
```

```
74                     // has been stopped, or (ii) there
75                     // are still tasks to be processed
76                     auto predicate = [this] ( ) -> bool {
77                         return  (stop_pool) ||
78                                !(tasks.empty());
79                     };
80
81                     // wait to be waken up on
82                     // aforementioned conditions
83                     cv.wait(unique_lock, predicate);
84
85                     // exit if thread pool stopped
86                     // and no tasks to be performed
87                     if (stop_pool && tasks.empty())
88                         return;
89
90                     // else extract task from queue
91                     task = std::move(tasks.front());
92                     tasks.pop();
93                     before_task_hook();
94                 } // here we release the lock
95
96                 // execute the task in parallel
97                 task();
98
99                 {   // adjust the thread counter
100                    std::lock_guard<std::mutex>
101                        lock_guard(mutex);
102                    after_task_hook();
103                } // here we release the lock
104            }
105        };
106
107        // initially spawn capacity many threads
108        for (uint64_t id = 0; id < capacity; id++)
109            threads.emplace_back(wait_loop);
110    }
```

Listing 4.25: Thread pool: constructor.

First, the member variables `stop_pool`, `active_threads`, and `capacity` are initialized in Lines 55–57. Second, we define the core wait loop which is performed by each thread in the pool with a closure in Line 60. Each thread probes in an infinite loop (Line 63) if it is signaled to wake-up (Line 83) using the aforementioned mutex and condition variable. The binary predicate in Line 76 ensures that no spurious wake-up occurs. A thread has to perform work in two cases: (i) the pool was signaled to stop, or (ii) there are still tasks in the queue to be processed. The condition in Line 87 terminates the thread if both the pool is stopped and the task queue is empty. In contrast, if there are still tasks in the queue, we remove a task from the queue and subsequently execute the helper function `before_task_hook`. The whole waiting procedure has to be enclosed by a `std::unique_lock` for two reasons: (i) a lock is needed to perform the conditional wait in Line 83, and (ii) the queue is not thread-safe, i.e. we have to exclude race condition while removing tasks. Third, the actual task is executed with no lock held (Line 97). In contrast, moving the task execution in the scope of the lock would result in implicit serialization of the to be processed jobs. Fourth, we execute the helper function `after_task_hook` in a locked scope to ensure the race condition-free decrement of the variable `active_threads`. Finally, we spawn `capacity` many threads and store them in the vector `threads`. The destructor in Listing 4.26 performs the necessary steps to stop the thread pool.

```
111     ~ThreadPool() {
112
113         { // acquire a scoped lock
114             std::lock_guard<std::mutex>
115                 lock_guard(mutex);
116
117             // and subsequently alter
118             // the global state to stop
119             stop_pool = true;
120         } // here we release the lock
121
122         // signal all threads
123         cv.notify_all();
124
125         // finally join all threads
126         for (auto& thread : threads)
127             thread.join();
128     }
```

Listing 4.26: Thread pool: destructor.

First, we acquire a scoped lock in Line 114 in order to allow for the safe manipulation of the shared global state `stop_pool` which we now set to `true`. Second, we notify all threads in the pool to wake-up. The corresponding logic in the constructor handles whether to stop the threads immediately or to process the remaining tasks in the queue before termination. Finally, we enforce a global barrier by joining all threads. Consequently, the destructor terminates if all threads have performed the return statement in Line 88. At the very end, we have to implement the enqueue function template in Listing 4.27.

```
129     template <
130         typename     Func,
131         typename ... Args,
132         typename Rtrn=typename std::result_of<Func(Args...)>::type>
133     auto enqueue(
134         Func &&      func,
135         Args && ... args) -> std::future<Rtrn> {
136
137         // create the task, get the future
138         // and wrap task in a shared pointer
139         auto task = make_task(func, args...);
140         auto future = task.get_future();
141         auto task_ptr = std::make_shared<decltype(task)>
142                         (std::move(task));
143
144         {  // lock the scope
145             std::lock_guard<std::mutex>
```

```
146             lock_guard(mutex);
147
148             // you cannot reuse pool after being stopped
149             if(stop_pool)
150                 throw std::runtime_error(
151                     "enqueue on stopped ThreadPool"
152                 );
153
154             // wrap the task in a generic void
155             // function void -> void
156             auto payload = [task_ptr] ( ) -> void {
157                 // basically call task()
158                 task_ptr->operator()();
159             };
160
161             // append the task to the queue
162             tasks.emplace(payload);
163         }
164
165         // tell one thread to wake-up
166         cv.notify_one();
167
168         return future;
169     }
170 };
171
172 // this #endif belongs to the include guard in Lines 1-2
173 #endif
```

Listing 4.27: Thread pool: enqueue tasks.

The `enqueue` function template is variadic, i.e. it accepts a function `func` with arbitrarily many arguments `args`. The third template parameter is used to determine the return value type of `make_task` which is `std::future<Rtrn>` where `Rtrn` is the return value type of `func`. Thus, you can use the same thread pool object to process tasks with distinct return value types.

First, we create a packaged task using our custom `make_task` factory and assign the associated future object to a variable `future` in Lines 139–140. Second, we wrap the packaged task with a shared pointer in Line 141 since we do not want to handle memory deallocation ourselves. Furthermore, a shared pointer guarantees persistence of the task object during its usage even if it leaves its corresponding scope of declaration. Third, we acquire a `std::unique_lock` to store a bulk task `payload` in the task queue `tasks` (Line 162). The lambda expression `payload` in Line 156 wraps the packaged task with a closure in order to allow for the storage of tasks with arbitrary return value types. Note that enqueuing should not be allowed if the pool is already stopped (Line 149). Fourth, we release the lock and notify a thread to wake-up. Finally, we return the future associated to the packaged task.

At the very end, let us discuss the member variable `active_threads`. You might have noticed that we do not actually need to count the number of active threads since we do not exploit it anywhere in the scheduling logic. Hence, you could simply remove it and its corresponding helper functions `before_task_hook` and `after_task_hook`. However, we will see in the next chapter how the number of active threads can be used to implement a work-sharing approach where threads in the pool can recursively submit tasks to the pool in order to distribute their work to idling threads.

4.7 ADDITIONAL EXERCISES

1. Provide a real-world example where the use of several threads executed on a single CPU core is still beneficial.
2. Consider two square matrices $A, B \in \mathbb{R}^{n\times n}$ each of width $n = 1024$. In the following we want to calculate their sum $C = A + B$ by the index-wise addition of entries $C_{ij} = A_{ij} + B_{ij}$ for all $i, j \in \{0, \ldots, n-1\}$. Implement a parallel program that uses p threads to compute the entries of C. Experiment with different static thread distribution patterns.
3. The Euler–Riemann zeta function $\zeta(s) = \sum_{n=1}^{\infty} n^{-s}$ is often used in natural science, notably in applied statistics and during the regularization of quantum field theoretical descriptions of the Casimir Effect.[2] An alternative formula for the computation of $\zeta(s)$ on real domains is given as follows:

$$\zeta(s) = 2^s \cdot \lim_{k\to\infty} \sum_{i=1}^{k} \sum_{j=1}^{k} \frac{(-1)^{i+1}}{(i+j)^s} \quad .$$

Thus, we can approximate $\zeta(s)$ up to the degree k if we omit the leading limit. The following code snippet implements this idea:

```
double Riemann_Zeta(double s, uint64_t k) {

    double result = 0.0;

    for (uint64_t i = 1; i < k; i++)
        for (uint64_t j = 1; j < k; j++)
            result += (2*(i&1)-1)/pow(i+j, s);

    return result*pow(2, s);
}
```

The asymptotic time complexity of a single call to `Riemann_Zeta` is obviously in $\mathcal{O}(k^2)$. Let us now investigate the approximation quality depending on the parameter k. We calculate the corresponding values of `Riemann_Zeta(x, k)` for all $k \in \{0, \ldots, n-1\}$ and write the result to a vector `X` of length `n`:

```
for (uint64_t k = 0; k < n; k++)
    X[k] = Riemann_Zeta(2, k);  // = pi^2/6
```

a. Parallelize this loop using the C++11 threads. Are there any data dependencies or shared variables?

[2] See http://en.wikipedia.org/wiki/Casimir_effect.

b. Discuss load balancing for different thread distribution patterns.
c. Verify your theoretical assumptions with experimental results.

4. Have a look at following code fragment:

```
#include <iostream>
#include <thread>

int main () {

    auto child = [] () -> void {
        std::cout << "child" << std::endl;
    };

    std::thread thread(child);
    thread.detach();

    std::cout << "parent" << std::endl;
}
```

In the majority of cases, the output of the program is exclusively "parent" since we reach the end of the main function before the print statement in the `child` closure. However, the use of `thread.join();` instead of `thread.detach();` results in the output "child" followed by "parent." Now assume, that we had to implement our own `join`-method using condition variables. We would proceed as follows:

a. We introduce a global Boolean variable `done = false`. Furthermore, we need a mutex `m` and a condition variable `c`.
b. After the print statement in `child` we set `done = true` in a locked scope and afterwards notify the condition variable `c`.
c. Our custom `join`-method performs a conditional wait in a locked scope as long as `done == false`.

Implement the described custom `join`-method. Try out different implementations: experiment with explicit locking of mutexes and scoped locks.

5. Revisit the aforementioned exercise. Now implement a custom `join`-method using one-shot synchronization as described in Section 4.5.
6. C++11 threads do not support the configuration of priorities. However, we can access the underlying POSIX thread via `std::thread::native_handle`. Implement a simple program that allows for the manipulation of C++11 thread priorities by manipulating the associated POSIX thread. Use one-shot synchronization or a conditional wait to suspend the thread before performing the changes.
7. Recall the dynamically scheduled block-cyclic distribution for the computation of the all-pairs distance matrix presented in Section 4.4. The proposed scheme dynamically picks chunks of fixed size one by another beginning at row zero. Reverse the indexing scheme in such a manner that the schedule begins with the chunks at the bottom of the matrix containing the longest rows. Do you observe an improvement in runtime? Explain the result.

REFERENCES

[1] Thomas H. Cormen, et al., Introduction to Algorithms, 3rd edition, The MIT Press, 2009, ISBN: 0262033844, 9780262033848.
[2] Ulrich Drepper, What every programmer should know about memory, http://people.redhat.com/drepper/cpumemory.pdf, 2007.
[3] Yann LeCun, Corinna Cortes, Christopher J.C. Burges, The MNIST database of handwritten digits, http://yann.lecun.com/exdb/mnist/ (visited on 01/12/2016).
[4] Scott Meyers, Effective Modern C++: 42 Specific Ways to Improve Your Use of C++11 and C++14, 1st edition, O'Reilly Media, Inc., 2014, ISBN: 1491903996, 9781491903995.
[5] Jakob Progsch, A simple yet versatile thread pool, https://github.com/progschj/ThreadPool (visited on 01/12/2017).
[6] Bjarne Stroustrup, The C++ Programming Language, 4st edition, Addison-Wesley Professional, 2013, ISBN: 0321563840, 9780321563842.

CHAPTER

ADVANCED C++11 MULTITHREADING

5

Abstract

The previous chapter introduced the basic concepts of multithreading using the C++11 threading API starting with basic spawn and join approaches, while finishing with non-trivial synchronization based on mutexes and condition variables. However, the major bottleneck of application performance is usually caused by contention for a shared resource. In case of mutex-based programming all participating threads usually try to acquire the same lock in parallel which effectively serializes the program for lightweight operations such as increment/decrement or updates of a single scalar value. Fortunately, modern CPUs provide dedicated commands that allow for the efficient execution of uninterruptible read-modify-write operations for scalar data types (so-called atomics). In this chapter you will learn how to improve your application's performance by enforcing race condition-free code without having to deal with the overhead of lock-based approaches.

Keywords

C++11, Multithreading, Atomic operation, Mutex, Parallel reduction, Compare and swap loop, Custom atomic functions, Thread pool, Tree traversal, Graph search, Knapsack problem, Acquire, Release, Memory barrier

CONTENTS

Parallel Programming. DOI: 10.1016/B978-0-12-849890-3.00005-8

5.1 LOCK-FREE PROGRAMMING (ATOMICS, COMPARE-AND-SWAP)

Up to this point, we have implemented algorithms that were either based on race condition-free access patterns such as dense matrix vector multiplication or alternatively resolved concurrent access to a shared resource by locking a mutex during a critical section as realized in our thread pool library. However, locks are slow: if we submit tons of lightweight jobs to our thread pool implementation from the previous chapter, we observe insufficient CPU utilization caused by the lock-based scheduling logic which effectively serializes the whole program. As a result, the impact of dynamic scheduling can only be neglected if the actual task takes reasonably more time to compute than the associated scheduling logic of the thread pool.

ATOMIC COUNTING

A practical workaround for CPU underutilization caused by locks would be the redesign of algorithms to not rely on mutexes. Unfortunately, it is not always possible to rewrite an implementation to be completely free of race conditions. As an example, the task queue of our thread pool implementation is manipulated by all threads in parallel and thus we have to enforce sequential execution during the insertion and removal of jobs. However, we have already seen examples in this chapter (see Section 4.4), where a single counter variable is concurrently manipulated by all threads. Increment of an integer is a lightweight operation and thus it is unreasonable to acquire an expensive lock for such a basic instruction.

Addressing this issue, C++11 introduces *atomic* data types that can be safely manipulated in a concurrent context without acquiring time-consuming locks. Note that the name atomic refers to the ancient Greek assumption that atoms are indivisible.[1] Modern processing units such as CPUs and CUDA-enabled GPUs feature efficient hardware-level instructions for the atomic manipulation of 32-bit and 64-bit integer types. Moreover, CPUs with x86_64 instruction set additionally provide support for atomic operations on data types consisting of 8, 16, or 128 bits. These hardware instructions allow for the atomic increment/decrement of a single variable, or the atomic exchange of two variables without interruption.

Let us verify the difference in execution time for the concurrent increment of a variable using both mutexes and atomics. Listing 5.1 shows the corresponding code fragment. The closure `lock_count` in Lines 16–24 realizes the traditional approach where the counter variable is manipulated within a locked scope in order to exclude race conditions. By contrast, the closure `atomic_count` in Lines 26–32 performs the concurrent increment using the atomic data type `std::atomic<uint64_t>`. Atomic types are defined in the `atomic` header file provided by C++11 (Line 5). Notably, the second closure looks more natural since it looks exactly like sequential code.

```
1 #include <iostream>
2 #include <cstdint>
3 #include <vector>
4 #include <thread>
5 #include <atomic>  // <- atomic data types
```

[1]The first experimental evidence was provided by Dalton in the 19th century. Nowadays, we know that this assumption is incorrect since atoms are assembled from leptons in the hull and quarks in the nucleus.

```
6   #include <mutex>
7   #include "../include/hpc_helpers.hpp"
8
9   int main( ) {
10
11      std::mutex mutex;
12      std::vector<std::thread> threads;
13      const uint64_t num_threads = 10;
14      const uint64_t num_iters = 100'000'000; // C++14 syntax
15
16      auto lock_count =
17          [&] (volatile uint64_t* counter,
18               const auto& id) -> void {
19
20          for (uint64_t i = id; i < num_iters; i += num_threads) {
21              std::lock_guard<std::mutex> lock_guard(mutex);
22              (*counter)++;
23          }
24      };
25
26      auto atomic_count =
27          [&] (volatile std::atomic<uint64_t>* counter,
28               const auto& id) -> void {
29
30          for (uint64_t i = id; i < num_iters; i += num_threads)
31              (*counter)++;
32      };
33
34      TIMERSTART(mutex_multithreaded)
35      uint64_t counter = 0;
36      threads.clear();
37      for (uint64_t id = 0; id < num_threads; id++)
38          threads.emplace_back(lock_count, &counter, id);
39      for (auto& thread : threads)
40          thread.join();
41      TIMERSTOP(mutex_multithreaded)
42
43      TIMERSTART(atomic_multithreaded)
44      std::atomic<uint64_t> atomic_counter(0);
45      threads.clear();
46      for (uint64_t id = 0; id < num_threads; id++)
47          threads.emplace_back(atomic_count, &atomic_counter, id);
48      for (auto& thread : threads)
49          thread.join();
50      TIMERSTOP(atomic_multithreaded)
51
52      std::cout << counter << " " << atomic_counter << std::endl;
53  }
```

Listing 5.1: Counting without race conditions.

The code can be compiled on a C++14 compliant compiler using the `-latomic` flag:

```
g++ -O2 -std=c++14 -pthread -latomic \
    atomic_count.cpp -o atomic_count
```

When executed on a Xeon E5-2683 v4 CPU with 10 threads we obtain the following execution times:

```
# elapsed time (mutex_multithreaded): 16.0775s
# elapsed time (atomic_multithreaded): 2.25695s
100000000 100000000
```

We can see that the use of atomics is approximately seven times more efficient than the traditional approach using mutexes and locks. Moreover, the atomic implementation is simpler in terms of code while computing the same result.

NON-FUNDAMENTAL ATOMIC DATA TYPES

C++11 provides support for a wide range of generic integer data types being 8, 16, 32, or 64 bits wide [3]. Moreover, 128-bit atomics with hardware support on x64_64 architectures can be defined using structs of the same width. Interestingly, we can even wrap structs and objects of different length with `std::atomic<T>`. In this case, the compiler treats those objects formally as atomic but realizes their concurrent manipulation with expensive locks. Hence, the atomic interface allows for the unified and race condition-free treatment of objects within the scope of the same API. As a result, your code remains correct even if the underlying hardware does not support the corresponding data type.

Listing 5.2 shows how to define atomic structs of length 24, 32, 48, 64, 80, and 128 bits. Note that the member function `is_lock_free()` can be used to determine whether the manipulation of the corresponding data type is accomplished with atomics or mutexes.

```
1  #include <iostream>
2  #include <atomic>
3
4  template <
5      typename x_value_t,
6      typename y_value_t,
7      typename z_value_t>
8  struct state_t {
9
10     x_value_t x;
11     y_value_t y;
12     z_value_t z;
13
14     // NOTE: no non-default constructor allowed
15 };
16
17 template <
18     typename R,
19     typename S,
20     typename T>
21 void status() { // report size and if lock-free
22
```

```
23      typedef std::atomic<state_t<R,S,T>> atomic_state_t;
24
25      std::cout << sizeof(atomic_state_t) << "\t"
26                << atomic_state_t().is_lock_free() << std::endl;
27  }
28
29  int main () {
30
31      std::cout << "size\tlock_free?" << std::endl;
32
33      // Let us have a look at the properties of distinct types
34      status<uint8_t,  uint8_t,  uint8_t >(); //  24 bit mutex
35      status<uint16_t, uint8_t,  uint8_t >(); //  32 bit atomic
36      status<uint16_t, uint16_t, uint8_t >(); //  48 bit mutex
37      status<uint32_t, uint16_t, uint16_t>(); //  64 bit atomic
38      status<uint32_t, uint32_t, uint16_t>(); //  80 bit mutex
39      status<uint64_t, uint32_t, uint32_t>(); // 128 bit atomic
40  }
```

Listing 5.2: Overview of atomic data types.

The output of the program reveals that the 32, 64, and 128 bit structs are atomics with hardware support, i.e. they are lock-free. The 24, 48, and 80 bit structs are not lock-free. In the latter case, the programmer has to decide whether to align the lock-based atomics in an appropriate manner such that they fit into a wider real atomic data type in order to benefit from hardware support. Alternatively, one could choose a more memory efficient implementation that uses expensive locks.

Finally, let us make an important remark: structs and objects that shall be wrapped by `std::atomic<T>` are not allowed to have a constructor. By contrast, fundamental data types can be initialized as usual. If you have to perform an explicit initialization of the member variables you can still implement a custom member function that has to be invoked on the to be wrapped object before passing it to `std::atomic`:

```
// instantiate object, initialize it and
// subsequently declare atomic type
T my_object;
my_object.init(arguments); // you have to implement init!
std::atomic<T> my_atomic_object;

// the assignment operator is equivalent
// to: my_atomic_object.store(my_object)
my_atomic_object = my_object;

// access a member variable after loading atomically
std::cout << my_atomic_object.load().member << std::endl;
```

Conceptually, you should treat atomics as a chunk of consecutively stored bytes that can be loaded and stored without interruption. Note that the increment operator acting directly on the atomic is only provided for generic integer types.

ATOMIC PARALLEL MAX-REDUCTION USING COMPARE-AND-SWAP

As mentioned before, we can only assume that our program computes the correct result when the concurrent manipulation of an atomic variable is realized in an indivisible step without interruption. In particular, this means that we could end up with incorrect results if we issue more than one operation. As an example, the following code fragment is incorrect since we cannot guarantee that the increment is performed on the most recent state:

```
// WARNING: this code fragment computes
//          incorrect results
std::atomic<uint64_t> atomic (0);

{ // two or more threads perform this
    // equivalent to value=atomic.load()
    value = atomic; // 1st operation

    // increment the integer
    value++;        // 2nd operation

    //equivalent to atomic.store(value)
    atomic = value; // 3rd operation
}
```

By contrast, a simple call to `atomic++;` is correct since the loading, incrementation and storing are issued within a single operation. Unfortunately, atomics feature only a limited set of predefined operations: generic integer atomics allow for the increment/decrement (`operator++`/`operator-`), and basic assignments on a value-level and bit-level such as `operator+=`, `operator-=`, `operator&=`, `operator|=`, and `operator^=`. Common operations like maximum or minimum are not supported out-of-the-box. This seems to be a hard constraint on first sight; however, we will see that we can implement virtually every function in an atomic fashion.

Every C++11 atomic data type feature a *compare-and-swap* (CAS) operation which can be used to implement arbitrary assignments. The corresponding method is either `compare_exchange_strong()` or `compare_exchange_weak()`. The latter is more efficient but might suffer from spurious fails similar to conditional waits while the first variant guarantees non-spurious behavior. Note that in the majority of cases we have to check whether a certain swap was successful in a loop and thus the more efficient weak variant is often the better choice. The corresponding CAS operation of an atomic `std::atomic<t> atomic` accepts two arguments:

1. a reference to a value that is expected to be stored in `atomic`, and
2. a value that shall be stored in `atomic` under the constraint that `atomic.load() == expected`:

```
atomic.compare_exchange_weak(T& expected, T desired);
```

CAS performs the following three steps in an atomic manner without interruption:

1. Compare a given value `expected` with the value stored in `atomic`.

2. If both values stored in `expected` and `atomic` coincide then set `atomic` to a given value `desired`, otherwise write the actual value stored in `atomic` to `expected`.
3. Return `true` if the swap in 2. was successful, otherwise return `false`.

Loosely speaking, we can exclude operations on outdated states by providing our perception of the presumed state by means of the value `expected`. The swap is not performed if our assumption turns out to be wrong. In this case, we have to handle the error explicitly by exemplary repeating the whole procedure with our new assumption `expected` which now corresponds to the actual value stored in `atomic`. Note that the weak variant can spuriously fail even if the condition of the swap is actually met. Consequently, CAS operations should always be performed in loops. The only exception to this rule is if solely one thread manipulates an atomic using `compare_exchange_strong`. Listing 5.3 shows how to atomically compute the maximum of a sequence in parallel.

```
1  #include <iostream>
2  #include <cstdint>
3  #include <vector>
4  #include <thread>
5  #include <atomic>
6  #include "../include/hpc_helpers.hpp"
7
8  int main( ) {
9
10     std::vector<std::thread> threads;
11     const uint64_t num_threads = 10;
12     const uint64_t num_iters = 100'000'000;
13
14     // WARNING: this closure produces incorrect results
15     auto false_max =
16         [&] (volatile std::atomic<uint64_t>* counter,
17              const auto& id) -> void {
18
19         for (uint64_t i = id; i < num_iters; i += num_threads)
20             if(i > *counter)
21                 *counter = i;
22     };
23
24     // Using a compare and swap-loop for correct results
25     auto correct_max =
26         [&] (volatile std::atomic<uint64_t>* counter,
27              const auto& id) -> void {
28
29         for (uint64_t i = id; i < num_iters; i += num_threads) {
30             auto previous = counter->load();
31             while (previous < i &&
32                 !counter->compare_exchange_weak(previous, i)) {}
33         }
34     };
35
36     TIMERSTART(incorrect_max)
37     std::atomic<uint64_t> false_counter(0);
38     threads.clear();
```

```
39      for (uint64_t id = 0; id < num_threads; id++)
40          threads.emplace_back(false_max, &false_counter, id);
41      for (auto& thread : threads)
42          thread.join();
43      TIMERSTOP(incorrect_max)
44
45      TIMERSTART(correct_max)
46      std::atomic<uint64_t> correct_counter(0);
47      threads.clear();
48      for (uint64_t id = 0; id < num_threads; id++)
49          threads.emplace_back(correct_max, &correct_counter, id);
50      for (auto& thread : threads)
51          thread.join();
52      TIMERSTOP(correct_max)
53
54      std::cout << false_counter << " "
55                << correct_counter << std::endl;
56  }
```

Listing 5.3: Atomically determine the maximum of a sequence.

The closure `false_max` in Lines 15–22 occasionally computes the incorrect result since the condition in Line 20 and the atomic store in Line 21 are two independent operations. Thus, two or more threads could have read the same value before performing the assignment in random order. The computed value tends to be less than the expected result 999,999,999. By contrast, the closure `correct_max` in Lines 25–34 computes the correct result using a CAS loop.

ARBITRARY ATOMIC OPERATIONS

In this subsection, we implement two atomic assignment operations that turn out to be quite useful in daily programming. The first one atomically computes a new value from the value stored in `atomic` and another variable `operand` by means of a given function and subsequently checks if a certain predicate is met. Consider that we want to compute the maximum of a sequence constrained on the subset of even numbers. Hence, the function is the maximum map and the predicate to check whether the result is even. Listing 5.4 shows the corresponding code fragment.

```
1   #include <iostream>
2   #include <cstdint>
3   #include <vector>
4   #include <thread>
5   #include <atomic>
6   #include "../include/hpc_helpers.hpp"
7
8   template <
9       typename atomc_t,
10      typename value_t,
11      typename funct_t,
12      typename predc_t>
13  value_t binary_atomic(
14      atomc_t& atomic,
```

```
15      const value_t& operand,
16      funct_t function,
17      predc_t predicate) {
18
19      value_t expect = atomic.load();
20      value_t target;
21
22      do {
23          // compute preliminary new value
24          target = function(expect, operand);
25
26          // immediately return if not fulfilling
27          // the given constraint for a valid result
28          if (!predicate(target))
29              return expect;
30
31      // try to atomically swap new and old values
32      } while (!atomic.compare_exchange_weak(expect, target));
33
34      // either new value if successful or the old
35      // value for unsuccessful swap attempts:
36      // in both cases it corresponds to atomic.load()
37      return expect;
38  }
39
40  int main( ) {
41
42      std::vector<std::thread> threads;
43      const uint64_t num_threads = 10;
44      const uint64_t num_iters = 100'000'000;
45
46      auto even_max =
47          [&] (volatile std::atomic<uint64_t>* counter,
48               const auto& id) -> void {
49
50          auto func = [] (const auto& lhs,
51                          const auto& rhs) {
52              return lhs > rhs ? lhs : rhs;
53          };
54
55          auto pred = [] (const auto& val) {
56              return val % 2 == 0;
57          };
58
59          for (uint64_t i = id; i < num_iters; i += num_threads)
60              binary_atomic(*counter, i, func, pred);
61      };
62
63      TIMERSTART(even_max)
64      std::atomic<uint64_t> even_counter(0);
65      for (uint64_t id = 0; id < num_threads; id++)
66          threads.emplace_back(even_max, &even_counter, id);
```

```
67     for (auto& thread : threads)
68         thread.join();
69     TIMERSTOP(even_max)
70
71     // 999999998 <- the biggest even number < 10^9
72     std::cout << even_counter << std::endl;
73 }
```

Listing 5.4: Arbitrary atomic operation based on binary conditions.

Let us talk about the code. The function template `binary_atomic` in Lines 8–38 generalizes the described approach to arbitrary functions and predicates. First, we load the actual value stored in `atomic` and declare a variable for the desired target value (Lines 19–20). Second, the function value of our expectation and the operand is computed in Line 24. Third, we immediately terminate the attempt to swap if the computed function value does not fulfill the given constraint (Lines 28–29). Otherwise, we iteratively try to swap the new value with the old one (Line 32). If the swap was not successful we perform another iteration of the loop until we either succeed or violate the constraint. Atomic assignment with ternary conditions of the form

```
predicate(atomic, operand) ? f_true (atomic, operand) :
                             f_false(atomic, operand) ;
```

can also be realized with CAS loops. In case of even maximum, `predicate` refers to the Boolean expression `operand > atomic && operand % 2 == 0`, the function `f_true` simply returns `operand`, and `f_false` the value stored in `atomic`. Listing 5.5 shows the corresponding code fragment. Note that this implementation is slightly slower than the first one since we rewrite the old value if we do not fulfill the constraint.

```
1  #include <iostream>
2  #include <cstdint>
3  #include <vector>
4  #include <thread>
5  #include <atomic>
6  #include "../include/hpc_helpers.hpp"
7
8  template <
9      typename atomc_t,
10     typename value_t,
11     typename funcp_t,
12     typename funcn_t,
13     typename predc_t>
14 value_t ternary_atomic(
15     atomc_t& atomic,
16     const value_t& operand,
17     funcp_t pos_function,
18     funcn_t neg_function,
19     predc_t predicate) {
20
21     value_t expect = atomic.load();
22     value_t target;
23
```

```
24        do {
25            // ternary block: pred ? pos_func : neg_func
26            if (predicate(expect, operand))
27                target = pos_function(expect, operand);
28            else
29                target = neg_function(expect, operand);
30
31        // try to atomically swap new and old values
32        } while (!atomic.compare_exchange_weak(expect, target));
33
34        // either new value if successful or the old
35        // value for unsuccessful swap attempts:
36        // in both cases it corresponds to atomic.load()
37        return expect;
38    }
39
40
41    int main( ) {
42
43        std::vector<std::thread> threads;
44        const uint64_t num_threads = 10;
45        const uint64_t num_iters = 100'000'000;
46
47        auto even_max =
48            [&] (volatile std::atomic<uint64_t>* counter,
49                 const auto& id) -> void {
50
51            auto pos_func = [] (const auto& lhs,
52                                const auto& rhs) {
53                return lhs;
54            };
55
56            auto neg_func = [] (const auto& lhs,
57                                const auto& rhs) {
58                return rhs;
59            };
60
61            auto pred = [] (const auto& lhs,
62                            const auto& rhs) {
63                return lhs > rhs && lhs % 2 == 0;
64            };
65
66            for (uint64_t i = id; i < num_iters; i += num_threads)
67                ternary_atomic(*counter, i, pos_func, neg_func, pred);
68        };
69
70        TIMERSTART(even_max)
71        std::atomic<uint64_t> even_counter(0);
72        for (uint64_t id = 0; id < num_threads; id++)
73            threads.emplace_back(even_max, &even_counter, id);
74        for (auto& thread : threads)
75            thread.join();
```

```
76      TIMERSTOP(even_max)
77
78      std::cout << even_counter << std::endl;
79  }
```

Listing 5.5: Arbitrary atomic operation based on ternary conditions.

Obviously, we can express complex atomic functions with non-trivial constraints in just a few lines of code. Let us make a final remark: this section focused on integer types but you can extend this approach to any other data type as long you can represent it as an integer on a bit-level. As an example, you could design a high precision floating-point data type with 128 bits by storing two 64 bit integers in a struct. Subsequently, the associated operations addition, subtraction, multiplication, division, square-root, and so forth have to be implemented by you using CAS loops. An interested reader might refer to [2] for an extensive discussion of double stacking techniques for the approximation of high precision floating-point values.

THE ABA PROBLEM

Atomics are fast and versatile. However, they have a slight disadvantage in comparison to mutex-based locks. If two or more threads try to acquire a lock, we can be sure that every thread will finally have acquired it after termination. Hence, each thread has performed its modification on the data. This is not necessarily the case when using CAS loops. Assume we want to implement atomic negation: a zero is rewritten to a one and vice versa. In this context, the following situation might occur:

1. Thread 0 reads the atomic with value 0 and attempts to change it to 1 but does not perform the CAS yet.
2. Thread 1 reads the atomic with value 0 and successfully changes it to 1.
3. Thread 1 reads the atomic with value 1 and successfully changes it to 0.
4. Thread 0 finally performs the CAS and succeeds since the atomic is in the expected state 0.

In Phase 4, Thread 0 cannot distinguish whether it has changed the state from Phase 1 or the state from Phase 3. The CAS in Phase 4 is blind for the changes performed in Phases 2 and 3. Consequently, the state cannot be used for synchronization purposes. By contrast, a lock-based approach would have performed Phases 1 and 4 without interruption since Thread 1 could not have acquired the lock in-between.

This observation is known as "ABA problem" in the literature. Fortunately, one can resolve this problem by introducing a dedicated counter variable which atomically counts how often a state was modified. Nevertheless, the payload and its corresponding counter have to be packed into a single atomic data type. A reasonable choice would be the following: we encode the binary payload in the most significant bit of an integer and use the remaining bits for the counting. Nevertheless, the ABA problem remains a crucial challenge during the design of efficient lock-free data structures such as dynamic arrays [1] and concurrent hash maps [4].

5.2 WORK-SHARING THREAD POOL (TREE TRAVERSAL)

This section discusses a work-sharing approach among threads. Assume we have a thread pool with a fixed capacity of $c > 1$ idling threads as discussed in Section 4.6. Subsequently, we submit a single task being executed by one thread. Hence, the utilization rate of our pool remains low at $\frac{1}{c}$ unless someone submits a new job. In order to improve the utilization rate of resources one could pursue the following scheduling strategy: each thread executing a task (initially one) shares parts of its work load with other idling threads as long as the number of active threads is smaller than the capacity c. In the following we extend our custom thread pool implementation from Section 4.6 to support the dynamic redistribution of work among threads.

USE CASE FOR A WORK-SHARING THREAD POOL

Before we start coding, let us have a look at a typical use case. Assume we want to traverse a binary tree with a recursive scheme as shown in Listing 5.6. At each node we perform some compute-intensive operation.

```
1  #include <iostream>
2  #include <cstdint>
3  #include "../include/hpc_helpers.hpp"
4
5  void waste_cycles(uint64_t num_cycles) {
6
7      volatile uint64_t counter = 0;
8      for (uint64_t i = 0; i < num_cycles; i++)
9          counter++;
10 }
11
12 void traverse(uint64_t node, uint64_t num_nodes) {
13
14     if (node < num_nodes) {
15
16         // do some work
17         waste_cycles(1<<15);
18
19         // visit your children more often!
20         traverse(2*node+1, num_nodes);
21         traverse(2*node+2, num_nodes);
22     }
23 }
24
25 int main() {
26
27     TIMERSTART(traverse)
28     traverse(0, 1<<20);
29     TIMERSTOP(traverse)
30 }
```

Listing 5.6: Sequential tree traversal using recursion.

Let us talk about the code. The recursive function `traverse` implicitly enumerates the `num_nodes` many nodes of the tree by calling itself for the left $(2 \cdot \text{node} + 1)$ and right child $(2 \cdot \text{node} + 2)$ until reaching the leaves. Moreover, we simulate the compute-heavy task by wasting cycles in a for-loop. Note that the keyword `volatile` prevents the compiler performing the (admittedly very efficient) substitution `counter += num_cycles;` in Lines 8–9. Finally, we call `traverse` in the `main` function for the root of the tree. A work-sharing parallel implementation could be implemented similarly to the code shown in Listing 5.7.

```
1  #include <iostream>
2  #include <cstdint>
3  #include "threadpool.hpp"
4  #include "../include/hpc_helpers.hpp"
5
6  ThreadPool TP(8); // 8 threads do the job
7
8  void waste_cycles(uint64_t num_cycles) {
9
10     volatile uint64_t counter = 0;
11     for (uint64_t i = 0; i < num_cycles; i++)
12         counter++;
13 }
14
15 void traverse(uint64_t node, uint64_t num_nodes) {
16
17     if (node < num_nodes) {
18
19         // do some work
20         waste_cycles(1<<15);
21
22         // try to execute the left branch using
23         // a potentially idling thread
24         TP.spawn(traverse, 2*node+1, num_nodes);
25
26         // execute the right branch sequentially
27         traverse(2*node+2, num_nodes);
28     }
29 }
30
31 int main() {
32
33     TIMERSTART(traverse)
34     TP.spawn(traverse, 0, 1<<20);
35     TP.wait_and_stop();
36     TIMERSTOP(traverse)
37 }
```

Listing 5.7: Parallel tree traversal using a recursive work-sharing scheme.

The code is almost identical to the sequential version. The only difference is that we try to delegate the computation of the left branches to another thread in the pool, i.e., at each node we can potentially split the work in half until there are no idling threads left. Finally, we solely submit the root node to the thread pool in order to initiate the parallel computation.

The actual implementation of this work-sharing thread pool performs the concurrent traversal of the whole tree in approximately 11 seconds on a Xeon E5-2683 v4 CPU using eight threads. By contrast, the sequential version runs for roughly 90 seconds. Hence, we observe linear speedup with only a minor modification of the code.

IMPLEMENTATION OF WORK-SHARING

In this subsection, we discuss how to extend our thread pool from Section 4.6. If you have not read this section before, we strongly advise to revisit the discussed code examples since we only present the modifications of the source code that are necessary to implement work-sharing.

Let us begin with the member variables of the thread pool. First, we have to query the number of executed tasks stored in `active_threads` and compare it to the capacity of the pool for each attempt to delegate work. This has to be accomplished without race conditions and thus we have to either protect the load of the variable `active_threads` with a slow lock or alternatively use an efficient atomic. The obvious choice is the latter. Second, we need another condition variable `cv_wait` that is used to synchronize the final `wait_and_stop()` call of the thread pool. This is necessary since our `spawn` method does not return a corresponding future object for synchronization. Hence, `wait_and_stop()` ensures that all tasks that have been submitted to the pool have terminated. Listing 5.8 shows the discussed modifications.

```
1  #ifndef THREADPOOL_HPP
2  #define THREADPOOL_HPP
3
4  #include <cstdint>
5  #include <future>
6  #include <vector>
7  #include <queue>
8  #include <thread>
9  #include <mutex>
10 #include <atomic>              // do not forget to include this
11 #include <condition_variable>
12
13 class ThreadPool {
14
15 private:
16
17     // storage for threads and tasks
18     std::vector<std::thread> threads;
19     std::queue<std::function<void(void)>> tasks;
20
21     // primitives for signaling
22     std::mutex mutex;
23     std::condition_variable cv, cv_wait;  // another cv
24
25     // the state of the thread, pool
26     bool stop_pool;
```

```
27  std::atomic<uint32_t> active_threads; // now atomic
28  const uint32_t capacity;
```

Listing 5.8: Initial portion of our work-sharing thread pool.

The modifications are simple: we include the atomic header in Line 10, rewrite the variable `active_threads` to be atomic in Line 27, and finally add another condition variable in Line 23.

At this point, we have to ensure that we correctly signal the `wait_and_stop()` method of the pool when all tasks have finished their computation. To achieve that we check after each successful execution of a task if the task queue is empty, and moreover if there are no tasks being executed by running threads. Listing 5.9 shows the necessary modifications in the function `after_task_hook()`. Note that the whole function is already enclosed by a lock and thus we do not need to acquire another one.

```
1   // will be executed after execution of a task
2   void after_task_hook() {
3       active_threads--;
4
5       // this synchronization step is new
6       if (active_threads == 0 && tasks.empty()) {
7           stop_pool = true;
8           cv_wait.notify_one();
9       }
10  }
```

Listing 5.9: Modifications of hook after task execution.

The implementation of `wait_and_stop()` is straightforward: we perform a conditional wait on the condition variable `cv_wait` (see Listing 5.10). The method is blocking the master thread until all tasks have finished. This step is important in order to avoid that we immediately reach the end of the `main` function which would result in the calling of the destructor of the pool.

```
1   // public member function
2   void wait_and_stop() {
3
4       // wait for pool being set to stop
5       std::unique_lock<std::mutex>
6           unique_lock(mutex);
7
8       auto predicate = [&] () -> bool {
9           return stop_pool;
10      };
11
12      cv_wait.wait(unique_lock, predicate);
13  }
```

Listing 5.10: New function for thread pool synchronization.

Finally, we have to implement the `spawn` method which delegates works to other idling threads. Listing 5.11 shows the corresponding code fragment.

```
1   // public member function template
2   template <
```

```
3        typename     Func,
4        typename ... Args>
5    void spawn(
6        Func &&      func,
7        Args && ... args) {
8
9        if (active_threads < capacity) // enqueue if idling
10           enqueue(func, args...);
11       else                           // process sequentially
12           func(args...);
13   }
```

Listing 5.11: New function for delegating work.

The variadic function template spawn accepts a function func with arbitrarily many arguments args. First, we check if the number of active threads is smaller than the capacity of the thread pool using the atomic variable active_threads. If we come to the conclusion that there are still threads idling, we submit a new task to the queue via enqueue. Otherwise, we execute the task sequentially in the calling thread.

Note that this code has a minor flaw. Consider that two threads perform the query if (active_threads < capacity) one after another in a short period of time before the issuing of enqueue. In this situation only one of the two tasks could be processed in case of a single idling thread; the other one would end up in the task queue without being processed immediately. Nevertheless, this behavior does not influence the correctness of the thread pool. A rigid workaround would be the use of a Double-Checked Locking Pattern (DCLP) where we first check the constraint atomically and subsequently repeat the check together with enqueue within a locked scope on success. Historically, DCLP was considered a broken antipattern until the release of C+11 because of a lacking memory model of previous C++ standards. An interested reader might refer to [5] for an extensive overview of the topic.

5.3 PARALLEL GRAPH SEARCH (BINARY KNAPSACK PROBLEM)

A parallel code is considered efficient if it exhibits linear speedup, i.e., the same speedup as the number of used processing units. In theory, linear speedup is the best speedup we can possibly achieve. However, we sometimes encounter *superlinear speedup* caused by superior caching properties. As an example, if we work on local copies of a huge matrix, we might benefit from more cache hits because of the better locality of processed data. Another example is a dual socket CPU which provides the double amount of caches in comparison to a single CPU. Those occurrences of superlinear speedup can be explained by overcoming resource limitations of the corresponding hardware architecture.

However, we can even encounter *spurious* superlinear speedup without overcoming resource limitations as we will show in this section. A representative class of algorithms that exhibits this behavior are branch-and-bound algorithms which recursively generate solution candidates while determining a globally optimal solution. The Traveling Salesman Problem, and the Binary Knapsack Problem are popular examples where a globally optimal state is determined from the set of all possible states by recursively traversing the space of implicitly enumerable solution candidates. In this section, we focus on the Knapsack problem.

Table 5.1 An example of a Knapsack problem with a capacity constraint of $c = 200$ consisting of $n = 4$ items sorted by their value density. The optimal solution $\mathcal{J}^* = \{1, 2\}$ among the $2^4 = 16$ solution candidates with an accumulated value of 196 and a weight of 185 does not include item 0 despite being the most valuable item per weight.

Item	Value	Weight	Value density
0	97	91	1.06593
1	98	92	1.06522
2	98	93	1.05376
3	92	91	1.01099

THE BINARY KNAPSACK PROBLEM

The *Binary Knapsack Problem* is an optimization task, where a thief tries to pack up to n objects $h \in \{0, \ldots, n-1\}$ of value $V[h]$ and weight $W[h]$ into a backpack of limited capacity c. For each object h, the thief may decide if he or she takes it or leaves it behind. The goal is to maximize the accumulated value of the chosen objects without violating the capacity constraint. Mathematically speaking, we are interested in the optimal subset $\mathcal{J}^* \subseteq \mathcal{I}$ among all index subsets $\mathcal{J} \subseteq \mathcal{I} := \{0, \ldots, n-1\}$ such that

$$\mathcal{J}^* := \underset{\mathcal{J} \subseteq \mathcal{I}}{\operatorname{argmax}} \sum_{j \in \mathcal{J}} V[j] \quad \text{subject to:} \quad \sum_{j \in \mathcal{J}} W[j] \leq c \quad . \tag{5.1}$$

An example is shown in Table 5.1.

For exclusively integral weights there exists a pseudo-polynomial algorithm, which solves this NP-complete problem in $\mathcal{O}(n \cdot c)$ time and space with the help of dynamic programming. However, we want to determine the optimal solution with a branch-and-bound algorithm that recursively explores the complete state space. Obviously, there are 2^n distinct possibilities to fill the backpack. Our program should be able to pack up to $n = 32$ objects such that we can store the container state in an unsigned integer variable. Moreover, we would like to remember the optimal accumulated value (global bound) of the objects in another global unsigned integer. Thus, the global state and accumulated value can be packed into an atomic integer being 64 bits wide. Using CAS-loops, we can concurrently update the global state in a lock-free manner.

The state space can be represented as a binary tree (see Fig. 5.1). At each level $h < n$ we decide either to pick item h or leave it behind. At each node we have to calculate the new accumulated value and weight of the candidate and afterwards compare it to the capacity c and global state `global_state`. If the accumulated weight of a candidate exceeds the capacity limit, the state including its corresponding subtree can safely be pruned since it is impossible to generate a valid solution from it. Moreover, we define a local upper bound at each node by overoptimistically estimating the value of the future states. Assume the initial index set is sorted in such a manner that the associated value density is non-increasing, i.e.

$$\frac{V[0]}{W[0]} \geq \frac{V[1]}{W[1]} \geq \cdots \geq \frac{V[h]}{W[h]} \geq \cdots \geq \frac{V[n-1]}{W[n-1]} \quad . \tag{5.2}$$

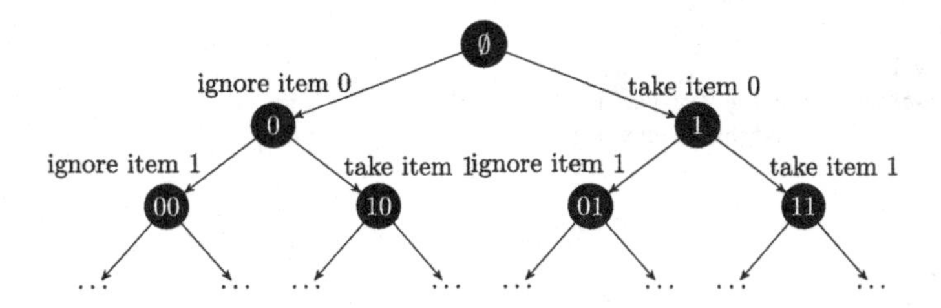

FIGURE 5.1

Visualization of the state space represented as binary tree. We start at the root node and consecutively decide whether to take item 0 (right branch) or leave it behind (left branch). Each leaf at the bottom of the full binary tree corresponds to a possible solution candidate.

At a node of height h we can upper-bound the accumulated value for the final candidate by greedily picking the next consecutive item of highest value density until we exceed the capacity. This upper bound must be greater than the optimal solution since we relaxed (ignored) the capacity constraint. If this upper bound has a smaller value than the global lower bound, which was initially generated by a valid solution, we can safely prune the candidate and the corresponding subtree since it is impossible to create a solution candidate with higher accumulated value than the global state's value.

SEQUENTIAL IMPLEMENTATION

Let us start coding. Listing 5.12 shows the header of our program. Initially, we define two auxiliary structs: `generic_tuple_t` (Lines 6–22) is used to store pairs of values and weights ($V[h]$, $W[h]$), and `state_t` (Lines 24–38) consists of the globally best-so-far observed value and a corresponding binary mask encoding the taken items. Note that `state_t` will later be wrapped by `std::atomic` and thus we are not allowed to implement a constructor. Furthermore, we abbreviate the used data types for convenience (Lines 41–45) and subsequently declare the global state, capacity of the backpack, number of items, and a vector for the storage of values and weights (Lines 48–51).

```
1  #include <algorithm>  // std::sort
2  #include <iostream>   // std::cout
3  #include <vector>     // std::vector
4  #include <random>     // std::uniform_int_distribution
5
6  template <
7      typename value_t_,
8      typename weight_t_>
9  struct generic_tuple_t {
10
11     value_t_  value;
12     weight_t_ weight;
13
14     // expose types
15     typedef value_t_ value_t;
16     typedef value_t_ weight_t;
17
```

```
18      generic_tuple_t(
19          value_t_  value_,
20          weight_t_ weight_) : value (value_ ),
21                               weight(weight_) {}
22  };
23
24  template <
25      typename bmask_t_,
26      typename value_t_>
27  struct state_t {
28
29      bmask_t_ bmask=0;
30      value_t_ value=0;
31
32      // expose template parameters
33      typedef bmask_t_ bmask_t;
34      typedef value_t_ value_t;
35
36      // non-default constructors are not allowed
37      // when wrapped with std::atomic<state_t>
38  };
39
40  // shortcuts for convenience
41  typedef uint64_t index_t;
42  typedef uint32_t bmask_t;
43  typedef uint32_t value_t;
44  typedef uint32_t weight_t;
45  typedef generic_tuple_t<value_t, weight_t> tuple_t;
46
47  // the global state encoding the mask and value
48  state_t<bmask_t, value_t> global_state;
49  const value_t capacity (1500);
50  const index_t num_items (32);
51  std::vector<tuple_t> tuples;
```

Listing 5.12: Initial portion of the Knapsack code.

Now we can proceed with the initialization of the tuples $(V[h], W[h])$ in Listing 5.13. We sample n pairs of values and weights from a uniform integer distribution in the range of 80 to 99 (inclusive). This is accomplished with a Mersenne Twister pseudo-random number generator from the standard library (Lines 14–20). Finally, the tuples are sorted in Lines 23–28 according to Eq. (5.2) such that the quotient of values and weights is non-increasing.

```
52  // initializes Knapsack problem
53  template <
54      typename tuple_t,
55      typename index_t>
56  void init_tuples(
57      std::vector<tuple_t>&  tuples,
58      index_t num_entries) {
59
60      // recover the types stored in tuple_t
```

```
61      typedef typename tuple_t::value_t  value_t;
62      typedef typename tuple_t::weight_t weight_t;
63
64      // C++11 random number generator
65      std::mt19937 engine(0); // mersenne twister
66      std::uniform_int_distribution<value_t>  rho_v(80, 100);
67      std::uniform_int_distribution<weight_t> rho_w(80, 100);
68
69      // generate pairs of values and weights
70      for (index_t index = 0; index < num_entries; index++)
71          tuples.emplace_back(rho_v(engine), rho_w(engine));
72
73      // sort two pairs by value/weight density
74      auto predicate = [] (const auto& lhs,
75                           const auto& rhs) -> bool {
76          return lhs.value*rhs.weight > rhs.value*lhs.weight;
77      };
78
79      std::sort(tuples.begin(), tuples.end(), predicate);
80  }
```

Listing 5.13: Initialization of values and weights.

The traversal logic of the binary tree is shown in Listing 5.14. Let us begin with the explanation of the core method `traverse` in Lines 115–147. First, we determine the corresponding bit in the binary mask at position `height` in Line 125. We pack the associated item in our bag if it is set to one. Consequently, we have to adjust the value and weight of our haul stored in `tuple` (Lines 126–127). In case we exceed the capacity of the backpack, we can terminate the recursive candidate generation since we cannot generate further valid solutions from the current state (Lines 130–131). Otherwise, we have found a valid solution which is subsequently checked versus the global state in Line 134 using the auxiliary method `sequential_update` in Lines 81–92. It compares our current solution candidate with the best global solution and performs an update in case of improvement.

The following step is optional, however, it may significantly speed up the traversal of the binary tree (in our case by one order-of-magnitude). We compute a relaxed solution of the Knapsack Problem which overestimates the actual achievable value of the haul by greedily packing the items starting from position `height+1` until we exceed the capacity (Lines 94–113). The computed value `dantzig_bound(height+1, tuple)` is most likely too high since it assumes that you are allowed to pack fractional items into your backpack. Nevertheless, you can exploit this overestimation to exclude a possible solution candidate if its value is smaller than the value of the best observed global solution. Note that we can realize the greedy packing strategy with a simple for-loop since we have already sorted the tuples according to their value density in the initialization step.

The final portion of the method `traverse` in Lines 143–147 recursively generates two new solution candidates: one that takes the item at position `height+1` and another one that leaves it behind. The traversal routine terminates if either all new solution candidates are pruned by the global bound in Line 134, the local bound in Lines 139–149, or alternatively we have reached the leaves of the binary tree. The worst-case traversal has to probe all 2^n item combinations.

```
81  template <
82      typename tuple_t,
83      typename bmask_t>
84  void sequential_update(
85      tuple_t tuple,
86      bmask_t bmask) {
87
88      if (global_state.value < tuple.value) {
89          global_state.value = tuple.value;
90          global_state.bmask = bmask;
91      }
92  }
93
94  template <
95      typename index_t,
96      typename tuple_t>
97  typename tuple_t::value_t dantzig_bound(
98      index_t height,
99      tuple_t tuple) {
100
101     auto predicate = [&] (const index_t& i) {
102         return i < num_items &&
103                tuple.weight < capacity;
104     };
105
106     // greedily pack items until backpack full
107     for (index_t i = height; predicate(i); i++) {
108         tuple.value  += tuples[i].value;
109         tuple.weight += tuples[i].weight;
110     }
111
112     return tuple.value;
113 }
114
115 template <
116     typename index_t,
117     typename tuple_t,
118     typename bmask_t>
119 void traverse(
120     index_t height,  // height of the binary tree
121     tuple_t tuple,   // weight and value up to height
122     bmask_t bmask) {  // binary mask up to height
123
124     // check whether item packed or not
125     const bool bit  = (bmask >> height) % 2;
126     tuple.weight += bit*tuples[height].weight;
127     tuple.value  += bit*tuples[height].value;
128
129     // check versus maximum capacity
130     if (tuple.weight > capacity)
131         return; // my backpack is full
```

```
132
133     // update global lower bound if needed
134     sequential_update(tuple, bmask);
135
136     // calculate local Danzig upper bound
137     // and compare with global upper bound
138     auto bsf = global_state.value;
139     if (dantzig_bound(height+1, tuple) < bsf)
140       return;
141
142     // if everything was fine generate new candidates
143     if (height+1 < num_items) {
144       traverse(height+1, tuple, bmask+(1<<(height+1)));
145       traverse(height+1, tuple, bmask);
146     }
147 }
```

Listing 5.14: Recursive traversal of the solution space.

Finally, we implement the actual computation in the `main` function. The corresponding code fragment is shown in Listing 5.15. First, we initialize the tuples in Line 151 by means of the aforementioned auxiliary function `init_tuples`. Second, we perform two recursive traversals of the left and right branch below the root node in Lines 154–155: one that takes the first item and another one that leaves it behind. The remaining code in Lines 160–165 reports the optimal value and prints the binary mask using a binary decomposition of the integer `bmask`.

```
148 int main () {
149
150     // initialize tuples with random values
151     init_tuples(tuples, num_items);
152
153     // traverse left and right branch
154     traverse(0, tuple_t(0, 0), 0);
155     traverse(0, tuple_t(0, 0), 1);
156
157     // report the final solution
158     std::cout << "value " << global_state.value << std::endl;
159
160     auto bmask = global_state.bmask;
161     for (index_t i = 0; i < num_items; i++) {
162         std::cout << bmask % 2 << " ";
163         bmask >>= 1;
164     }
165     std::cout << std::endl;
166 }
```

Listing 5.15: Main function of the Knapsack program.

The sequential program executed on a Xeon E5-2683 v4 CPU obtains the optimal solution in approximately 2.4 seconds:

```
value 1571
11111111111111100010001000000000
```

Note that it takes more than 22 seconds to compute the same result if we remove the optional local bound in Lines 138–140. The discussed techniques are also applicable to the Traveling Salesman Problem: the global state stores the length of the currently shortest valid tour together with the sequence of visited cities.

PARALLEL IMPLEMENTATION

A parallelization of the described algorithm is simple. Let $p = 2^{h+1}$ be the number of available processing units for some integer $h \geq 0$, then we can concurrently process each node in level h of the binary tree and their associated branches with a dedicated thread. As an example, set $h = 0$, then we can independently compute the left and right branch below the root node using $p = 2$ threads. Moreover, a work-sharing strategy as proposed in the previous section is conceivable: we delegate the left branches of the nodes to idling threads in a thread pool. Unfortunately, the relatively expensive scheduling logic of a thread pool dwarfs the benefits of work-sharing in case of the lightweight traversal scheme used above.

Furthermore, we have to check the code for potential race conditions. The binary mask `bmask`, as well as the corresponding accumulated value and weight `tuple`, is exclusively copied by value. Thus, we can rule out concurrent access to shared resources during the recursive invocation of branches. Nevertheless, the global state is simultaneously updated by distinct threads which might cause incorrect results. Consequently, the struct `global_state` in Line 48 should be wrapped by `std::atomic`:

```
std::atomic<state_t<bmask_t, value_t>> global_state;
```

We further have to modify the auxiliary function `sequential_update` in Lines 81–92 to support lock-free updates based on CAS-loops (see Listing 5.16).

```
1  template <
2      typename tuple_t,
3      typename bmask_t>
4  void atomic_update(
5      tuple_t tuple,
6      bmask_t bmask) {
7
8      typedef typename tuple_t::value_t value_t;
9
10     auto g_state = global_state.load();
11     auto l_value = tuple.value;
12     state_t<bmask_t, value_t> target;
13
14     do {
15
16         // exit if solution is not optimal
17         if (g_state.value > l_value)
18             return;
19
20         // construct the desired target
21         target.value = l_value;
22         target.bmask = bmask;
23
```

```
24    } while (!global_state.compare_exchange_weak(g_state, target));
25 }
```

Listing 5.16: Lock-free update of the global state.

First, a copy of the global state is stored atomically in the variable `g_state`. Second, we consecutively try to atomically swap the local state `target` (consisting of `bmask` and `tuple.value`) with the global state in case the current solution has a higher accumulated value than the global solution. Finally, we use our thread pool from the previous section to compute both branches below the root node (see Listing 5.17).

```
1  #include "threadpool.hpp"
2
3  int main () {
4
5      ThreadPool TP(2); // 2 threads are sufficient
6
7      // initialize tuples with random values
8      init_tuples(tuples, num_items);
9
10     // traverse left and right branch
11     TP.spawn(traverse<index_t, tuple_t, bmask_t>,
12              0, tuple_t(0, 0), 0);
13     TP.spawn(traverse<index_t, tuple_t, bmask_t>,
14              0, tuple_t(0, 0), 1);
15
16     // wait for all tasks to be finished
17     TP.wait_and_stop();
18
19     // report the final solution
20     auto g_state = global_state.load();
21     std::cout << "value " << g_state.value << std::endl;
22
23     auto bmask = g_state.bmask;
24     for (index_t i = 0; i < num_items; i++) {
25         std::cout << bmask % 2 << " ";
26         bmask >>= 1;
27     }
28     std::cout << std::endl;
29 }
```

Listing 5.17: Main function for the concurrent computation of the Knapsack Problem.

The parallel program executed on a Xeon E5-2683 v4 CPU using two threads obtains the optimal solution in less than 0.6 seconds:

```
value 1571
11111111111111100010001000000000
```

This is somehow surprising because the single-threaded implementation needed approximately 2.4 seconds to compute the same result: this is four times slower than our concurrent version using two threads. The observed superlinear speedup of 4 can be explained by the concurrent traversal scheme of the parallel implementation.

Starting from the root node, the sequential version is forced to find a reasonable solution in the left branch (leave item 0 behind) before traversing the right branch (take item 0). However, item 0 has the highest value density and thus it is highly probable that the global state stores a suboptimal solution during the traversal of the left branch. As a result, the local Dantzig bound exhibits almost no pruning power, i.e., it is very unlikely that we can rule out solution candidates in an early level of the tree. By contrast, the concurrent traversal scheme explores the binary tree in parallel. While thread 0 observes the same suboptimal solutions in the left branch, thread 1 simultaneously discovers high quality solutions in the right branch. As a result, the global state stores an almost optimal solution within a few steps which can be used to excessively prune worse solutions in the left branch. Loosely speaking, the concurrent traversal is a non-trivial mixture of depth-first-search and breadth-first-search which allows for a more efficient pruning of solution candidates in contrast to the traditional sequential algorithm using a pure depth-first-search based scheme.

This observation can be generalized to other algorithms that exhaustively probe candidates in order to find an optimal solution. Those occurrences of spurious superlinear speedup happen if two or more threads share a global state that can be exploited to exclude large portions of the state space. A trivial example is concurrent search of a node in a graph. If a thread discovers a node with a desired property it can simply signal other threads to stop by atomically writing a Boolean variable `is_found=true;`.

You might argue that we are actually not seeing any superlinear speedup since there exists at least one sequential algorithm that traverses the data structure in exactly the same order as the concurrent one. We could even find a sequential algorithm that needs less steps in total than the concurrent implementation. In fact, we can demystify the remarkable speedup of our Knapsack implementation by interchanging the two traversal calls in the main function of the sequential algorithm.

5.4 OUTLOOK

Now that we have come to the end of this chapter, we want to mention a few things that have not been discussed in detail. The multithreading API of the C++11 dialect is usually said to be the major contribution of this release. Atomic functions are perceived to be a fancy feature on top which allows for the implementation of platform-independent applications in a lock-free manner. The latter, however, is actually the fundamental contribution since atomic operations may exhibit a vastly different behavior on distinct hardware architectures. C++11 atomics guarantee more than the plain fact that each read-modify-operation is performed without interruption. They additionally imply memory order constraints that have to be enforced on every hardware architecture with a corresponding C++11 compiler. As an example, some architecture such as ARM allow for the reordering of instructions stated in the binary code of your application while Intel CPUs have relatively strict ordering guarantees. Let us demonstrate that with a code example:

```
payload_t payload;
std::atomic<uint8_t> flag;
...
payload = some_payload; // set the payload
flag.store(1);          // signal others to start
```

Here we set an arbitrary instance of a payload struct to a certain value. This could be a video sequence or any other data to be processed by one or more threads in parallel. The atomic variable `flag` is set to one in order to signal potentially waiting threads that they can start their computation. On the receiver side the code would look similar to this:

```
while(!flag.load());      // busy wait
process(payload);         // read the payload
```

In both cases, a naive compiler cannot see any dependency between the payload struct and the atomic flag and thus could safely assume that setting/reading the payload are interchangeable. Even when the compiler keeps the correct order during code generation, an ARM CPU might reorder the corresponding instructions which would lead to incorrect code. Hence, atomic operations have to guarantee that the sequential consistency of our program is preserved. C++11 performs this with a dedicated memory model that acts as naively expected in default settings but can also be relaxed to less restrictive behavior. While the fine-grained tuning of the memory model can have significant impact on application's performance on weakly ordered architectures such as ARM, it is only barely noticeable on Intel x64_64 CPUs that exhibit a relatively strong memory order. As a result, we omit the details of the different memory ordering modes and stick to the default behavior.

Assume we acquire (lock) a mutex and release (unlock) it right after a critical section. While it is not very smart from a performance point of view, it is totally valid to move any operations issued before the lock inside the critical section unless it breaks sequential consistency (you cannot move the statement `x=1;` after `y=x;` without violating dependencies):

```
std::mutex mutex;

statement_that_could_be_moved_below();

// |||||||||||||||||||||||||||||||||
// vvvvvvvvvvvvvvvvvvvvvvvvvvvvvvvvv
mutex.lock();         // <-- acquire

// critical code section goes here

mutex.unlock();       // <-- release
// ^^^^^^^^^^^^^^^^^^^^^^^^^^^^^^^^^
// |||||||||||||||||||||||||||||||||

statement_that_could_be_moved_above();
```

A similar statement is true for any instruction issued after the unlock: we could move it inside the locked scope without harming the correctness of the program. By contrast, it should be not allowed to move any instruction issued inside the locked scope below the release or above the acquire operation since this would defeat the purpose of a mutex. So to speak, both acquiring and releasing a mutex act as a one-way memory barrier with the following properties:

1. No memory operation issued after an acquire operation may be moved above the acquire operation.
2. No memory operation issued before a release operation may be moved below the release operation.

The described acquire–release semantic is also applicable to atomics. An atomic store in default memory order (`std::memory_order_seq_cst`) acts like a release. The payload is modified before setting the flag to one and thus has to be visible to other threads at the point of the release. An atomic load in default memory order acts like acquire: we can be sure that reading the payload is issued after the load. This behavior is crucial to ensure correct results. Hence, you can always assume for default memory ordering that (i) nothing written before an atomic store will ever be moved below the store, and (ii) nothing read after an atomic load will be moved above the load.

Concluding, atomic operations are more than uninterruptible read-modify-store instructions: they guarantee additional synchronization properties between stores and loads to the same variable when using default memory order. Nevertheless, C++11 allows for the relaxation of those synchronization properties which enables fine-grained tuning of the memory barrier properties. As an example, the weakest memory order `std::memory_order_relaxed` strips off any synchronization or ordering properties which results in a plain atomic that only guarantees stores and loads without interruption. This might result in potentially erroneous code with unpredictable synchronization behavior. Hence, we strongly advise to completely avoid the use of non-default memory order settings unless you develop a high performance application for hardware architectures with a weak memory model such as ARM.

5.5 ADDITIONAL EXERCISES

1. Write an efficient parallel program that computes the maximum element in an array of n floats. Initially, partition the array into equally sized portions and concurrently perform a local max-reduction on each fragment. Subsequently, merge the partial results using either locks or atomics. Compare the execution times. Which approach performs better?
2. An open-addressing hash map is a data structure that allows for the efficient storage and querying of key–value pairs (under certain conditions even in constant time). In the following, we assume that both the key and the value are 32-bit unsigned integers. The n keys and values are either stored in distinct locations as struct of arrays or interleaved as array of structs. For the sake of simplicity we choose the latter configuration $(key_0, val_0, key_1, val_1, \ldots, key_{n-1}, val_{n-1})$. All keys are initialized with a placeholder $\emptyset := 2^{32} - 1$ indicating that the corresponding slot is empty. The insertion and querying of a key–value pair (*key*, *val*) is accomplished as follows:
   a. We hash the key with an appropriate hash function, e.g. the collision-free Murmur hash integer finalizer:

```
uint32_t fmix32 (uint32_t key){
    key ^= key >> 16;
    key *= 0x85ebca6b;
    key ^= key >> 13;
    key *= 0xc2b2ae35;
    key ^= key >> 16;

    return key;
}
```

   and subsequently compute the corresponding slot as `fmix32(key) % n`. Note that the modulo operation might introduce collisions, i.e., several keys map onto the same slot. Values of existing keys are overwritten.

b. We probe if the key in the corresponding slot is the same or the empty placeholder and write both the key and the value in case of a success. Otherwise, we linearly probe in the neighborhood if we can find the same key or an empty slot. The key–value pair is inserted in the first valid slot next to the original position.
c. A query is accomplished analogous. We compute the position of the slot and probe to the right until we find the desired key or encounter an empty placeholder. In the latter case, we can guarantee that the key is not stored in the hash map.

Implement a sequential version of an open-addressing hash map using a linear probing scheme as described above. Thoroughly check your implementation for errors. Subsequently, design a lock-based parallel version that guarantees insertion without race conditions. Can you improve the insertion performance by using more than one lock? Finally, implement a lock-free version performing the insertion with CAS-loops. Compare the lock-based and lock-free implementation in terms of code complexity and performance.

3. Revisit the aforementioned exercise. Robin Hood hashing (RHH) is slightly modified variant of an open-addressing hash map [6]. We augment each key–value pair with an additional counter that stores the number of hops performed during the linear probing phase. Note that RHH can also be used with other probing schemes such as quadratic or chaotic probing. A "rich" key–value–count triple has performed less hops during the insertion than a "poor" one. RHH introduces the following modification of the probing scheme: whenever a poor triple encounters a rich triple during probing, the poor triple is stored at the position of the rich one. Afterwards, the probing is continued with the rich triple until it finds an empty slot or an even richer triple. RHH performs better for higher load factors (number of stored elements/capacity) since it automatically equalizes the probing lengths during insertion.
   a. Can you exploit the counter stored in a triple in order to accelerate the querying step?
   b. Implement a lock-free version of a Robin Hood hash map that uses 128 bits for the key–value–counter triple.
   c. Extend your implementation to support key, values, and counters of arbitrary width, e.g. 20 bits for the key, 20 bits for the value, and 24 bits for the counter.
4. Revisit the rules of the game Boggle in Section 4.6. Find a text file containing English words by querying the expression "english dictionary text file" in a search engine of your choice. Subsequently store the words in a hash set `std::unordered_set<std::string>` and implement a constant time function `bool is_word(std::string word)` for querying English words. Proceed as follows:
   a. Implement a sequential backtracking algorithm that recursively enumerates all valid path in a 4×4 Boggle board. Prevent cells from being visited twice by passing a binary mask of already visited cells: a 16 bit integer should suffice to encode the whole board. Store valid solutions in a dedicated vector of strings.
   b. Parallelize your sequential implementation by spawning a thread for each of the 16 cells. Is your code free of race conditions? If not make it so.
   c. Introduce a global state that stores the longest valid word being observed so far. Use a mutex to allow for its safe manipulation in a concurrent context. Can you achieve the same in a lock-free manner?

REFERENCES

[1] Damian Dechev, Peter Pirkelbauer, Bjarne Stroustrup, Lock-free dynamically resizable arrays, in: Mariam Momenzadeh, Alexander A. Shvartsman (Eds.), Principles of Distributed Systems: 10th International Conference, Proceedings, OPODIS 2006, Bordeaux, France, December 12–15, 2006, Springer Berlin Heidelberg, Berlin, Heidelberg, 2006, pp. 142–156.

[2] Yozo Hida, Xiaoye S. Li, David H. Bailey, Quad-Double Arithmetic: Algorithms, Implementation, and Application, 2000.

[3] C++11 standard (ISO/IEC 14882:2011), C++ reference of atomic data types, http://en.cppreference.com/w/cpp/atomic/atomic (visited on 01/12/2017).

[4] Tobias Maier, Peter Sanders, Roman Dementiev, Concurrent hash tables: fast and general?(!), in: Proceedings of the 21st ACM SIGPLAN Symposium on Principles and Practice of Parallel Programming, PPoPP '16, ACM, Barcelona, Spain, ISBN 978-1-4503-4092-2, 2016, pp. 34:1–34:2, http://doi.acm.org/10.1145/2851141.2851188.

[5] Jeff Preshing, Double-checked locking is fixed in C++11, http://preshing.com/20130930/double-checked-locking-is-fixed-in-cpp11/ (visited on 01/12/2016).

[6] Sebastian Sylvan, Robin Hood Hashing should be your default Hash Table implementation, https://www.sebastiansylvan.com/post/robin-hood-hashing-should-be-your-default-hash-table-implementation/ (visited on 01/12/2016).

CHAPTER

OPENMP

6

Abstract

OpenMP is an application programming interface (API) for platform-independent shared-memory parallel programming in C, C++, and Fortran. It pursues a semi-automatic parallelization approach which allows for the augmentation of sequential code using simple compiler directives in contrast to the explicit programming model of C++11 multithreading which involves the manual spawning, joining, and synchronization of threads. As an example, complex schedules, task parallelism, the convenient parallelization of loops and reductions can be realized with just a few additional lines of code. Despite its simplicity, the parallelization efficiency of OpenMP-augmented CPU code is often as good as handcrafted solutions using C++11 threads. Hence, OpenMP is ideally suited to investigate the parallelization potential of sequential code or to speed up trivially parallelizable algorithms within a short period of time. While targeting the same hardware architectures as with explicit multithreading, it simplifies a program's structure by providing ready-to-use building blocks. As a result, the programmer can focus on logical and abstract issues of the code.
This chapter teaches you how to parallelize a variety of scientific applications. We start with the concurrent processing of embarrassingly parallel tasks using loops. Subsequently, we discuss parallel reductions being the basic building block of many core algorithms such as counting, summation of values, or the streamed computation of extrema within a sequence. Moreover, we demonstrate how you can handle load imbalance caused by skewed work distributions using static and dynamic schedules. Another important aspect is task parallelism over implicitly enumerable sets such as graphs and trees. We conclude this chapter with a discussion of semi-automatic vectorization techniques.

Keywords

OpenMP, Compiler directive, Pragma, Race condition, Computation-to-communication ratio, Synchronization, Variable sharing, Privatization, SPMD, Nearest neighbor classifier, Mutex, Atomic operations, Static scheduling, Dynamic scheduling, Block distribution, Cyclic distribution, Custom reduction operator, Task parallelism, Tree traversal, Vectorization, SIMD, Data dependency, Offloading

CONTENTS

Parallel Programming. DOI: 10.1016/B978-0-12-849890-3.00006-X

6.1 INTRODUCTION TO OPENMP (HELLO WORLD)

OpenMP is an application programming interface (API) for platform-independent shared-memory parallel programming in C, C++, and Fortran on multiple different platforms. As an API, it provides a high-level abstraction layer over low-level multithreading primitives as the ones described in Chapter 4. Its programming model and interface are portable to different compilers and hardware architectures, scalable over an arbitrary number of CPU cores, and flexible in terms of writing compact and expressive source code. Hence, it is applicable to a variety of parallel applications executed in diverse scenarios ranging from desktop PCs with a few CPU cores to compute nodes in supercomputers utilizing up to hundreds of cores.

A BRIEF HISTORY OF OPENMP

The OpenMP API specifications are produced and published by the OpenMP Architecture Review Board (ARB), which consists of a group of major computer hardware and software vendors, as well as major parallel computing user facilities. Historically, the first specification released was OpenMP for Fortran 1.0 in October 1997, followed by OpenMP for C and C++ 1.0 in October 1998. Significant milestones for the C and C++ version are version 2.0 in March 2002, version 2.5 in May 2005, version 3.0 in May 2008, version 3.1 in July 2011, version 4.0 in July 2013, and version 4.5 from November 2015 [2,3]. Specifications of corresponding releases are publicly available on the respective website of the OpenMP ARB: http://www.openmp.org/specifications. Throughout the rest of this chapter we use OpenMP version 2.0 as baseline since it is supported by the majority of compilers: whenever features from newer versions are introduced, we explicitly mention it in the text.

BASICS

The basic philosophy of OpenMP is to augment sequential code by using special comment-like compiler directives – so-called **pragmas** – to give hints to the compiler on how the code can be parallelized. Pragmas can be used to exploit compiler-specific functionality. If the compiler does not support the corresponding feature it will simply ignore the pragma. Note that GCC supports the compiler flag `-Wunknown-pragmas` which reports unknown pragmas in your code.

By just adding appropriate pragmas, OpenMP can be easily used to parallelize existing sequential code. Written software can be compiled and executed with a single or multiple threads. Whether the compiler should include support for the OpenMP API is specified by the compiler flag `-fopenmp`. If we compile the code without this flag, we end up with sequential code. This behavior allows for the convenient deployment of sequential code that can be augmented with optional support for parallel execution.

The default number of threads used is usually determined at runtime to be equal to the number of logical CPU cores seen by the operating system. It can also be modified by setting the environment variable `OMP_NUM_THREADS`. Moreover, the explicit number of utilized threads can be further specified in the source code using the command `set_num_threads()` or dedicated clauses. The group of threads that is currently executing the program is called a *team*.

In this chapter, we investigate the parallelization potential of programs that have already been discussed in the previous chapter on multithreading using the C++11 threading API. The workflow is as follows: we start with a sequential program, analyze data dependencies and race conditions, and finally augment the code with OpenMP pragmas.

Now we get right into coding. Let us start with a basic *hello world* example (see Listing 6.1). It is almost identical to the sequential *hello world* program written in C++: the only difference is an additional `#pragma omp parallel` in Line 5 which tells the compiler that the following scope (only covering the print statement in Line 6) is executed in parallel.

```
1  #include <iostream>
2
3  int main() {
```

```
4       // run the statement after the pragma in all threads
5       #pragma omp parallel
6       std::cout << "Hello world!" << std::endl;
7   }
```

Listing 6.1: OpenMP hello world.

This code example can be compiled using GCC version 5.2 which supports OpenMP up to version 4.0:

```
g++ -O2 -std=c++14 -fopenmp hello_world.cpp -o hello_world
```

When run, we observe as many "Hello World" lines as there are CPU cores on the executing machine. You may set the environment variable `OMP_NUM_THREADS` before running the program in order to control the number of threads used:

```
OMP_NUM_THREADS=2 ./hello_world
Hello world!
Hello world!
```

Note that the output may not be perfectly separated by line breaks as in this example because all threads write concurrently to the standard output file descriptor which might result in cluttered output.

The `#pragma omp parallel` statement is translated by the compiler to the spawning and joining of a new team of threads and the execution of the statements in the following scope. The statement following the pragma does not need to be just a single line of code like in the example. It can also be a structured block of code within curly braces. Refer to Fig. 4.1 in Chapter 4 for more details on the behavior of thread spawning and joining.

The size of the team can be controlled at runtime, as we have already seen, but also by the programmer at the time of writing: changing Line 5 in the example above to `#pragma omp parallel num_threads(2)` will always create a team of just two threads regardless of other facts.

If a program needs to be aware of its OpenMP parameters, e.g. the default number of threads or the inherent identifier of a thread, a runtime library can be used to gather this information. An additional code example that shows this feature is listed below:

```
1   #include <iostream>
2   #include <omp.h>
3
4   int main() {
5       // run the block after the pragma in four threads
6       #pragma omp parallel num_threads(4)
7       {
8           int i = omp_get_thread_num();
9           int n = omp_get_num_threads();
10          std::cout << "Hello world from thread "
11                    << i << " of " << n << std::endl;
12      }
13  }
```

Listing 6.2: OpenMP hello world on steroids.

Again, we compile the aforementioned code fragment using

```
g++ -O2 -std=c++14 -fopenmp hello_world.cpp -o hello_world
```

and execute it on the command line

```
./hello_world
Hello world from thread 1 of 4
Hello world from thread 0 of 4
Hello world from thread 2 of 4
Hello world from thread 3 of 4
```

As expected, the threads are executed in arbitrary order. Thus, you will most likely observe shuffled or cluttered output. Note that the code in this example cannot be trivially compiled without OpenMP support since it explicitly requires OpenMP library functions defined in the `omp.h` header file. However, you can always mask portions of your code that explicitly depend on OpenMP using preprocessor macros

```
#if defined(_OPENMP)
#include <omp.h>
#endif
...
#if defined(_OPENMP)
// code explicitly depending on omp.h
#else
// refactored code not depending on omp.h
#endif
```

since the macro `_OPENMP` is automatically defined when activating the `-fopenmp` compiler flag.

6.2 THE `parallel for` DIRECTIVE (BASIC LINEAR ALGEBRA)

One of the most common concurrent approaches for *work sharing* is loop parallelization. Using OpenMP, it is trivial to execute a for-loop without data dependencies in parallel. In most cases, a composite statement of the `parallel` and `for` directives is used:

```
#pragma omp parallel for
for (...) {
    ...
}
```

But keep in mind that this is just a shortcut notation for the more explicit form:

```
#pragma omp parallel
{
    #pragma omp for
    for (...) {
        ...
    }
}
```

Let us discuss the aforementioned code fragment in more detail. The compiler directive `#pragma omp for` in front of the for-loop is crucial: each executed thread would process the whole for-loop independently if we omit it. Hence, computation would be redundant resulting in no runtime improvement or even performance degradation. The directive `#pragma omp for` acts as splitting primitive that partitions the set of n indices into chunks of approximate size n/p where p is the number of utilized threads.

Consequently, you should refer to `#pragma omp for` as `#pragma omp split_for` in your mind's eye. This has important implications on the syntax of the for-loop: the compiler has to determine the total amount of iterations as closed mathematical expression at compile time. As a result, we are not allowed to manipulate the iterator index or its bounds in the body of the for-loop. Further restrictions follow directly from this:

1. The loop iteration variable must be an integer up to OpenMP version 2.0. Unsigned integers are supported beginning from version 3.0.
2. The loop control parameters must be the same for all threads.
3. It is not allowed to branch (goto) out of a loop. This includes early-exits using `break`.

The good news is that the compiler will tell you if your for-loop violates one of the aforementioned constraints. However, algorithms that heavily rely on early-exits within for-loops have to be completely rewritten to be compliant with concurrent computation. Another construct for work sharing that is not bound by these restrictions is presented in Section 6.6.

Finally, we want to make an important remark: OpenMP does not magically transform your sequential for-loop into correct parallel code. You have to ensure by means of theoretical considerations that the computed result does not depend on the execution order of the loop iterations. This is trivially the case if there are no data dependencies among threads. Concluding, `#pragma omp parallel for` performs two steps: (i) splitting the for-loop indices evenly into chunks, and (ii) subsequently processing each chunk independently, no matter if the result is correct from a sequential point of view. Note that each thread can access any variable or array declared outside of the `parallel` scope since OpenMP is a framework operating on shared memory architectures. Hence, you have to guarantee that your code is free of race conditions.

VECTOR ADDITION

Let us illustrate the theoretical considerations from the previous subsection with an example of vector addition. We compute the sum of two integer-valued vectors $z = x + y$ where $x, y, z \in \mathbb{N}_0^n$ in coordinate representation:

$$z[i] = x[i] + y[i] \quad \text{for all } i \in \{0, \ldots, n-1\} \quad . \tag{6.1}$$

The vectors x and y are initialized with arbitrary values. Here we chose $x[i] = i$ and $y[i] = n - i$ such that $z[i] = n$ for all $i \in \{0, \ldots, n-1\}$. Furthermore, we store each of the three vectors in a `std::vector<uint64_t>` container. The corresponding source code is shown in Listing 6.3.

```
1  #include <iostream>
2  #include <cstdint>
3  #include <vector>
4
5  // hpc_helpers contains the TIMERSTART and TIMERSTOP macros
6  // and the no_init_t template that disables implicit type
7  // initialization
8  #include "../include/hpc_helpers.hpp"
9
```

```
10  int main() {
11      // memory allocation for the three vectors x, y, and z using
12      // the no_init_t template as a wrapper for the actual type
13      TIMERSTART(alloc)
14      const uint64_t num_entries = 1UL << 30;
15      std::vector<no_init_t<uint64_t>> x(num_entries);
16      std::vector<no_init_t<uint64_t>> y(num_entries);
17      std::vector<no_init_t<uint64_t>> z(num_entries);
18      TIMERSTOP(alloc)
19
20      // manually initialize the input vectors x and y
21      TIMERSTART(init)
22      #pragma omp parallel for
23      for (uint64_t i = 0; i < num_entries; i++) {
24          x[i] = i;
25          y[i] = num_entries - i;
26      }
27      TIMERSTOP(init)
28
29      // compute x + y = z sequentially
30      TIMERSTART(add_seq)
31      for (uint64_t i = 0; i < num_entries; i++)
32          z[i] = x[i] + y[i];
33      TIMERSTOP(add_seq)
34
35      // compute x + y = z in parallel
36      TIMERSTART(add)
37      #pragma omp parallel for
38      for (uint64_t i = 0; i < num_entries; i++)
39          z[i] = x[i] + y[i];
40      TIMERSTOP(add)
41
42      // check if summation is correct
43      TIMERSTART(check)
44      #pragma omp parallel for
45      for (uint64_t i = 0; i < num_entries; i++)
46          if (z[i] - num_entries)
47              std::cout << "error at position "
48                        << i << std::endl;
49      TIMERSTOP(check)
50  }
```

Listing 6.3: OpenMP Vector Addition.

In Line 8, we include the `hpc_helpers.hpp` header file which is distributed with this book. It contains the `no_init_t` wrapper type that disables `std::vector`'s implicit variable initialization of plain old data types, as shown in Listing 4.9 in the previous chapter. This avoids redundant initialization of the vectors during their declaration and a second time during our handcrafted initialization step in Lines 23–26 which is now performed in parallel. The macros `TIMERSTART` and `TIMERSTOP` are also defined in the header file allowing for convenient time measurement.

All threes phases, the initialization (Lines 23–26), the actual computation of the addition (Lines 38–39), and the final check (Lines 45–48) are embarrassingly parallel since there are no data dependencies between the slots of the vector. We have included an additional sequential computation of the vector sum in Lines 31–32. When we execute the program on a dual socket 32-core Intel Xeon CPU E5-2683 v4 @ 2.10GHz machine using eight threads, we obtain the following runtimes:

```
OMP_NUM_THREADS=8 ./vector_add
# elapsed time (alloc): 2.4933e-05s
# elapsed time (init): 0.888266s
# elapsed time (add_seq): 4.17046s
# elapsed time (add): 0.68106s
# elapsed time (check): 0.256372s
```

We observe a speedup of roughly 5 using eight threads. Unfortunately, the program is memory bound since we perform almost no computation besides a single addition. Consequently, we cannot expect significantly higher speedups when using more cores. The parallel runtime for 64 threads accounts for approximately 0.45 seconds which corresponds to a disastrous parallelization efficiency of roughly 15%. In general, it is not advisable to parallelize linear time algorithms with lightweight computation over a huge number of cores. However, we also discuss examples with a better compute-to-memory-access ratio in this chapter.

Implicit Synchronization

The aforementioned code has a slight disadvantage: we spawn the threads at the beginning of each `#pragma omp parallel for` directive and join them afterwards. It would be sufficient to spawn them once and synchronize the threads between the individual phases. Many OpenMP directives such as `#pragma omp for` support implicit synchronization, i.e. idling threads block the execution at the end of the loop body until completion.

```
#pragma omp parallel
{   // <- spawning of threads

    #pragma omp for
    for (uint64_t i = 0; i < num_entries; i++) {
        x[i] = i;
        y[i] = num_entries - i;
    }

    // <- implicit barrier

    #pragma omp for
    for (uint64_t i = 0; i < num_entries; i++)
        z[i] = x[i] + y[i];

    // <- another implicit barrier

    #pragma omp for
    for (uint64_t i = 0; i < num_entries; i++)
        if (z[i] - num_entries)
            std::cout << "error at position "
                      << i << std::endl;
```

```
} // <- joining of threads
// <- final barrier of the parallel scope
```

This behavior may be disadvantageous if we want to process two or more independent loops that do not need to wait for each other. The `nowait` clause can be used to explicitly remove the barriers usually resulting in a slight improvement of runtime:

```
#pragma omp parallel
{
    #pragma omp for nowait
    for (...) { ... }

    // <- here is no barrier

    #pragma omp for
    for (...) { ... }
}
```

VARIABLE SHARING AND PRIVATIZATION

Up to now, we have exclusively accessed variables and array entries in the scope of `#pragma omp parallel` without race conditions. Let us have a look at the following program which is both valid C++ and C99 code. It prints all index combinations (i, j) for $i, j \in \{0, 1, 2, 3\}$ to the command line.

```
#include <stdio.h>

int main () {

    #pragma omp parallel for
    for (int i = 0; i < 4; i++)
       for (int j = 0; j < 4; j++)
           printf("%d %d\n", i, j);
}
```

Declaring Private Variables

However, initial declarations of for-loop iterator variables of the form `int i = 0` are forbidden in the C90 dialect of the C programming language. The corresponding program has to be rewritten to

```
#include <stdio.h>

int main () {

    int i, j; // must be declared before the loops

    // #pragma omp parallel for produces incorrect
    // results due to race condition on variable j
    for (i = 0; i < 4; i++)
       for (j = 0; j < 4; j++)
           printf("%d %d\n", i, j);
}
```

Unfortunately, we cannot simply add `#pragma omp parallel for` in front of the first loop since each thread now concurrently manipulates the globally declared variable `j` which is not local to the threads anymore. Hence, we have to give the compiler a hint, that each thread should have its own copy of `j`. This can be achieved with the `private` clause. The corresponding directive reads now

```
#pragma omp parallel for private(j)
```

Note that you never have to care about the iterator variable of the loop directly following the `#pragma omp for` directive (here `i`). The globally declared variable `j` is not modified in the parallel scope since threads operate on thread-local versions, i.e. its value remains the same no matter if its privatized thread-local variants are modified.

Initialization of Privatized Variables

Privatized variables are not initialized in the parallel scope and thus have to be manually assigned a value as done with the statement `j = 0` in the inner loop of our example. Otherwise you end up with undefined behavior like in the following example:

```
#include <stdio.h>

int main () {

    int i = 1;

    // each thread declares its own i
    // but leaves it uninitialized
    #pragma omp parallel private(i)
    {
        // WARNING: i is not initialized here!
        printf("%d\n", i); // could be anything
    }
}
```

However, you can copy the value of the globally declared variable to the thread-local version using the clause `firstprivate`:

```
#include <stdio.h>

int main () {

    int i = 1;

    // each thread declares its own i
    // and sets it to i = 1
    #pragma omp parallel firstprivate(i)
    {
        printf("%d\n", i); // i == 1
    }
}
```

Nevertheless, modifications of privatized thread-local variables in the parallel scope still leave the globally declared variable unchanged.

Capturing Private Variables

If you want to access the thread-local values of a privatized variable from outside of the parallel scope, you can write them back to a globally declared auxiliary array. Moreover, this example demonstrates how to use OpenMP in a Single Program Multiple Data (SPMD) fashion where each thread executes the same program on different input (the thread identifier `j` in our case):

```
#include <omp.h>

int main () {

    // maximum number of threads and auxiliary memory
    const int num = omp_get_max_threads();
    int * aux = new int[num];

    int i = 1; // we pass this via copy by value

    #pragma omp parallel firstprivate(i) num_threads(num)
    {
        // get the thread identifier j
        const int j = omp_get_thread_num();
        i += j;         // any arbitrary function f(i, j)
        aux[j] = i;     // write i back to global scope
    }

    delete [] aux; // aux stores the values [1, 2, 3, ...]
}
```

We cannot return a unique value of a privatized variable since each thread assigns a potentially different value to it. As a result, one needs to define a criterion for the selected value. Note that more sophisticated approaches, which combine all values, so-called *reductions*, are discussed in the next section. In the special case of loops we can return the value of a privatized variable using the `lastprivate` clause. It copies the current value of the thread-local variable processed in the last iteration back to the global variable. This value is unique if the value of the thread-local variable does not depend on the execution order of iterations, e.g. when its value exclusively depends on the iteration index. Let us demonstrate this with a basic example:

```
int main () {
    int i;

    #pragma omp parallel for lastprivate(i) num_threads(16)
    for (int j = 0; j < 16; j++)
        i = j;

    // now, i has the value 15 since j == 15 in the last
    // iteration; execution order of threads is irrelevant
}
```

Final Remarks on Variable Privatization/Sharing

The clauses `firstprivate` and `lastprivate` can be combined realizing both the copying of values into the parallel scope, and the subsequent assignment of the thread-local value stemming from the last iteration to the global variable. Although the described value passing mechanism is convenient you will most likely never need it since C++ allows for the initialization of iterator variables within for-loop headers which automatically ensures thread-local memory allocation. Moreover, the transfer of values from thread-local memory to global memory should always be realized in an explicit fashion using dedicated memory such as the helper array `aux`. Concluding, let us make a few statements on variable sharing:

- Variables not being privatized are shared by default. However, the redundant clause `shared` could be used to explicitly emphasize this in the code.
- More than one variable can be privatized/shared by providing comma-separated lists, e.g. `private(i, j)`.
- It is not advisable to privatize raw pointers to arrays since only the address itself is privatized. Moreover, `private(data_ptr)` will cause a segmentation fault due to missing initialization: use `firstprivate` instead.
- If you explicitly need dedicated thread-local arrays simply allocate and free them in the parallel scope.

OpenMP can be combined with value passing mechanisms from other libraries as well. Feel free to mix pragmas with promises and futures from C++11, or lock-free data structures provided by the Boost library [1].

MATRIX VECTOR MULTIPLICATION

Let us have a look at a slightly more advanced real-life application that we have already discussed in the previous chapter on C++11 multithreading: Dense Matrix Vector Multiplication (DMV). Let $A \in \mathbb{R}^{m \times n}$ be a real-valued matrix of shape $m \times n$ and $x \in \mathbb{R}^n$ be a vector in n dimensions, then A linearly maps x from an n-dimensional vector space to an m-dimensional vector space (i.e. parallel lines are mapped onto parallel lines). The entries of A shall be denoted as A_{ij} where the index i enumerates the rows and j refers to its columns. The product $b := A \cdot x$ can be written in coordinate form as follows:

$$b_i = \sum_{j=0}^{n-1} A_{ij} \cdot x_j \quad \text{for all} \quad i \in \{0, \dots, m-1\} \quad . \tag{6.2}$$

The sum over j can be evaluated independently for each fixed index i resulting in $m \cdot n$ additions overall. Hence, it is advisable to parallelize the matrix vector product over the $\mathcal{O}(m)$ outer indices enumerated by i. Another approach would parallelize the inner sum over j using a parallel reduction over $\mathcal{O}(\log(n))$ steps. The latter, however, involves non-trivial synchronization after each reduction step which would unnecessarily complicate this example. The corresponding sequential and parallel implementation is shown in Listing 6.4.

```
1  #include <iostream>
2  #include <cstdint>
3  #include <vector>
4
5  // hpc_helpers contains the TIMERSTART and TIMERSTOP macros
```

```
6   // and the no_init_t template that disables implicit type
7   // initialization
8   #include "../include/hpc_helpers.hpp"
9
10  template <typename value_t,
11            typename index_t>
12  void init(std::vector<value_t>& A,
13            std::vector<value_t>& x,
14            index_t m,
15            index_t n) {
16
17      for (index_t row = 0; row < m; row++)
18          for (index_t col = 0; col < n; col++)
19              A[row*n+col] = row >= col ? 1 : 0;
20
21      for (index_t col = 0; col < n; col++)
22          x[col] = col;
23  }
24
25  template <typename value_t,
26            typename index_t>
27  void mult(std::vector<value_t>& A,
28            std::vector<value_t>& x,
29            std::vector<value_t>& b,
30            index_t m,
31            index_t n,
32            bool parallel) {
33
34      #pragma omp parallel for if(parallel)
35      for (index_t row = 0; row < m; row++) {
36          value_t accum = value_t(0);
37          for (index_t col = 0; col < n; col++)
38              accum += A[row*n+col]*x[col];
39          b[row] = accum;
40      }
41  }
42
43  int main() {
44      const uint64_t n = 1UL << 15;
45      const uint64_t m = 1UL << 15;
46
47      TIMERSTART(overall)
48      // memory allocation for the three vectors x, y, and z
49      // with the no_init_t template as a wrapper for the actual type
50      TIMERSTART(alloc)
51      std::vector<no_init_t<uint64_t>> A(m*n);
52      std::vector<no_init_t<uint64_t>> x(n);
53      std::vector<no_init_t<uint64_t>> b(m);
54      TIMERSTOP(alloc)
55
56      // manually initialize the input matrix A and vector x
57      TIMERSTART(init)
```

```
58      init(A, x, m, n);
59      TIMERSTOP(init)
60
61      // compute A * x = b sequentially three times
62      for (uint64_t k = 0; k < 3; k++) {
63          TIMERSTART(mult_seq)
64          mult(A, x, b, m, n, false);
65          TIMERSTOP(mult_seq)
66      }
67      // compute A * x = b in parallel three times
68      for (uint64_t k = 0; k < 3; k++) {
69          TIMERSTART(mult_par)
70          mult(A, x, b, m, n, true);
71          TIMERSTOP(mult_par)
72      }
73      TIMERSTOP(overall)
74
75      // check if (last) result is correct
76      for (uint64_t index = 0; index < m; index++)
77          if (b[index] != index*(index+1)/2)
78              std::cout << "error at position " << index
79                        << " " << b[index] << std::endl;
80  }
```

Listing 6.4: OpenMP Matrix Vector Multiplication.

Let us discuss the source code. First, we include necessary headers from the standard library and a custom `hpc_helpers.hpp` header file for the convenient time measurement which is distributed with this book. The function template `init` in Line 12 fills the matrix A with ones below the diagonal and with zeros otherwise, simulating prefix summation. The vector $x = (0, 1, 2, \dots)$ is initialized with ascending integers. Consequently, we expect the entries b_i of $b = A \cdot x$ to be the partial sums from zero to i. Second, the function template `sequential_mult` in Line 27 processes the actual matrix vector product by consecutively computing scalar products of the i-th row of A with the vector x. Third, we allocate the storage for the matrix A and the vectors x and b within the `main` function (see Lines 51–53). Afterwards, we initialize them using `init` in Line 58 and finally execute the matrix vector product sequentially (Line 64) and in parallel (Line 70). We run each computational variant (sequential and parallel) three times in order to obtain robust runtimes.

You might have noticed that the only "parallel" code in the aforementioned Listing is the compiler directive `#pragma omp parallel for if(parallel)` in Line 32. Moreover, the pragma can be masked using an `if` clause which accepts a Boolean argument. Hence, we can conveniently switch between sequential and parallel execution mode by altering the last parameter `parallel` of the function `mult` in Line 32. The code can be compiled with a C++14 compliant compiler (here GCC 5.2):

```
g++ -O2 -std=c++14 -fopenmp \
    matrix_vector.cpp -o matrix_vector
```

When we execute the program on a dual socket 32-core Intel Xeon CPU E5-2683 v4 @ 2.10GHz machine using eight threads, we obtain the following runtimes:

```
OMP_NUM_THREADS=8 ./matrix_vector
```

```
# elapsed time (alloc): 2.3118e-05s
# elapsed time (init): 2.89897s
# elapsed time (mult_seq): 1.2596s
# elapsed time (mult_seq): 1.25956s
# elapsed time (mult_seq): 1.24509s
# elapsed time (mult_par): 0.273558s
# elapsed time (mult_par): 0.250817s
# elapsed time (mult_par): 0.249425s
# elapsed time (overall): 7.43739s
```

The resulting speedup of roughly 5 and the corresponding parallelization efficiency of 63% are comparable to the handcrafted parallel code for DMV using C++11 threads from Section 4.3. This is quite remarkable considering the fact that we have added only a single line of code to implement parallelization. Note that similarly to vector addition this algorithm is also memory-bound and thus we cannot expect good scalability over a huge amount of CPU cores: an issue that we will address in the next section.

6.3 BASIC PARALLEL REDUCTIONS (NEAREST-NEIGHBOR CLASSIFIER)

This section shows how to handle situations where a shared accumulator variable is correctly updated in a parallel context, e.g. the concurrent summation of all entries stored in an array. Later in this chapter, we will generalize this technique to work with any associative binary reduction map. Moreover, we investigate the impact of parallelization granularity on the overall runtime of our programs. This is accomplished by studying the parallel performance of a one-nearest-neighbor (1NN) classifier trained on the popular MNIST dataset consisting of tens of thousands of handwritten digits.

ONE-NEAREST-NEIGHBOR CLASSIFICATION

One-nearest-neighbor classifiers [4] are among the most versatile classifiers in the field of machine learning. Their only prerequisite is a pairwise distance measure $dist(X_{\text{test}}^{(i)}, X_{\text{train}}^{(j)})$ assigning a notion of similarity between two objects $X_{\text{test}}^{(i)}$ and $X_{\text{train}}^{(j)}$. 1NN classifiers often provide reasonable classification accuracy and are easy to implement. The goal of classification is to correctly predict the class labels $Y_{\text{test}}^{(i)}$ for all objects $X_{\text{test}}^{(i)}$ stemming from a test set of unlabeled objects by assigning the corresponding label $Y_{\text{train}}^{(j^*)}$ of the nearest instance $X_{\text{train}}^{(j^*)}$ from a labeled training set. The label assignment step for a fixed index i in the test set reads as follows:

$$Y_{\text{test}}^{(i)} \leftarrow Y_{\text{train}}^{(j^*)} \quad \text{where} \quad j^* = \operatorname*{argmin}_j dist(X_{\text{test}}^{(i)}, X_{\text{train}}^{(j)}) \quad . \tag{6.3}$$

Loosely speaking, you maintain a collection of labeled objects, let us say dogs and cats, and a suitable measure how to compare them to other unlabeled objects. If someone hands you a new animal and asks you whether it is a dog or a cat, you simply compare it to each animal in your collection and determine the most similar one: let us assume it is a cat. Finally, one assigns the label of the nearest instance to the unseen object – in our case it would have been classified as a cat. Notably, this approach allows for the implicit definition of a cat without ever being explicit about what it needs to be a cat. Instead, we learn the concept of distinguishing dogs and cats by showing the classifier representatives of both classes.

The described procedure has several advantages. A 1NN classifier needs no time-consuming training phase or a complex mathematical modeling of the prediction step. Training is equivalent to maintaining a collection of labeled objects and determining a suitable notion of similarity (which is basically free). Moreover, we do not have to perform an expensive grid search over parameters if the similarity measure is free of external dependencies which is the case for basic measures such as Euclidean Distance, Cross Correlation, or Mutual Information.

Unfortunately, this procedure also presents drawbacks. It is not always straightforward to determine a measure which captures a suitable notion of similarity. However, we will see that Euclidean Distance is a good choice for approximately aligned handwritten digits. Another disadvantage is the high computational cost of the prediction/inference step. It involves a brute force comparison of the objects to be classified with all instances in the training set. Although there exist spatial data structures that allow for the acceleration of the linear search step such as Kd-trees or the excessive use of lower bounds, we often end up with no significant improvement in runtime due to the curse of dimensionality [7]. Note that the dimension of an image is equivalent to the amount of pixels: querying spatial data structures usually degenerates to linear search for $d \gg 10$. Hence, we limit ourselves to basic linear search to evaluate Eq. (6.3) in our example.

THE MNIST DATASET OF HANDWRITTEN DIGITS

In the following, we construct a 1NN classifier over a training set of $n = 55{,}000$ handwritten digits of the MNIST dataset [6]. MNIST consists of overall 65,000 images stored as gray-scale arrays of shape 28×28 with their corresponding labels ranging from zero to nine. The remaining $m = 10{,}000$ images are used as test set for subsequent prediction and evaluation of classification performance. Fig. 6.1 depicts twenty typical samples. For our purpose we interpret each of the images as a flattened vector consisting of $d = 784$ intensity values. Assume that we store the n images in the training set as a data matrix $D^{\text{train}}_{jk} = X^{(j)}_{\text{train}}[k]$ of shape $n \times d$ where j denotes the n image indices and the index k enumerates the d pixels of each image. The data matrix of the test set $D^{\text{test}}_{ik} = X^{(i)}_{\text{test}}[k]$ has shape $m \times d$ and stores m images of the same dimension.

THEORETICAL ASPECTS OF ALL-PAIRS DISTANCE COMPUTATION

In the following, we are interested in the all-pairs distance matrix Δ of shape $m \times n$ using squared Euclidean Distance as similarity measure:

$$\Delta_{ij} = dist\left(X^{(i)}_{\text{test}}, X^{(j)}_{\text{train}}\right) = \sum_{k=0}^{d-1} \left(X^{(i)}_{\text{test}}[k] - X^{(j)}_{\text{train}}[k]\right)^2 \quad , \tag{6.4}$$

where $i \in \{0, \ldots, m-1\}$ and $j \in \{0, \ldots, n-1\}$. We observe that the time complexity of the algorithm $\mathcal{O}(m \cdot d \cdot n)$ is almost three orders of magnitude higher than its memory complexity being $\mathcal{O}(m \cdot n)$ since the number of pixels per image $d = 784$ is reasonably high. Hence, we expect this program to scale significantly better than the previously discussed vector addition and matrix vector multiplication

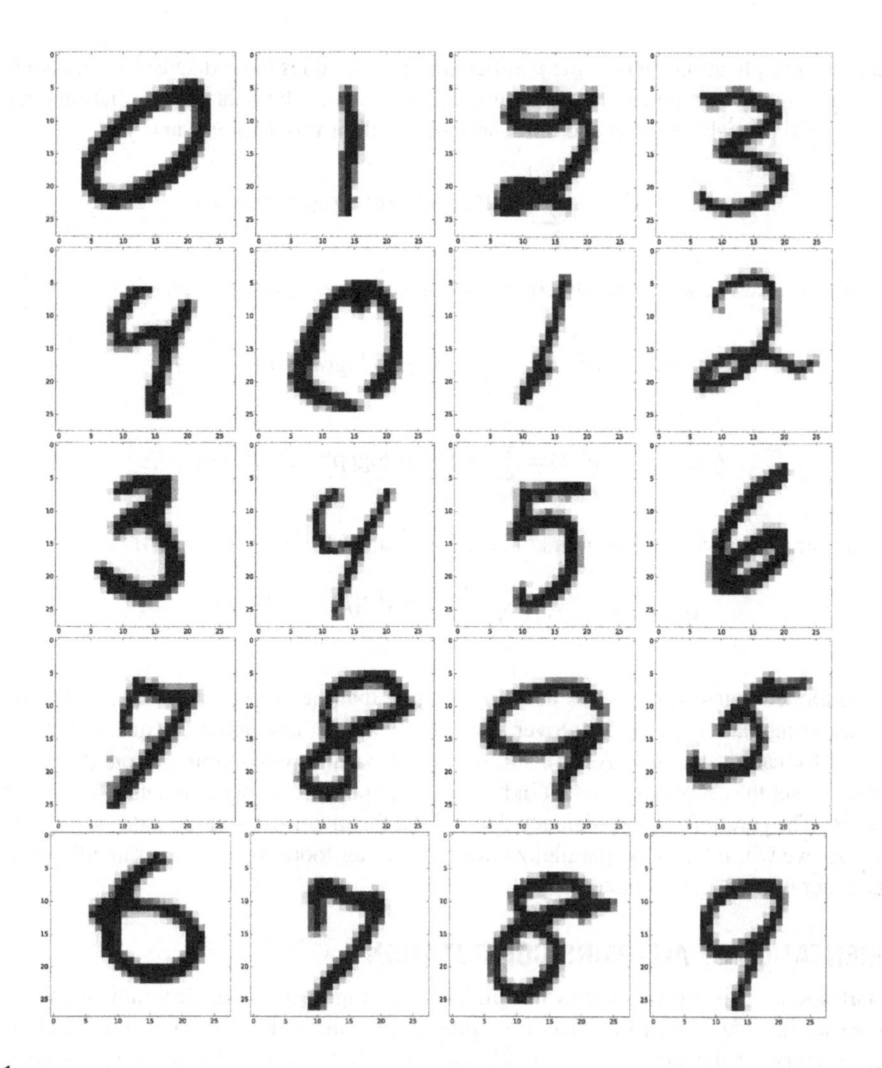

FIGURE 6.1

Twenty exemplary samples from the MNIST dataset consisting of 65,000 handwritten digits. The images of shape 28×28 are embedded as plain vectors in $\mathbb{R}^{784}$.

examples. Furthermore, if we rewrite Eq. (6.4) by expanding it in terms of the binomial theorem

$$\Delta_{ij} = \sum_{k=0}^{d-1} \left(X_{\text{test}}^{(i)}[k]\right)^2 - 2\sum_{k=0}^{d-1} \left(X_{\text{test}}^{(i)}[k] \cdot X_{\text{train}}^{(j)}[k]\right) + \sum_{k=0}^{d-1} \left(X_{\text{train}}^{(j)}[k]\right)^2 \quad . \tag{6.5}$$

We observe a decomposition into two self-interaction terms (first and last) which only depend on either the index i or j and a third mixing term (middle) depending on both i and j which is actually dense matrix matrix multiplication. Hence, the parallelization techniques to be discussed can also be applied to other incrementally accumulated similarity measures such as the Pearson correlation coefficient of two z-normalized (vanishing mean and unit variance) random variables $x^{(i)}$ and $y^{(j)}$

$$\rho(x^{(i)}, y^{(j)}) = \sum_{k=0}^{d-1} x^{(i)}[k] \cdot y^{(j)}[k] \quad \text{(middle term)} \quad , \tag{6.6}$$

cross-entropy and Kullback–Leibler divergence of probability vectors $p^{(i)}$ and $q^{(j)}$

$$H(p^{(i)}, q^{(j)}) = -\sum_{k=0}^{d-1} p^{(i)}[k] \cdot \log(q^{(j)}[k]) \quad ,$$

$$KLD(p^{(i)} || q^{(j)}) = \sum_{k=0}^{d-1} p^{(i)}[k] \cdot \log(p^{(i)}[k]/q^{(j)}[k]) \quad , \tag{6.7}$$

and the Hamming distance between pairs of strings $s^{(i)}$ and $t^{(j)}$ of fixed length d

$$Ham(s^{(i)}, t^{(j)}) = \sum_{k=0}^{d-1} \begin{cases} 0 & \text{if } s^{(i)}[k] == t^{(j)}[k] \\ 1 & \text{else} \end{cases} \quad . \tag{6.8}$$

All aforementioned pairwise measures have in common that they can be computed independently for a fixed index combination (i, j). Moreover, their final value is computed as sum over the individual contributions for each value of k. As a result, we have basically two options for parallelization: either we parallelize over the set of independent indices (i, j) or perform a concurrent sum-reduction over the inner index k. The parallelization efficiency depends on the dimensions of the data matrices m, n, d. In the following, we will refer to the parallelization of the outer loops as "coarse-grained" parallelization and to the inner one as "fine-grained."

IMPLEMENTATION OF ALL-PAIRS COMPUTATION

Let us start coding. Listing 6.5 shows the initial code fragment of our program. At the beginning (Lines 1–3) we include standard headers providing print statements, numeric limits of plain old data types, and vectors for the storage of data. We further include the `hpc_helpers.hpp` header (Line 8) for convenient time measurement and `binary_IO.hpp` (Line 11) which provides methods facilitating the reading and writing of binary data from or to the file system. The implementation of the all-pairs computation is straightforward: we loop over all index combinations (i, j) in Lines 24–25 and finally accumulate the contributions of the measure for each dimension k in Line 27. Note that we store both data matrices $D^{\text{test}}_{ik} = X^{(i)}_{\text{test}}[k]$ and $D^{\text{train}}_{jk} = X^{(j)}_{\text{train}}[k]$ in row-major order, i.e. the index k enumerating the d dimension alters the least significant bits of an entry's position in linear memory. This access pattern is more cache-friendly than the naive index ordering of ordinary matrix matrix multiplication

$$C_{ij} = \sum_{k=0}^{d-1} A_{ik} \cdot B_{kj} \quad \text{for all } i, j \quad , \tag{6.9}$$

which suffers from bad caching behavior for the second factor B_{kj} as shown in Section 3.3.

```
1  #include <iostream>   // std::cout
2  #include <limits>     // std::numeric_limits
3  #include <vector>     // std::vector
4
5  // hpc_helpers contains the TIMERSTART and TIMERSTOP macros
6  // and the no_init_t template that disables implicit type
7  // initialization
8  #include "../include/hpc_helpers.hpp"
9  // binary_IO contains the load_binary function to load
10 // and store binary data from and to a file
11 #include "../include/binary_IO.hpp"
12
13 template <typename value_t,
14           typename index_t>
15 void all_vs_all(value_t* test,
16                 value_t* train,
17                 value_t* delta,
18                 index_t num_test,
19                 index_t num_train,
20                 index_t num_features,
21                 bool parallel) {
22
23     #pragma omp parallel for collapse(2) if(parallel)
24     for (index_t i = 0; i < num_test; i++)
25         for (index_t j = 0; j < num_train; j++) {
26             value_t accum = value_t(0);
27             for (index_t k = 0; k < num_features; k++) {
28                 const value_t residue = test [i*num_features+k]
29                                       - train[j*num_features+k];
30                 accum += residue*residue;
31             }
32             delta[i*num_train+j] = accum;
33         }
34 }
```

Listing 6.5: OpenMP Nearest-Neighbor Classification: initial code fragment.

Here, we have chosen a coarse-grained parallelization strategy which concurrently performs the computation for distinct index combinations (i, j). The corresponding compiler directive can be found in Line 23. You may have noticed that we have introduced a novel clause `collapse(2)` to further specify the `for` splitting primitive. It collapses the following two for-loops as if they were written as one entity using a hyper-index h:

```
#pragma omp parallel for
for (index_t h = 0; h < num_test*num_train; h++) {
    const index_t i = h / num_test;
    const index_t j = h % num_test;
}
```

This might become handy if the number of samples in the test set is smaller than the amount of existing physical cores of your CPU. The naive directive `#pragma omp parallel for` applied to the outer loop would then result in inefficient parallel code caused by thread undersubscription.

Problems Emerging During Fine-Grained Parallelization

Alternatively, we could choose a fine-grained parallelization scheme which concurrently sums up the $d = 784$ pixels of the residual image. Unfortunately, we cannot choose the following combination of compiler directives since it introduces a race condition on the variable `accum`

```
// CAUTION: this code is erroneous!
for (index_t i = 0; i < num_test; i++)
   for (index_t j = 0; j < num_train; j++) {
       value_t accum = some_initial_value;
       #pragma omp parallel for
       for (index_t k = 0; k < num_features; k++)
           // introducing a race condition on accum
           accum += some_value;
   }
```

This code most probably computes an incorrect result because the load modify–store operation `accum+=some_value` is not atomic. A fast and dirty but inefficient hack would be the atomic increment of `accum`:

```
// CAUTION: this code is correct but inefficient
for (index_t i = 0; i < num_test; i++)
   for (index_t j = 0; j < num_train; j++) {
       value_t accum = some_initial_value;
       #pragma omp parallel for
       for (index_t k = 0; k < num_features; k++)
           // sanitizing the race condition using atomics
           #pragma omp atomic
           accum += some_value;
   }
```

or the even slower mutex-based variant which effectively serializes our program

```
// CAUTION: this code is correct but dead-slow
for (index_t i = 0; i < num_test; i++)
   for (index_t j = 0; j < num_train; j++) {
       value_t accum = some_initial_value;
       #pragma omp parallel for
       for (index_t k = 0; k < num_features; k++)
           // sanitizing the race condition using locks
           #pragma omp critical
           accum += some_value;
   }
```

None of the three aforementioned solutions is satisfying from a parallel point of view since we either compute an incorrect result or significantly degrade the performance by increasing contention for a shared resource: being an atomic variable in the second and a mutex in the third example. Note that the lock-based approach using `#pragma omp critical` is always applicable while `#pragma omp atomic` can only be used for operations with dedicated hardware support such as +=, *=, -=, /=, &=, ^=, |=, «=, »=, as well as all combinations of pre/post in/decrement operators ++, - acting on a single scalar variable.

Parallel Block-Wise Reduction

Conceptually, we already know how to resolve the race condition in an efficient manner: privatized variables. A viable computation pattern would be the following: each thread declares a thread-local variant of accum initialized with a default value (zero for addition) and subsequently performs the summation independently over its corresponding iterations of the loop in a sequential manner. Hence, each thread adds approximately n/p numbers where n is the number of iterations and p denotes the number of threads. Finally, we have to sum the partial results computed by each thread and add the overall result on top of the current value of the global variable accum. OpenMP provides a specialized reduction clause performing the described procedure. So to speak, reduction is a combination of the clause private with default initialization and subsequent merging of partial results. Now, the code reads as follows:

```
for (index_t i = 0; i < num_test; i++)
    for (index_t j = 0; j < num_train; j++) {
        value_t accum = some_initial_value;
        #pragma omp parallel for reduction(+:accum)
        for (index_t k = 0; k < num_features; k++)
            accum += some_value;
    }
```

In general, the clause reduction(operator:variable,...) accepts a list of one or more pairs each consisting of a reduction operator and a corresponding variable: operator:variable. OpenMP features out-of-the-box support for common operators such as addition (+ and the redundant -), multiplication (*), extreme values (min and max), as well as associative bitwise and logical operators (&, |, ^, &&, and ||). Note that you can even define custom associative operators with individual initialization and complex pairwise combination rules. This topic is discussed in Section 6.5 on advanced reductions in OpenMP.

PARALLEL LABEL PREDICTION

Now that we have cached the all-pairs distance information in the matrix Δ, we can proceed with the label assignment step which evaluates Eq. (6.3) for each image $X^{(i)}_{\text{test}}$ in the test set. For a fixed index i we scan through all n distance scores Δ_{ij} and determine the index j^* of the nearest image $X^{(j^*)}_{\text{train}}$ in the training set. Subsequently, we set the predicted class label to $Y^{(j^*)}_{\text{train}}$. In our case, we compare the (actually unknown) ground truth label $Y^{(i)}_{\text{test}}$ with the predicted label in order to estimate the quality of the classifier. Classification accuracy is a popular measure defined as quotient of the amount of correctly classified and the total number of classified objects. Hence, we simply have to count in parallel how many labels have been classified correctly and divide it by the number of images in the test set m. Note that the labels of the 10-digit classes are stored in one-hot encoding for the MNIST dataset, i.e. each label is represented by a binary vector of length 10 which is zero everywhere except at one position. As an example, the digit 3 is encoded as $(0, 0, 0, 1, 0, 0, 0, 0, 0, 0)$. Admittedly, this is a quite memory-inefficient representation carrying redundant information: a simple scalar 3 would have been sufficient. However, we will actually need the one-hot encoded label representation for another classifier discussed in this chapter and thus we stick to it. The corresponding source code is shown in Listing 6.6.

```
35  template <typename label_t,
36            typename value_t,
37            typename index_t>
38  value_t accuracy(label_t* label_test,
39                   label_t* label_train,
40                   value_t* delta,
41                   index_t num_test,
42                   index_t num_train,
43                   index_t num_classes,
44                   bool parallel) {
45
46      index_t counter = index_t(0);
47
48      #pragma omp parallel for reduction(+:counter) if(parallel)
49      for (index_t i = 0; i < num_test; i++) {
50
51          // the initial distance is float::max
52          // the initial index j_star is some dummy value
53          value_t bsf = std::numeric_limits<value_t>::max();
54          index_t jst = std::numeric_limits<index_t>::max();
55
56          // find training sample with smallest distance
57          for (index_t j = 0; j < num_train; j++) {
58              const value_t value = delta[i*num_train+j];
59              if (value < bsf) {
60                  bsf = value;
61                  jst = j;
62              }
63          }
64
65          // compare predicted label with original label
66          bool match = true;
67          for (index_t k = 0; k < num_classes; k++)
68              match &&= label_test [i   *num_classes+k] ==
69                        label_train[jst*num_classes+k];
70
71          counter += match;
72      }
73
74      return value_t(counter)/value_t(num_test);
75  }
```

Listing 6.6: OpenMP Nearest-Neighbor Classification: inference step.

Here, we have chosen a coarse-grained parallelization approach by augmenting the outer loop over the images in the test set (see Line 48). Initially, we determine the index j^* of the nearest neighbor for a fixed index i by searching for the smallest similarity score in the corresponding row i of the all-pairs distance matrix Δ_{ij} (see Lines 57–63). Afterwards, we compare if the ground truth and predicted label vector are identical. Technically, both inner loops can also be parallelized using a fine-grained scheme: the first loop (Line 57) is an argmin-reduction (no explicit OpenMP support) which is a minor modification of min-reduction, and the second loop could be parallelized using the directive `#pragma omp parallel for reduction(&&:match)`.

PERFORMANCE EVALUATION

The `main` function of our program combines the two discussed methods `all_vs_all` and `classify` (see Listing 6.7). After memory allocation and reading the images from disk (Lines 93–95), we subsequently compute the all-pairs distance matrix (Lines 103–107), and finally perform classification (Lines 111–116). Both functions accept as last parameter a Boolean parameter `parallel` which is determined in Line 38 from the command line arguments of our program. The overall classification accuracy of roughly 96.8 percent is a respectable result for a 10-class problem. Note that a dummy classifier which assigns labels by plain guessing would have performed at 10 percent.

```
76  int main(int argc, char* argv[]) {
77
78      // run parallelized when any command line argument is given
79      const bool parallel = argc > 1;
80
81      std::cout << "running "
82                << (parallel ? "in parallel" : "sequentially")
83                << std::endl;
84
85      // the shape of the data matrices
86      const uint64_t num_features = 28*28;
87      const uint64_t num_classes = 10;
88      const uint64_t num_entries = 65000;
89      const uint64_t num_train = 55000;
90      const uint64_t num_test = num_entries-num_train;
91
92      // memory for the data matrices and all-pairs matrix
93      std::vector<float> input(num_entries*num_features);
94      std::vector<float> label(num_entries*num_classes);
95      std::vector<float> delta(num_test*num_train);
96
97      // get the images and labels from disk
98      load_binary(input.data(), input.size(), "./data/X.bin");
99      load_binary(label.data(), label.size(), "./data/Y.bin");
100
101     TIMERSTART(all_vs_all)
102     const uint64_t inp_off = num_train * num_features;
103     all_vs_all(input.data() + inp_off,
104                input.data(),
105                delta.data(),
106                num_test, num_train,
107                num_features, parallel);
108     TIMERSTOP(all_vs_all)
109
110     TIMERSTART(classify)
111     const uint64_t lbl_off = num_train * num_classes;
112     auto acc = accuracy(label.data() + lbl_off,
113                         label.data(),
```

```
114                        delta.data(),
115                        num_test, num_train,
116                        num_classes, parallel);
117     TIMERSTOP(classify)
118
119     // fraction of labels assigned correctly in test set: 0.9677
120     std::cout << "test accuracy: " << acc << std::endl;
121 }
```

Listing 6.7: OpenMP Nearest Neighbor Classification: main function.

When we sequentially execute the program on a dual socket 32-core Intel Xeon CPU E5-2683 v4 @ 2.10GHz machine using one thread, we obtain the following runtimes:

```
./1NN
running sequentially
# elapsed time (all_vs_all): 446.91s
# elapsed time (classify): 0.735543s
test accuracy: 0.9677
```

Obviously, the major portion of runtime is spent on the all-pairs distance computation. The classification step itself is negligible. As a result, we focus on different parallelization strategies of the function `all_vs_all`. The coarse-grained parallelization of the outer loop using 32 physical cores and 32 hyperthreaded cores achieves a speedup of approximately 39 which corresponds to 120% of the expected linear speedup of 32. The classification step cannot benefit from parallelization by the same extent since it is heavily memory-bound (linear search in an array).

```
./1NN parallel
running in parallel
# elapsed time (all_vs_all): 11.4908s
# elapsed time (classify): 0.0514747s
test accuracy: 0.9677
```

In contrast, the fine-grained parallelization of the inner loop exhibits disastrous scaling properties:

```
for (index_t i = 0; i < num_test; i++)
    for (index_t j = 0; j < num_train; j++) {
        value_t accum = value_t(0);
        #pragma omp parallel for reduction(+:accum)
        for (index_t k = 0; k < num_features; k++)
            ...
    }
```

In this case, the computation of the matrix Δ takes several hours – a degradation of the sequential runtime of more than one order-of-magnitude. This massive difference in the performance can be explained as follows: The coarse-grained parallelization over the index combinations (i, j) performs $m \cdot n = 10{,}000 \cdot 65{,}000$ sequential reductions in parallel. In contrast, the fine-grained variant sequentially computes $m \cdot n$ parallel reductions. However, the inner loop accumulates only $d = 784$ values, i.e., in case of 64 threads each chunk covers only 12.25 indices – that is, less than the length of a cache line (16 for `float` on our machine). Besides that, all threads are spawned and joined after each sequential iteration of the outer loops which causes massive overhead.

Let us conclude what we have seen. OpenMP supports parallel reductions in order to allow for the concurrent and race condition-free accumulation of values by means of an associative and commutative binary operation. Nevertheless, the granularity of parallelization has to be chosen carefully. Otherwise you might even encounter performance degradation despite using more than one thread. As a rule of thumb we advise to first target the outer loops of an algorithm for coarse-grained parallelization. The considerable overhead of spawning and joining threads might render fine-grained schemes computationally intractable as demonstrated in our example. Note that the opposite is true for GPU parallelizations using the CUDA programming language as we will see in the next chapter.

6.4 SCHEDULING OF IMBALANCED LOOPS (INNER PRODUCTS)

All-pairs distance computation between objects in a test set and training set is similar to dense matrix matrix multiplication $C = A \cdot B$ of two matrices A and B. Sometimes one is interested in the inner product $C = A \cdot A^T$ of a single matrix A. The resulting product C is a symmetric and positive-definite matrix with non-negative eigenvalues. Matrices of this form are ubiquitous in natural science, e.g. moment of inertia tensors in classical mechanics, Gramian matrices in differential geometry, covariance matrices in stochastics, and construction of non-linear support vector machines and other spectral approaches (kernel machines, spectral clustering, etc.) in machine learning.

LOAD IMBALANCE CAUSED BY SYMMETRY

You might argue that $C = A \cdot A^T$ is just a special case of $C = A \cdot B$ for $B = A^T$; however, the resulting matrix C exhibits a symmetry in the entries C_{ij} since $C^T = (A \cdot A^T)^T = A \cdot A^T = C$. Hence, the indices of the expression C_{ij} can be swapped without changing its actual value. As a result, it is sufficient to compute only the contributions either below or above the diagonal for $i \leq j$ or $i \geq j$, respectively. Unfortunately, this introduces a load imbalance issue during parallelization using a simple `#pragma omp parallel for` directive for the outer loop. Because of the triangular structure of C some threads have to compute more entries than others.

We have already discussed techniques in Section 4.3 and Section 4.4 which allow for the balancing of heavily skewed work distributions: static and dynamic schedules of for-loops. In detail, we have partitioned the set of m rows of the resulting matrix $C = A \cdot A^T$ into batches of fixed chunk size $1 \leq c \leq \lceil m/p \rceil$ and subsequently processed the $\lceil m/c \rceil$ batches one after another using a fixed number of threads p. In the static case a predetermined number of batches is assigned to each of the p threads. Note that the special case $c = 1$ is referred to as *cyclic* and the special case $c = \lceil m/p \rceil$ as *block* distribution. In contrast, dynamic scheduling consecutively chooses the next idling thread for an unprocessed batch.

The actual implementation of the corresponding static or dynamic block-cyclic distribution scheme can be cumbersome and error-prone since (i) m might not be a multiple of p in general, and (ii) the need for a global mutex or atomic variable in case of dynamic scheduling. As a result, you end up with nested loops using non-trivial indexing (see Listing 4.16) which heavily decreases code readability.

The good news is that OpenMP features out-of-the-box support for static and dynamic block-cyclic distributions and a further scheduling mode called "guided." Moreover, you can choose an automatic schedule “auto” or the schedule of the runtime environment “runtime.” The scheduling modes can be briefly summarized as follows:

- `static`: All iterations are divided into roughly m/c many chunks each performing c sequential iterations of the overall m indices. Those batches of chunk size c are then distributed to the threads of the team in a round-robin fashion (block-cyclic distribution). If a thread has completed the computation of its assigned chunks, it will be idling until the remaining threads finish their work. The chunk size c defaults to $c \approx m/p$ if not defined (pure block distribution).
- `dynamic`: All iterations are again divided into equally sized chunks and distributed one by another to the executing threads as soon as they are a thread waiting for work. Therefore, idling can only occur when every thread is executing its last batch. The chunk size defaults to 1 if not specified (pure cyclic distribution).
- `guided`: All iterations are divided into chunks of decreasing size (up to the minimal size of a user-defined chunk size) and the batches are dispatched one by another like in the `dynamic` schedule. If the chunk size is not specified, it defaults to m/p.
- `auto`: The scheduling decision for one of the above modes is made by the compiler and/or the runtime system.
- `runtime`: The scheduling decision for one of the above kinds is determined by the runtime system according to the environment variable `OMP_SCHEDULE`.

The `static` and `dynamic` schedules are usually sufficient for everyday use. The `guided` mode is infrequently used in applications with heavily skewed work distributions. The last two modes somehow defeat the purpose of explicitly specifying the work distribution pattern since we lose the control to either the operating system or a third party that sets the environment variables. Hence, we concentrate on the first two schedules in our examples. The syntax of a scheduled loop in OpenMP is as easy as

```
#pragma omp for schedule(mode,chunk_size)
```

where `mode` is one of the aforementioned options and `chunk_size` controls the batch size c.

IMPLEMENTATION OF INNER PRODUCT COMPUTATION

Let us start coding. Listing 6.8 shows the initial code fragment consisting of the header portion and a function for the computation of the inner product.

```
1   #include <iostream>   // std::cout
2   #include <vector>     // std::vector
3
4   // hpc_helpers contains the TIMERSTART and TIMERSTOP macros
5   #include "../include/hpc_helpers.hpp"
6   // binary_IO contains the load_binary function to load
7   // and store binary data from and to a file
8   #include "../include/binary_IO.hpp"
9
10  // we will change this mode later
11  #define MODE static
12
```

```
13 template <typename value_t,
14           typename index_t>
15 void inner_product(value_t * data,
16                    value_t * delta,
17                    index_t num_entries,
18                    index_t num_features,
19                    bool    parallel) {
20
21     #pragma omp parallel for schedule(MODE) if(parallel)
22     for (index_t i = 0; i < num_entries; i++)
23         for (index_t j = i; j < num_entries; j++) {
24             value_t accum = value_t(0);
25             for (index_t k = 0; k < num_features; k++)
26                 accum += data[i*num_features+k] *
27                          data[j*num_features+k];
28             delta[i*num_entries+j] =
29             delta[j*num_entries+i] = accum;
30         }
31 }
```

Listing 6.8: OpenMP Inner Product: initial code fragment.

For clarity, we have defined a placeholder `MODE` for the scheduling parameters in Line 11. The corresponding compiler directive in Line 21 issues the parallelization of the outer loop of the inner product computation. Again, we use the `if(parallel)` clause to switch between sequential and parallel execution. Note that the clause `collapse(2)` from the previous section cannot be applied here since the range of the inner loop over j now depends on the index i. The accumulation of the innermost loop over k computes the scalar product of the i-th and j-th rows of the input matrix stored in `float* data`. The remaining `main` function is shown in Listing 6.9.

```
32 int main(int argc, char* argv[]) {
33     // run parallelized when any command line argument given
34     const bool parallel = argc > 1;
35
36     std::cout << "running "
37               << (parallel ? "in parallel" : "sequentially")
38               << std::endl;
39
40     // the shape of the data matrix
41     const uint64_t num_features = 28*28;
42     const uint64_t num_entries = 65000;
43
44     TIMERSTART(alloc)
45     // memory for the data matrix and inner product matrix
46     std::vector<float> input(num_entries*num_features);
47     std::vector<float> delta(num_entries*num_entries);
48     TIMERSTOP(alloc)
49
50     TIMERSTART(read_data)
51     // get the images from disk
52     load_binary(input.data(), input.size(), "./data/X.bin");
53     TIMERSTOP(read_data)
```

```
54
55      TIMERSTART(inner_product)
56      inner_product(input.data(), delta.data(),
57                    num_entries, num_features, parallel);
58      TIMERSTOP(inner_product)
59  }
```

Listing 6.9: OpenMP Inner Product: main function.

The code of the `main` function is straightforward. Initially, we declare the shape of the data matrix (Lines 41–42) and allocate memory for the data matrix and the inner product matrix (Lines 46–47). Subsequently, the MNIST images are loaded from disk in Line 52. Finally, the actual computation is performed in Lines 56–57.

PERFORMANCE EVALUATION

We investigate the sequential and parallel execution times on a dual socket 32-core Intel Xeon CPU E5-2683 v4 @ 2.10GHz machine using 64 threads for the following scheduling modes

1. SPB: `#define MODE static` (static pure block)
2. SPC: `#define MODE static,1` (static pure cyclic)
3. SBC: `#define MODE static,32` (static block-cyclic for $c = 32$)
4. DPC: `#define MODE dynamic` (dynamic pure cyclic)
5. DBC: `#define MODE dynamic,32` (dynamic block-cyclic for $c = 32$)

The sequential version takes roughly 30 minutes to compute all $(65{,}000^2 + 65{,}000)/2 = 2{,}112{,}532{,}500$ pairwise inner products over $d = 784$ pixels. The parallel execution times for the aforementioned scheduling modes using 64 threads are listed below:

Performance	SPB	SPC	SBC	DPC	DBC
Time in s	71.0	35.4	36.5	33.8	34.1
Speedup	26.4	53.9	51.3	55.4	55.0

As expected, the static pure block distribution performs worst since the majority of threads have already finished their computation while some are still processing the compute-heavy chunks associated with small indices of the outer loop. Note that a dynamic pure block distribution does not offer any advantage over the static one as the distribution of the iterations is exactly the same (all iterations are distributed at the beginning). In contrast, the remaining static and dynamic block-cyclic distributions with small chunk sizes are almost twice as fast. Although it is not explicitly shown here, the parallel performance degrades for increasing values of the chunk size as already discussed in detail in Section 4.4. Concluding, we achieve almost linear speedup in the number of software threads and superlinear speedup in the number of physical cores just by adding a single directive to a for-loop.

6.5 ADVANCED REDUCTIONS (SOFTMAX REGRESSION/AVX REDUCTIONS)

In this section we discuss code examples that are based on advanced reduction techniques that have been introduced in OpenMP versions 4.0 and 4.5. The first technique is parallel reduction over more than one or a few reduction variables. We investigate the parallelization potential of a softmax regression classifier that concurrently accumulates the gradients of its loss function for thousands of variables at once. During the second part of this section, we focus on the declaration of custom reduction operations. As an example, we show how to employ OpenMP in combination with manual vectorization techniques such as SSE or AVX intrinsics. Finally, we discuss common pitfalls you might encounter when using OpenMP reductions.

SOFTMAX REGRESSION CLASSIFIER OVER MNIST

In previous sections we have discussed the parallelization potential of one-nearest-neighbor (1NN) classification: a distance-based approach for label prediction. Its major disadvantage is the time-consuming linear search over all n training samples for each of the m test samples during the inference step. The resulting asymptotic time complexity $\mathcal{O}(m \cdot n)$ becomes intractable in terms of computational cost and memory consumption for huge image databases. Moreover, 1NN classifiers cannot deal with redundancy or feature correlation between the individual samples of the training set. As a result, the classifier grows linearly with the number of added samples to the training set no matter how (dis)similar the novel instances are.

Nowadays, the state-of-the-art solutions for image classification are deep neural networks. They are usually designed as a cascade of feature extraction filters and combinatorial layers which reason about the presence or absence of those features. As an example, a cat is a cat and not a dog if it has (presence of a feature) fluffy triangular ears and vertical pupils (the features) but does not exhibit (absence of a feature) circular pupils (the feature) like a dog. Neural networks automatically determine a set of discriminative features and combine their distribution of occurrence to complex hypotheses.

In this section, we design and implement a basic two-layer neural network that feeds handwritten images from the MNIST dataset to an input layer consisting of $d = 28 \times 28 = 784$ neurons and determines a class activation score for each of the $c = 10$ classes in the output layer. Hence, our neural network is a non-linear function f_θ mapping m vectors $x^{(i)} \in \mathbb{R}^d$ to m vectors $y^{(i)} \in \mathbb{R}^c$ where θ is a set of trainable parameters. Let $\theta = (W, b)$ be a dense matrix W of shape $c \times d$ and b a dense vector of length c, then $z^{(i)} = W \cdot x^{(i)} + b \in \mathbb{R}^c$ is a scaled and translated variant of the m original input vectors $x^{(i)}$ from the test set. The resulting vector $z^{(i)}$ stores c real-valued entries which correspond to the set of c distinct class labels:

$$z_j^{(i)} = \sum_{k=0}^{d-1} W_{jk} \cdot x_k^{(i)} + b_j \quad \text{for all } i \in \{0, \dots, m-1\}, j \in \{0, \dots, c-1\} \quad . \tag{6.10}$$

In the following, we interpret the j-th entry of a vector $z_j^{(i)}$ as the evidence that the i-th input image $x^{(i)}$ is associated with the j-th class label. High values correspond to high evidence and vice versa. The MNIST label vectors $y^{(i)}$ store the class assignment information in a one-hot representation, i.e. they are zero everywhere except at one position. As an example, a 3 is encoded as $(0, 0, 0, 1, 0, 0, 0, 0, 0, 0)$. Hence, we need a normalization procedure that maps the highest value in the evidence vector $z^{(i)}$

to (almost) one and the remaining ones to (almost) zero. This can be accomplished using a softmax activation function:

$$y_j^{(i)} = \text{softmax}(z^{(i)})_j = \frac{\exp(z_j^{(i)} - \mu)}{\sum\limits_{j=0}^{c-1} \exp(z_j^{(i)} - \mu)} \quad \text{for any } \mu \in \mathbb{R} \quad . \tag{6.11}$$

The exponential amplification of the entries $z_j^{(i)}$ and subsequent normalization effectively selects the greatest value and sets it to one – the remaining entries are close to zero. Note that the actual value of the softmax function is independent of the choice of the parameter μ; however, it is usually chosen to be the maximum of the evidence scores in order to guarantee numerical stability of floating-point computation. Now, the overall feedforward pass of our two-layer neural network reads as

$$y_j^{(i)} = \text{softmax}(W \cdot x^{(i)} + b)_j \quad . \tag{6.12}$$

From a mathematical perspective, all we have to do is to multiply the input vectors with a matrix W, translate the result with a vector b, and finally pass the result to the softmax activation function. From a machine learning point of view the entries of W and b are trainable parameters that have to be adjusted in such a manner that the predicted values from the forward pass agree with the one-hot encoding of the label vectors. For the time being, we assume that someone provided us the optimal parameters $\theta = (W, b)$. Later, we will discuss how to automatically determine their optimal values. Fig. 6.2 depicts the graph topology of our basic neural network.

Implementation of the Prediction Step

The implementation of the inference step is shown in Listing 6.10. We include the usual header files for time measurement, reading of binary data from disk, and standard imports in Lines 1–6. The `softmax_regression` function template computes the forward pass for a single input image x and writes the result to the output array y. Initially, we compute the affine map $z = W \cdot x + b$ in the loop in Lines 19–24. Afterwards, the maximum evidence $\mu = \max(z_j)$ is determined using a max-reduction (Lines 30–31). Now, we can compute the exponential amplification of the evidence vector in Lines 34–35. Subsequently, the normalization term $Z = \sum_j \exp(z_j - \mu)$ is processed using a sum-reduction in Lines 38–39. Finally, we normalize the amplified evidence by multiplying the individual entries with $1/Z$ (see Lines 42–43).

```
1  #include "../include/hpc_helpers.hpp"  // timers
2  #include "../include/binary_IO.hpp"    // load images
3
4  #include <limits>    // numerical limits of data types
5  #include <vector>    // std::vector
6  #include <cmath>     // std::max
7
8  template <
9      typename value_t,
10     typename index_t>
11 void softmax_regression(
12     value_t * input,
```

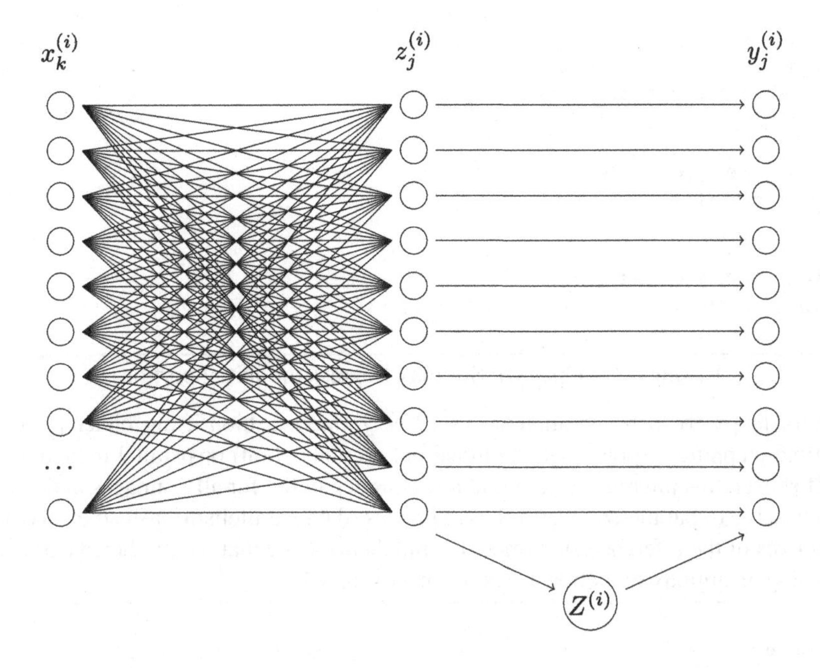

FIGURE 6.2

Schematic overview of the topology of our neural network for MNIST classification. The $d = 784$ input values of $x^{(i)}$ are forwarded to the next layer computing the intermediate result $z^{(i)} = W \cdot x^{(i)} + b$. Subsequently, the normalization of the softmax activation $Z^{(i)} = \sum_j \exp(z_j^{(i)} - \mu)$ is calculated and subsequently passed to the softmax layer which applies pointwise normalization to the evidence vector $z^{(i)}$.

```
13      value_t * output,
14      value_t * weights,
15      value_t * bias,
16      index_t   n_input,
17      index_t   n_output) {
18
19      for (index_t j = 0; j < n_output; j++) {
20          value_t accum = value_t(0);
21          for (index_t k = 0; k < n_input; k++)
22              accum += weights[j*n_input+k]*input[k];
23          output[j] = accum + bias[j];
24      }
25
26      value_t norm = value_t(0);
27      value_t mu = std::numeric_limits<value_t>::lowest();
28
29      // compute mu = max(z_j)
30      for (index_t index = 0; index < n_output; index++)
31          mu = std::max(mu, output[index]);
32
```

```
    // compute exp(z_j-mu)
    for (index_t j = 0; j < n_output; j++)
        output[j] = std::exp(output[j]-mu);

    // compute Z = sum_j exp(z_j)
    for (index_t j = 0; j < n_output; j++)
        norm += output[j];

    // compute y_j = exp(z_j)/Z
    for (index_t j = 0; j < n_output; j++)
        output[j] /= norm;
}
```

Listing 6.10: OpenMP Softmax Regression: initial portion.

Each of the for-loops could be parallelized in a fine-grained fashion using either plain concurrent for-loops (affine mapping, exponential amplification, normalization) or parallel reduction (max/sum-reduction). However, this inference step is embarrassingly parallel for all m images in the test set. As a result, it is advisable to parallelize over different images (data parallelism) instead of over the computational primitives of the inference step (model parallelism). Note that the predicted class label can be determined using an argmax-reduction as shown in Listing 6.11.

```
template <
    typename value_t,
    typename index_t>
index_t argmax(
    value_t * neurons,
    index_t   n_units) {

    index_t arg = 0;
    value_t max = std::numeric_limits<value_t>::lowest();

    for (index_t j = 0; j < n_units; j++) {
        const value_t val = neurons[j];
        if (val > max) {
            arg = j;
            max = val;
        }
    }

    return arg;
}
```

Listing 6.11: OpenMP Softmax Regression: argmax for the readout layer.

The computation of the classification accuracy is similar to the approach performed in the 1NN classifier (see Listing 6.12). We predict the class label for each image i in the test set and compare it to the (actually unknown) ground truth label. The coarse-grained parallelization of this phase is straightforward: we declare a counter variable and concurrently count the amount of correctly predicted labels by augmenting the outer loop with `#pragma omp parallel for reduction(+: counter)`. Note that it is sufficient to use a pure block distribution (which is the default schedule) here since all regression calls take approximately the same time to compute.

```
65  template <
66      typename value_t,
67      typename index_t>
68  value_t accuracy(
69      value_t * input,
70      value_t * label,
71      value_t * weights,
72      value_t * bias,
73      index_t   num_entries,
74      index_t   num_features,
75      index_t   num_classes) {
76
77      index_t counter = index_t(0);
78
79      #pragma omp parallel for reduction(+: counter)
80      for (index_t i= 0; i < num_entries; i++) {
81
82          value_t output[num_classes];
83          const uint64_t input_off = i*num_features;
84          const uint64_t label_off = i*num_classes;
85
86          softmax_regression(input+input_off, output, weights,
87                             bias, num_features, num_classes);
88
89          counter +=  argmax(output, num_classes) ==
90                      argmax(label+label_off, num_classes);
91      }
92
93      return value_t(counter)/value_t(num_entries);
94  }
```

Listing 6.12: OpenMP Softmax Regression: accuracy evaluation.

Gradient Descent-Based Parameter Optimization

The classification performance of our softmax regression classifier is mainly influenced by the choice of the weight matrix W and the bias vector b. In the following, we demonstrate how to properly determine a suitable parameter set $\theta = (W, b)$ in a parallel manner. This can be achieved by iteratively updating the parameters based on a gradient descent scheme that minimizes the mismatch of predicted and ground truth labels. Assume that $\bar{y}_j^{(i)}$ are ground truth label entries and $y_j^{(i)}(\theta)$ denote the entries of our predicted label vector. Categorical cross-entropy $H\big(\bar{y}^{(i)}, y^{(i)}(\theta)\big)$ is a popular measure that determines the agreement between two probability vectors

$$H(\bar{y}^{(i)}, y^{(i)}(\theta)) = -\sum_{j=0}^{c-1} \bar{y}_j^{(i)} \cdot \log\big(y_j^{(i)}(\theta)\big) \quad \text{for all } i \in \{0, \dots, m-1\} \quad . \tag{6.13}$$

The aforementioned expression vanishes for two identical label vectors since $1 \cdot \log 1 = 0$ and $\lim_{\lambda \to 0} \lambda \cdot \log \lambda = 0$. The overall loss function $L(\theta) = L(W, b)$ measures the model quality depending on the parameters $\theta = (W, b)$ by averaging the n individual cross-entropy contributions over the input images in the training set. Note that training is performed on training data and inference on test data. This distinction is crucial in order to avoid information leakage which would result in suboptimal generalization properties due to overfitting the test data.

$$L(\theta) = \frac{1}{n}\sum_{i=0}^{n-1} H\big(\bar{y}^{(i)}, y^{(i)}(\theta)\big) = -\frac{1}{n}\sum_{i=0}^{n-1}\sum_{j=0}^{c-1} \bar{y}_j^{(i)} \cdot \log\big(y_j^{(i)}(\theta)\big) \quad . \tag{6.14}$$

Our goal is to minimize the non-negative loss function $L(\theta)$ since $L(\theta) = 0$ implies perfect label agreement. A naive approach would incrementally sample a vast amount of random weight matrices and bias vectors, and finally keep the best observed solution. A less naive approach would perform an "informed" update procedure which slightly modifies a given parameter pair $\theta = (W, b)$ in a recursive manner. Gradient descent is a basic technique to compute incremental updates of the parameters by iteratively moving the parameters $\theta \mapsto \theta - \epsilon \nabla L(\theta)$ in the opposite direction of the gradient $\nabla L(\theta)$ of the loss function. The described update procedure converges for a suitably choice of the learning rate $\epsilon > 0$ to a local minimum or saddle point of the loss function which might not be globally optimal but sufficient in our case. For the sake of simplicity, we save you from agonizing gradient computations and postulate the final result:

$$\begin{aligned}
\Delta W_{jk} &:= \frac{\partial L(W,b)}{\partial W_{jk}} = \frac{1}{n}\sum_{i=0}^{n-1}\big(\text{softmax}(W \cdot x^{(i)} + b)_j - \bar{y}_j^{(i)}\big) \cdot x_k^{(i)} \\
\Delta b_j &:= \frac{\partial L(W,b)}{\partial b_j} = \frac{1}{n}\sum_{i=0}^{n-1}\big(\text{softmax}(W \cdot x^{(i)} + b)_j - \bar{y}_j^{(i)}\big)
\end{aligned} \tag{6.15}$$

From a computational point of view the update procedure is a sum-reduction over the index i enumerating the images in the training set. We start with a zeroed weight matrix ΔW and a zeroed bias vector Δb and add the contributions for each image $x^{(i)}$. Finally, we iteratively adjust the weight and bias until convergence:

$$W_{jk} \mapsto W_{jk} - \epsilon \Delta W_{jk} \quad \text{and} \quad b_j \mapsto b_j - \epsilon \Delta b_j \quad . \tag{6.16}$$

From a parallel programming perspective we have basically two options to perform the updates:

1. The naive approach would sequentially spawn a team of threads for each of the $c \times d$ entries of W and c entries of b which subsequently perform a parallel reduction of a single variable over the image index i. As a consequence, we would spawn $c \times (d + 1) \times w$ many teams where w is the number of iterations.
2. A better approach would spawn a team of threads once and perform $c \times d$ parallel reductions for W and c parallel reductions for b over the image index i. We keep the same team of threads until convergence.

The latter seems to be the obvious solution. However, OpenMP up to version 4.0 only supports parallel reductions over a few variables which have to be known at compile time. Remember that you have to specify the reduction variables one by another in the `reduction` clause. This becomes quite inconvenient for thousands of variables as in our case. Moreover, variables cannot be enumerated with indices which is mandatory in our application. The good news is that OpenMP in version 4.5 finally supports parallel reductions over arrays containing an arbitrary number of reduction variables. The syntax is as simple as

```
#pragma omp for reduction(operation:array[lower:length])
```

where `operation` is a predefined or user-defined reduction identifier, `array` is a pointer to linear memory, and `lower` as well as `length` specify the range of indices being privatized during reduction. Note that so-called *array sections* also support multidimensional arrays but unfortunately do not work out-of-the-box with containers from the standard library.

Implementation of the Training Step

The implementation of the training step is shown in Listing 6.13. Right after memory allocation for the gradients in Lines 110–111, we immediately spawn a team of threads in Line 114 which stays persistent during the whole process of updating all parameters over `num_iters` iterations. Afterwards, we overwrite the corresponding memory with zeros (Lines 118–125). You might have noticed a novel directive `single` in Line 118 which tells the compiler to execute the following scope with only one thread. This becomes handy if it is not beneficial to parallelize a loop over only a few (here $c = 10$) indices. The loop starting in Line 131 enumerates the images and is augmented with reduction clauses that support accumulation over the weights in W and bias variables in b. The body of the loop contains the computation of the gradients ΔW and Δb. The parameters are subsequently adjusted in Lines 164–172. Finally, we free the memory for the gradients.

```
95  template <
96      typename value_t,
97      typename index_t>
98  void train(
99      value_t * input,
100     value_t * label,
101     value_t * weights,
102     value_t * bias,
103     index_t   num_entries,
104     index_t   num_features,
105     index_t   num_classes,
106     index_t   num_iters=32,
107     value_t   epsilon=1E-1) {
108
109     // allocate memory for the gradients
110     value_t * grad_bias    = new value_t[num_classes];
111     value_t * grad_weights = new value_t[num_features*num_classes];
112
113     // spawn the team of threads once
114     #pragma omp parallel
115     for (uint64_t iter = 0; iter < num_iters; iter++){
116
117         // zero the gradients
118         #pragma omp single
119         for (index_t j = 0; j < num_classes; j++)
```

```
120             grad_bias[j] = value_t(0);
121 
122         #pragma omp for collapse(2)
123         for (index_t j = 0; j < num_classes; j++)
124             for (index_t k = 0; k < num_features; k++)
125                 grad_weights[j*num_features+k] = value_t(0);
126 
127         // compute softmax contributions
128         #pragma omp for \
129             reduction(+:grad_bias[0:num_classes]) \
130             reduction(+:grad_weights[0:num_classes*num_features])
131         for (index_t i = 0; i < num_entries; i++) {
132 
133             const index_t inp_off = i*num_features;
134             const index_t out_off = i*num_classes;
135 
136             value_t * output = new value_t[num_classes];
137             softmax_regression(input+inp_off,
138                                output,
139                                weights,
140                                bias,
141                                num_features,
142                                num_classes);
143 
144             for (index_t j = 0; j < num_classes; j++) {
145 
146                 const index_t out_ind = out_off+j;
147                 const value_t lbl_res = output[j]-label[out_ind];
148 
149                 grad_bias[j] += lbl_res;
150 
151                 const index_t wgt_off = j*num_features;
152                 for (index_t k = 0; k < num_features; k++) {
153 
154                     const index_t wgt_ind = wgt_off+k;
155                     const index_t inp_ind = inp_off+k;
156                     grad_weights[wgt_ind] +=lbl_res*input[inp_ind];
157                 }
158             }
159             delete [] output;
160         }
161 
162         // adjust bias vector
163         #pragma omp single
164         for (index_t j = 0; j < num_classes; j++)
165             bias[j] -= epsilon*grad_bias[j]/num_entries;
166 
167         // adjust weight matrix
168         #pragma omp for collapse(2)
169         for (index_t j = 0; j < num_classes; j++)
170             for (index_t k = 0; k < num_features; k++)
171                 weights[j*num_features+k] -= epsilon*
```

```
172                         grad_weights[j*num_features+k]/num_entries;
173         }
174
175         delete [] grad_bias;
176         delete [] grad_weights;
177 }
```

Listing 6.13: OpenMP Softmax Regression: parallel training.

Performance Evaluation

The main function of our softmax regression classifier is shown in Listing 6.14. After reading the images and class labels from disk in Lines 190–191, we alternate the training step acting on the first $n = 55{,}000$ images and an accuracy evaluation step on the last $m = 10{,}000$ images of the MNIST dataset in an infinite loop.

```
178 int main() {
179
180     const uint64_t num_features = 28*28;
181     const uint64_t num_classes = 10;
182     const uint64_t num_entries = 65000;
183
184     std::vector<float> input(num_entries*num_features);
185     std::vector<float> label(num_entries*num_classes);
186
187     std::vector<float> weights(num_classes*num_features);
188     std::vector<float> bias(num_classes);
189
190     load_binary(input.data(), input.size(), "./data/X.bin");
191     load_binary(label.data(), label.size(), "./data/Y.bin");
192
193     while(true) {
194
195         TIMERSTART(training)
196         train(input.data(),
197               label.data(),
198               weights.data(),
199               bias.data(),
200               55000UL,
201               num_features,
202               num_classes);
203         TIMERSTOP(training)
204
205         const uint64_t off_inp = 55000*num_features;
206         const uint64_t off_lbl = 55000*num_classes;
207
208         TIMERSTART(accuracy)
209         auto acc = accuracy(input.data()+off_inp,
210                             label.data()+off_lbl,
211                             weights.data(),
212                             bias.data(),
213                             10000UL,
```

```
214                                num_features,
215                                num_classes);
216             TIMERSTOP(accuracy)
217
218             std::cout << "accuracy_test: " << acc << std::endl;
219         }
220     }
```

Listing 6.14: OpenMP Softmax Regression: main function.

The code can be compiled with an OpenMP 4.5 compliant compiler (e.g. GCC 6):

```
g++-6 -O2 -std=c++14 -fopenmp softmax.cpp -o softmax
```

The single-threaded variant (compiled without `-fopenmp` flag) executed on a dual socket 32-core Intel Xeon CPU E5-2683 v4 @ 2.10GHz takes roughly 23 seconds to perform 32 updates of the whole parameter set $\theta = (W, b)$ over $n = 55{,}000$ images. In contrast, the time needed for the prediction of $m = 10{,}000$ image labels is less than 100 ms. The parallel version utilizing 64 threads achieves a speedup of approximately 40. Both the sequential and the parallel variants compute the same parameter set (W, b) resulting in a classification accuracy of 92.5%.

```
# elapsed time (training): 0.589874s
# elapsed time (accuracy): 0.00157976s
accuracy_test: 0.836
...
# elapsed time (training): 0.583625s
# elapsed time (accuracy): 0.00178474s
accuracy_test: 0.9251
...
```

Note that the program can be stopped after a sufficient amount of iterations by pressing `CTRL+C`.

CUSTOM REDUCTION OPERATOR

OpenMP 2.0 features out-of-the-box support for custom reduction maps such as addition, multiplication, and bitwise operators over plain old data types. OpenMP 3.1 added further support for minimum and maximum reduction. This is sufficient for the majority of applications. However, you might encounter code that uses arbitrary precision data types or AVX vector registers. The reduction variable could be a private member of an object manipulated by getter and setter functions. In this case, none of the already discussed mechanisms of OpenMP can be used for parallel reduction. Fortunately, OpenMP 4.0 introduces a `declare reduction` directive that allows for the definition of custom reduction operators.

Theoretical Background of Parallel Reduction

Before we can dive into programming, we have to briefly revisit the mathematical properties of reduction operations. Parallel reduction can be performed on different graph topologies such as linear chains, binary trees or hypercubes. The standard text book example reduces $n = 2^k$ elements in a binary tree of height k as shown in Fig. 6.3. Let $\cdot \circ \cdot$ be a binary operation that combines two values; then eight

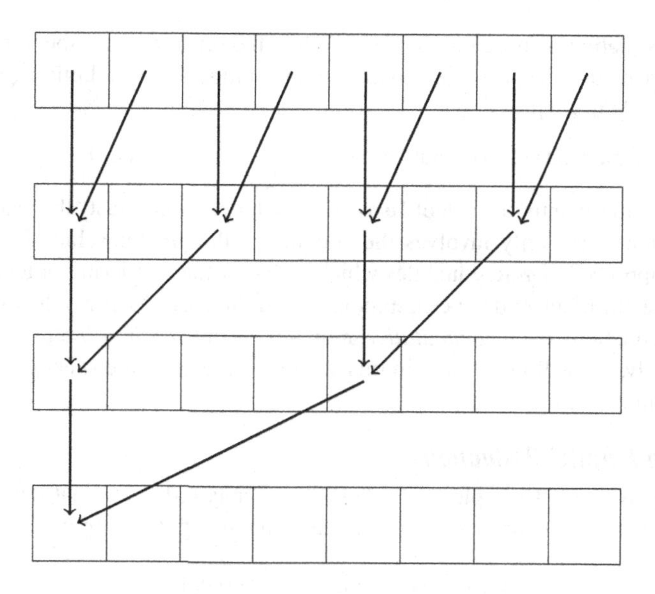

FIGURE 6.3

An example for parallel reduction over an array of size 8. In each of the $\log_2(8) = 3$ iterations, k we combine two values with a stride of $8/2^{k+1}$. The final result is stored at position zero.

values $a_0, \ldots, a_7$ are reduced within a tree topology as follows:

$$\big((a_0 \circ a_1) \circ (a_2 \circ a_3)\big) \circ \big((a_4 \circ a_5) \circ (a_6 \circ a_7)\big) \quad . \tag{6.17}$$

In contrast, the sequential reduction traverses a linear topology:

$$\left(\left(\left(\left((a_0 \circ a_1) \circ a_2\right) \circ a_3\right) \circ a_4\right) \circ a_5\right) \circ \ldots \quad . \tag{6.18}$$

In order to ensure that both computation schemes calculate the same result, we have to guarantee that for every choice of three values $a, b, c \in M$ their composition does not depend on the order of parentheses:

$$(a \circ b) \circ c = a \circ (b \circ c) \quad \text{for all } a, b, c \in M \quad . \tag{6.19}$$

This mathematical property is called *associativity*. When dealing with parallel sum-reduction over arrays whose length is not a power of two, we typically round up to the next power of two and fill the remaining slots with zeros. Mathematically speaking, we need a unique *neutral element* e such that $a \circ e = e \circ a = a$ for all choices of $a \in M$. Furthermore, we assume that the composition of every two elements $a \circ b$ is again an element in the set M. The whole mathematical structure $(M, \circ)$ consisting of the set $M = \{e, \ldots\}$ and an associative binary operation is called *monoid*. Concluding, parallel reduction computes the same result when dealing with monoids.

In practice, the situation is more complicated. While the set of real numbers with addition $(\mathbb{R}, +)$ is a monoid, the set of double precision floating-point numbers with addition (`double`, $+$) is not a monoid. As an example the expression

```
double(double(1UL<<53)+1.0)-double(1UL<<53) // should be 1
```

evaluates zero since the mantissa of `double` is 52 bits long. In contrast, the pair (`double`, max) is a monoid since composition only involves the copying of the greater value. To complicate matters further, OpenMP supports dynamic schedules which allow for the reordering of loop iterations. Hence, the result should be invariant under permutations of the indices. As a result, OpenMP exclusively supports parallel reductions over *commutative* monoids where $a \circ b = b \circ a$ for every pair of two values in M. Actually, we will later show in this section that even static schedules compute incorrect results on non-commutative monoids.

Declaring Custom Parallel Reductions

Let us start coding. We want to implement a sign-aware max-reduction that determines the element with the greatest modulus in an array of signed integers v while preserving its sign:

$$\text{absmax}(v) = v\Big[\underset{0 \leq i < n}{\text{argmax}}(|v[i]|)\Big] \quad . \tag{6.20}$$

As an example, $\text{absmax}([+1, -4]) = -4$ and $\text{absmax}([+5, -4]) = +5$. The associated binary operation is both associative and commutative since it inherits the mathematical properties of max-reduction. Listing 6.15 shows the corresponding OpenMP implementation.

```
1  #include <iostream>
2
3  template <typename value_t>
4  struct binop {
5      constexpr static value_t neutral = 0;
6
7      value_t operator()(
8          const value_t& lhs,
9          const value_t& rhs) const {
10
11         const value_t ying = std::abs(lhs);
12         const value_t yang = std::abs(rhs);
13
14         return ying > yang ? lhs : rhs;
15     }
16 };
17
18 int main () {
19     const uint64_t num_iters = 1UL << 20;
20     int64_t result = binop<int64_t>::neutral;
21
22     #pragma omp declare reduction(custom_op : int64_t : \
23     omp_out = binop<int64_t>()(omp_out, omp_in))        \
24     initializer (omp_priv=binop<int64_t>::neutral)
25
```

```
26      # pragma omp parallel for reduction(custom_op:result)
27      for (uint64_t i = 0; i < num_iters; i++)
28          result = binop<int64_t>()(result, i&1 ? -i : i);
29
30      std::cout << result << std::endl;
31  }
```

Listing 6.15: OpenMP Custom Reduction Operator.

Initially, we declare a functor struct in Lines 3–16 that consists of the neutral element of our monoid ($e = 0$) and the binary reduction map $\cdot \circ \cdot$. Subsequently, we declare the user-defined reduction in Lines 22–24 using a `#pragma omp declare reduction` directive. The syntax is straightforward. First, we provide the name of our operation `custom_op` followed by the data type `int64_t`. Subsequently, we have to provide an implementation how two values `omp_out` and `omp_in` shall be combined. Afterwards, we have to declare the neutral element of the monoid using the `initializer` clause. The loop in Line 27 finally computes the absmax-reduction over a sequence of sign-alternating integers. We can also mix reductions over AVX registers with OpenMP as shown in Listing 6.16.

```
1   #include <iostream>     // std::cout
2   #include <cstdint>      // uint64_t
3   #include <cmath>        // INFINITY
4   #include <random>       // random
5   #include <immintrin.h>  // AVX intrinsics
6
7   // (AVX _m256, max) monoid
8   struct avxop {
9
10      __m256 neutral;
11
12      avxop() : neutral(_mm256_set1_ps(-INFINITY)) {}
13
14      __m256 operator()(
15          const __m256& lhs,
16          const __m256& rhs) const {
17
18          return _mm256_max_ps(lhs, rhs);
19      }
20  };
21
22  // fill data with random numbers
23  void init(float * data, uint64_t length) {
24
25      std::mt19937 engine(42);
26      std::uniform_real_distribution<float> density(-1L<<28, 1L<<28);
27
28      for (uint64_t i = 0; i < length; i++)
29          data[i] = density(engine);
30  }
31
32  // computes v[0]+v[1]+v[2]+v[3]
33  inline float hmax_sse3(__m128 v) {
```

```
34     __m128 shuf = _mm_movehdup_ps(v);
35     __m128 maxs = _mm_max_ps(v, shuf);
36     shuf        = _mm_movehl_ps(shuf, maxs);
37     maxs        = _mm_max_ss(maxs, shuf);
38     return        _mm_cvtss_f32(maxs);
39 }
40
41 // computes v[0]+v[1]+v[2]+v[3]+...+v[7]
42 inline float hmax_avx(__m256 v) {
43     __m128 lo = _mm256_castps256_ps128(v);
44     __m128 hi = _mm256_extractf128_ps(v, 1);
45            lo = _mm_max_ps(lo, hi);
46     return hmax_sse3(lo);
47 }
48
49 int main () {
50
51     // allocate memory and fill it with random numbers
52     const uint64_t num_entries = 1UL << 28;
53     const uint64_t num_bytes = num_entries*sizeof(float);
54     auto data = static_cast<float*>(_mm_malloc(num_bytes , 32));
55     init(data, num_entries);
56
57     // declare a max-reduction over AVX registers
58     #pragma omp declare reduction(avx_max : __m256 :  \
59     omp_out = avxop()(omp_out, omp_in))               \
60     initializer (omp_priv=avxop().neutral)
61
62     auto result = avxop().neutral;
63
64     // use our new reduction operation
65     #pragma omp parallel for reduction(avx_max:result)
66     for (uint64_t i = 0; i < num_entries; i += 8)
67         result = avxop()(result, _mm256_load_ps(data+i));
68
69     // horizontal max over resulting register
70     std::cout << hmax_avx(result) << std::endl;
71
72     _mm_free(data);
73 }
```

Listing 6.16: Mixing OpenMP and AVX.

Initially, we declare our monoid with the neutral element $e = (-\infty, \ldots, -\infty)$ in Lines 8–20 and a binary combiner which vertically determines the maximum of two AVX registers `lhs` and `rhs`, i.e.

$$\big(\max(lhs[0], rhs[0]), \max(lhs[1], rhs[1]), \ldots, \max(lhs[7], rhs[7])\big) \quad . \tag{6.21}$$

In contrast, the functions `hmax_avx` (Lines 42–46) and `hmax_sse3` (Lines 33–39) compute the horizontal maximum of a single AVX or SSE register, i.e. the maximum of all entries $\max(m[0], m[1], m[2], \ldots)$. For further details on vectorization refer to Section 3.2 of this book. The declaration of the custom max-reduction in Lines 58–60 simply uses the operation and neutral element of our monoid. Subsequently, we loop over the entries of the array data with a stride of 8, again using the operation from the struct (see Lines 65–67). The final result is the horizontal maximum over the eight entries of the AVX register result.

OPENMP REDUCTIONS UNDER THE HOOD

After having revisited parallel reductions in detail you might still have some open questions about how OpenMP actually transforms your sequential code into parallel code. In particular, you might wonder how OpenMP detects the portion in the body of the loop that performs the reduction. As an example, this or a similar code fragment can often be found in code samples:

```
#pragma omp parallel for reduction(+:x)
for (int i = 0; i < n; i++)
    x -= some_value;
```

You could also use - as reduction operator (which is actually redundant to +). But how does OpenMP isolate the update step x-= some_value? The discomforting answer is that OpenMP does not detect the update at all! The compiler treats the body of the for-loop like this:

```
#pragma omp parallel for reduction(+:x)
for (int i = 0; i < n; i++)
    x = some_expression_involving_x_or_not(x);
```

As a result, the modification of x could also be hidden behind an opaque function call. This is a comprehensible decision from the point of view of a compiler developer. Unfortunately, this means that you have to ensure that all updates of x are compatible with the operation defined in the reduction clause. The overall execution flow of a reduction can be summarized as follows:

1. Spawn a team of threads and determine the set of iterations that each thread j has to perform.
2. Each thread declares a privatized variant x_j of the reduction variable x initialized with the neutral element e of the corresponding monoid.
3. All threads perform their iterations no matter whether or how they involve an update of the privatized variable x_j.
4. The result is computed as sequential reduction over the (local) partial results x_j and the global variable x. Finally, the result is written back to x.

Let us check that behavior with an admittedly useless but valid OpenMP fragment:

```
uint64_t x = 5;
#pragma omp parallel for reduction(min:x) num_threads(2)
for (uint64_t i = 0; i < 10; i++)
    x += 1;
std::cout << x << std::endl;
```

What do you think is the final value of x? Let us evaluate the aforementioned phases step by step. The initial value of x is irrelevant for the time being since it is modified exclusively in Phase 4. The OpenMP directive does not state an explicit schedule and thus we can assume that it is schedule(static). Hence, a team of two threads will be processing $10/2 = 5$ iterations each. Subsequently, each thread declares a privatized variable initialized with the neutral element which is the greatest unsigned integer $e = 3^{32} - 1$. Now, each thread performs five increments on their privatized variable resulting in four after overflow. The final value of x is then determined as the minimum of the partial results 4, 4 and the global value 5. Consequently, the integer 4 is printed to the command line.

Let us elaborate on commutativity of the reduction map. We have mentioned before that due to the unpredictable scheduling we cannot safely assume that iterations are processed in order. Moreover, the final reduction of partial results in Phase 4 does not necessarily need to be performed in order. We can check this with an associative but non-commutative reduction map such as string concatenation:

```
std::string result {"SIMON SAYS: "};
std::vector<std::string> data {"p", "a", "r", "a", "l", "l",
                               "e", "l", " ", "p", "r", "o",
                               "g", "r", "a", "m", "m", "i",
                               "n", "g", " ", "i", "s", " ",
                               "f", "u", "n", "!"};

#pragma omp declare reduction(op : std::string :             \
    omp_out = omp_out+omp_in)                                \
    initializer (omp_priv=std::string(""))

#pragma omp parallel for reduction(op:result) num_threads(2)
for (uint64_t i = 0; i < data.size(); i++)
    result = result+data[i];

std::cout << result << std::endl;
```

We expect the output to be `SIMON SAYS: parallel programming is fun!`. However, if we execute this program several times, we occasionally observe `SIMON SAYS: amming is fun!parallel progr`. Obviously, the partial results are reduced in potentially permuted order even when using static schedules. Concluding, commutativity of the reduction map is mandatory for OpenMP reductions.

At the very end, let us make an important statement on implicit barriers: as mentioned before, OpenMP allows for the removal of implicit barriers after loops using the `nowait` clause. This is a potentially dangerous approach when using parallel reductions since the final value of the reduction has to be written back to the global variable x in Phase 4. Hence, we cannot guarantee that x has the correct values unless we explicitly state a barrier using `#pragma omp barrier`:

```
#pragma omp parallel
{
    uint64_t x = 0;
    #pragma omp for reduction(+:x) nowait
    for (uint64_t i = 0; i < 1024; i++)
        x++;

    // x could be anything between 0 and 1024

    #pragma omp barrier
    std::cout << x << std::endl;
}
```

6.6 TASK PARALLELISM (TREE TRAVERSAL)

Up to now, all presented OpenMP directives implemented a pretty static approach to parallelization. For example, they only work on loops with an iteration count known at least at runtime (mostly written as for-loops), and not for recursive algorithms or handling linked list (mostly written as while-loops). To handle such use cases, there are OpenMP task directives, which are the closest mechanism to the explicit thread control and execution management explained previously in Chapter 4. They are available since OpenMP version 3.0 [9].

Tasks are basically parallel execution blocks that will be put into an execution queue and scheduled for execution by the current team of threads whenever they have idle capacities. Thus, the primarily visible aspect is task generation within the program, while task execution is performed "in the background," hidden from the programmer. The `task` statements must occur within a `parallel` region to actually be executed by the corresponding team of threads. The `task` directive itself does not have an implicit barrier – there is either an implicit barrier at the end of the `parallel` region, or the programmer can set an explicit barrier for just the current task queue using `#pragma omp taskwait`, or by using the more general `#pragma omp barrier`.

The programmer has to pay attention that task creation is not accidentally performed in parallel without intent. This can be avoided by wrapping the task generation parts in `#pragma omp single [nowait]` clauses. A `task` directive can also contain any of the previously known data sharing attributes such as `shared`, `private`, and `firstprivate`. Since OpenMP version 4.0, data dependencies between different task executions can be specified using the `depends` modifier.

TREE TRAVERSAL

In this example, we are going to use a (perfectly balanced) binary tree structure and perform some computationally intensive operations on the values of the nodes.

There are different ways to express such a tree data structure. One simple way is to just use an array and exploit the mathematical relations between indices of parents and children in each level of the tree. A tree of depth m can contain $n = 2^m - 1$ nodes. If the array representation assigns nodes in each level in a linear fashion, e.g. the root node in level 0 is at position 0, its two children in level 1 are at positions 1 and 2, and so on, there are simple formulas to calculate the indices of parents and children: $\text{parent}(i) = \frac{i-1}{2}$, $\text{left\_child}(i) = 2i + 1$, $\text{right\_child}(i) = 2(i + 1)$. Another way to express the tree structure is to store references (e.g. pointers) to children (and/or parents, depending on usage) within each node.

In this example, both approaches are present, since it is easier to initialize the nodes using an array and then attach the references to the child nodes.

```
1 #include <iostream>
2 #include <cstdint>
3 #include <cmath>
4
5 // hpc_helpers contains the TIMERSTART and TIMERSTOP macros
6 #include "../include/hpc_helpers.hpp"
7
8 #define VALUE_T double
9
```

```
10 template <typename value_t>
11 class Node {
12 public:
13   value_t value_;
14   Node<value_t>* left_;
15   Node<value_t>* right_;
16
17   Node() {}
18   Node(value_t value) : value_(value) {}
19   Node(value_t value, Node<value_t>* left, Node<value_t>* right)
20   : value_(value), left_(left), right_(right) {}
21
22   // inorder tree traversal, calling func with each value
23   // again, we are using the if clause of the omp pragma
24   // to be able to conditionally enable OpenMP
25   void traverse(auto func, bool parallel=false) {
26     if (this->left_) {
27       #pragma omp task if(parallel)
28       this->left_->traverse(func, parallel);
29     }
30     func(this->value_);
31     if (this->right_) {
32       #pragma omp task if(parallel)
33       this->right_->traverse(func, parallel);
34     }
35   }
36   // inorder tree traversal, printing each value
37   void traverse() {
38     traverse([](auto &val){std::cout << val << std::endl;});
39   }
40 };
41
42 int main() {
43   // height of a perfectly balanced binary tree
44   const uint64_t m = 15;
45   // number of elements in the perfectly balanced binary tree
46   const uint64_t n = (1UL << m) - 1;
47   // number of iterations within each task
48   const uint64_t iterations = 1UL << 12;
49
50   TIMERSTART(overall)
51
52   TIMERSTART(alloc)
53   Node<VALUE_T> nodes[n];
54   TIMERSTOP(alloc)
55
56   TIMERSTART(init)
57   for (uint64_t i = (n-1); i > 0; --i) {
58     if (i > ((n/2)-1)) {
59       // bottommost row, no children
60       nodes[i] = Node<VALUE_T>(i);
61     } else {
```

```
62         // not the bottommost row
63         // left child is 2*i+1, right child is 2*i+2
64         // parent is (i-1)/2
65         nodes[i] = Node<VALUE_T>(i, &nodes[2*i+1], &nodes[2*i+2]);
66       }
67     }
68     // root node
69     Node<VALUE_T> tree = Node<VALUE_T>(0, &nodes[1], &nodes[2]);
70     nodes[0] = tree;
71     TIMERSTOP(init)
72 
73     // TIMERSTART(print)
74     // tree.traverse();
75     // TIMERSTOP(print)
76 
77     TIMERSTART(sum)
78     VALUE_T sum = 0;
79     // this is a lambda closure capturing the variable sum from its
80     // outer scope by using the [&] syntax
81     tree.traverse([&](auto &val){sum += val;});
82     std::cout << (n*(n-1)/2) << " ?= " << sum << std::endl;
83     TIMERSTOP(sum)
84 
85     // auto func = [](auto &val){std::cout << val << std::endl;};
86     auto func = [](auto &val){
87       for (uint64_t i = 0; i < iterations; ++i)
88         val = std::pow(val, 1.1);
89     };
90 
91     TIMERSTART(sequential)
92     tree.traverse(func);
93     TIMERSTOP(sequential)
94 
95     TIMERSTART(parallel)
96     #pragma omp parallel
97     {
98       #pragma omp single
99       {
100        #pragma omp task
101        tree.traverse(func, true);
102      }
103    } // implicit barrier here
104    TIMERSTOP(parallel)
105 
106    TIMERSTOP(overall)
107  }
```

Listing 6.17: OpenMP Tasking Tree Traversal.

In Lines 10–40, we define a class template that represents one node of the tree. For convenience, all its members are defined as public, although they would be private in production code. Moreover, there are three constructors for a more terse programming style in this example. The `traverse` function in Lines 25–35 takes a function reference which is subsequently called with the value of each node. Also, the function again accepts an optional Boolean argument to toggle OpenMP parallelization at runtime while keeping the code example small. In the basic sequential example, we simply perform a so-called "inorder" tree traversal, which means that the operation on the node itself is executed after traversing the left child and before traversing the right child. During parallel execution this order cannot be guaranteed (of course, we could do it by using a barrier which defeats the purpose of parallelization).

The initialization in Lines 57–67 creates an array representation of the tree starting at the back (the lowest level of the tree). This simplifies the treatment of leaf nodes (no children) and further ensures that for the nodes in the upper levels, children are already defined and can be referenced using the simple index calculations mentioned above. The main calculation function consists of a compute-heavy loop of power operations in Lines 85–89. This is not a very useful calculation per se, but it is intensive enough to hide task maintenance overhead behind computation. Finally, we determine the speedup between sequential (Line 92) and parallel (Lines 96–103) version:

```
./tree
# elapsed time (alloc): 3.6e-08s
# elapsed time (init): 0.000370741s
536821761 ?= 5.36822e+08
# elapsed time (sum): 0.00177787s
# elapsed time (sequential): 2.2161s
# elapsed time (parallel): 1.1723s
# elapsed time (overall): 3.39083s
```

GENERATING TASKS IN A LOOP

If many tasks need to be created in a loop, the `taskloop` directive, might come in handy with some special features. It is written directly before a (possibly nested) for-loop and creates a task for each loop execution. Its `grainsize` and `num_tasks` clauses can be used to control the number of iterations per task (comparable to chunk size) or the total number of tasks to create. Since we still have the array representation of the tree, we can employ a task-loop on it:

```
105  TIMERSTART(taskloop)
106  #pragma omp parallel
107  {
108    #pragma omp single
109    {
110      #pragma omp taskloop
111      for (uint64_t i = 0; i < n; ++i) {
112        func(nodes[i].value_);
113      }
114    }
115  } // implicit barrier here
116  TIMERSTOP(taskloop)
```

Listing 6.18: OpenMP Taskloop Tree Traversal.

But while this makes the computation a little bit faster than the sequential version, it is not quite as fast as plain tasking:

```
./tree
# elapsed time (alloc): 3.5e-08s
# elapsed time (init): 0.000356017s
536821761 ?= 5.36822e+08
# elapsed time (sum): 0.00207262s
# elapsed time (sequential): 2.22064s
# elapsed time (parallel): 1.08154s
# elapsed time (taskloop): 1.96788s
# elapsed time (overall): 5.27283s
```

6.7 SIMD VECTORIZATION (VECTOR ADDITION)

As previously explained in Section 3.2, modern microprocessors and accelerators support, besides parallelization using multiple independent computational units (MIMD – Multiple Instruction Multiple Data), feature-specific processor instructions that perform a (basic) operation on a vector registers instead of on just a single scalar value (SIMD – Single Instruction Multiple Data). This enables the processor to inherently exploit the locality of the data. More information on vectorization primitives can be found in Section 3.2, which includes several programming examples demonstrating the explicit usage of intrinsics (e.g. Listing 3.2).

The traditional approach to SIMD vectorization is the augmentation of the code with assembly-style code fragments. This could be completely avoided if the compiler was able to determine with absolute certainty that it could auto-vectorize parts of the code without the occurrence of side effects. Unfortunately, fully unsupervised auto-vectorization tends to produce suboptimal code (although different compilers continuously improve support) and thus is infrequently used nowadays. Implementations with manually handcrafted (hard-wired) vectorization usually perform significantly better. The main reason for this discrepancy is that hard-wired implementations exploit additional mathematical properties such as associativity and commutativity of operations that cannot be derived directly from plain sequential code. As an example, you know how to exploit associativity to implement parallel or SIMD-aware reduction: the compiler however only sees a linear chain consisting of dependent operations.

Fortunately, OpenMP version 4.0 provides support for dedicated directives that help the compiler to spot vectorization patterns [5]. Basically, we tell the compiler that additional mathematical constraints are fulfilled in order to allow for effective vectorization. Let us reuse the example from Listing 6.3 and add vectorization (see Line 37):

```
1  #include <iostream>
2  #include <cstdint>
3  #include <vector>
4
5  // hpc_helpers contains the TIMERSTART and TIMERSTOP macros
6  // and the no_init_t template that disables implicit type
7  // initialization
8  #include "../include/hpc_helpers.hpp"
```

```
9
10  int main() {
11      // memory allocation for the three vectors x, y, and z
12      // with the no_init_t template as a wrapper for the actual type
13      TIMERSTART(alloc)
14      const uint64_t num_entries = 1UL << 30;
15      std::vector<no_init_t<uint64_t>> x(num_entries);
16      std::vector<no_init_t<uint64_t>> y(num_entries);
17      std::vector<no_init_t<uint64_t>> z(num_entries);
18      TIMERSTOP(alloc)
19
20      // manually initialize the input vectors x and y
21      TIMERSTART(init)
22      #pragma omp parallel for
23      for (uint64_t i = 0; i < num_entries; i++) {
24          x[i] = i;
25          y[i] = num_entries - i;
26      }
27      TIMERSTOP(init)
28
29      // compute x + y = z sequentially
30      TIMERSTART(add_seq)
31      for (uint64_t i = 0; i < num_entries; i++)
32          z[i] = x[i] + y[i];
33      TIMERSTOP(add_seq)
34
35      // compute x + y = z vectorized
36      TIMERSTART(add_vec)
37      #pragma omp simd
38      for (uint64_t i = 0; i < num_entries; i++)
39          z[i] = x[i] + y[i];
40      TIMERSTOP(add_vec)
41
42      // compute x + y = z in parallel
43      TIMERSTART(add_par)
44      #pragma omp parallel for
45      for (uint64_t i = 0; i < num_entries; i++)
46          z[i] = x[i] + y[i];
47      TIMERSTOP(add_par)
48
49      // compute x + y = z in parallel *and* vectorized
50      TIMERSTART(add)
51      #pragma omp parallel for simd
52      for (uint64_t i = 0; i < num_entries; i++)
53          z[i] = x[i] + y[i];
54      TIMERSTOP(add)
55
56      // check if summation is correct
57      TIMERSTART(check)
58      #pragma omp parallel for
59      for (uint64_t i = 0; i < num_entries; i++)
60          if (z[i] - num_entries)
```

```
61            std::cout << "error at position "
62                      << i << std::endl;
63        TIMERSTOP(check)
64  }
```

Listing 6.19: OpenMP Vector Addition with SIMD instructions.

We observe that the vectorized loop execution already gets us some speedup:

```
./vector_add
# elapsed time (alloc): 2.0813e-05s
# elapsed time (init): 0.683577s
# elapsed time (add_seq): 4.18215s
# elapsed time (add_vec): 2.85035s
# elapsed time (add_par): 0.474374s
# elapsed time (add): 0.460693s
# elapsed time (check): 0.166859s
```

In Line 51, we combine parallel and vectorized execution, which is also possible.

DATA DEPENDENCIES

One particular precondition where vectorization can be used very effectively while other speedup techniques come short is when there are data dependencies. This is not without caveats though: for this to work, the distance between dependent fields of the array in question has to be *bigger* than the chunk size of the vector operation. As an example, a loop like the following

```
for (i = 10; i < N; ++i)
    a[i] = a[i-10] + b;
```

has a dependency span of 10 and can be effectively vectorized for vector instructions of sizes 2, 4, or 8 but not for vector operations of size 12 or bigger. The safe width can be explicitly specified using the `safelen` clause:

```
#pragma omp simd safelen(10)
for (i = 10; i < N; ++i)
    a[i] = a[i-10] + b;
```

VECTORIZATION-AWARE FUNCTIONS

OpenMP can also vectorize complete functions [10]. This is realized by creating different version of the specified functions for different vector instruction sizes. As an example, assume a function definition as follows:

```
#pragma omp declare simd
int do(int a, int b) {
    return a + b;
}
```

then it can be called from within a SIMD loop and will benefit from the same vector instructions as if it would have been inlined:

```
#pragma omp simd
for (i = 0; i < N; ++i) {
    c[i] = do(a[i], b[i]);
}
```

The compiler creates an equivalent of the function `do` that, instead of taking a single scalar `int` value, accepts a vector register that can be processed by a vector instruction.

6.8 OUTLOOK

Now that we have reached the end of this chapter, we want to mention a few more advanced OpenMP subjects that have not been discussed. Some of them are too specific for a textbook while others need dedicated hardware and software support for massively parallel accelerators which at the time of writing is not included in default binary distributions of mainstream compilers.

A noteworthy OpenMP 4.0 feature is the `proc_bind` clause in combination with the environment variable `OMP_PLACES` which allows for the fine-grained control of thread affinities. In particular, this is crucial when developing code for multi-socket CPUs. The plain use of directives such as `#pragma omp parallel for`, which completely ignore the underlying topology of CPU cores and associated caches, might result in mediocre scaling properties of your code over more than one CPU socket.

We have discussed only one OpenMP 4.5 feature in this chapter: the so-called *array sections*. They allow for the reduction of arrays consisting of thousands of reduction variables at once. However, a major difference between OpenMP 4.0 and OpenMP 4.5 is the possibility to offload code sections to target devices. A target could be an Intel Xeon Phi many-core accelerator board or a CUDA-enabled GPU. Unfortunately, at the time of writing this involves a non-trivial recompilation of the GCC using vendor-specific binaries for code generation on Xeon Phi boards and CUDA-enabled GPUs. The vision of auto-generated code that processes loops over thousands of GPU cores using only a few pragmas sounds pleasant. The discomforting truth is that it is in fact possible to write platform portable OpenMP code that can be compiled for traditional CPUs architectures and GPUs, but the resulting code might not be performance portable. You usually end up writing architecture-specific pragmas which are enclosed by if-then-else macros that choose the corresponding optimization for each architecture at compile time. Even worse, many memory layout and scheduling optimizations for CUDA-enabled GPUs are completely non-transparent for a developer who is unfamiliar with the programming of massively parallel accelerators. Concluding, target offloading is a useful technique for an experienced CUDA developer performing rapid code prototyping. The next chapter will teach you to become one.

6.9 ADDITIONAL EXERCISES

1. Below you will find four sequential codes. Augment the code snippets with OpenMP pragmas if applicable. You have to make sure that no dependencies are broken and thus the algorithm still computes correct result. If the algorithm is parallelizable, briefly discuss your parallelization strategy. Otherwise, explain why it cannot be parallelized. Justify your claims.

(a) (Dense Matrix Multiplication)

```
1  // computes the product of an M x L matrix A
2  // with an L x N matrix B
3  for (i = 0; i < M; i++)
4      for (j = 0; j < N; j++) {
5          value_t sum = 0;
6          for (k = 0; k < L; k++)
7              sum += A[i*L+k] * B[k*N+j];
8          C[i*N+j] = sum;
9      }
```

(b) (Pseudo-Polynomial Knapsack Using Dynamic Programming)

```
1
2  #define AT(i,j)  ((i)*(C+1)+(j))
3  #define MAX(x,y) ((x)<(y)?(y):(x))
4
5  // w and v are __constant__ and __non-negative__
6  // arrays of length N, m is an array of length
7  // (C+1)*(N+1) initialized with zeros
8  for (i = 1; i < N+1; i++)
9      for (j = 0; j < C+1; j++)
10         if (w[i-1] <= j)
11             m[AT(i,j)] = MAX(m[AT(i-1,j)],
12                                  m[AT(i-1,j-w[i-1])]+v[i-1]);
13         else
14             m[AT(i,j)] = m[AT(i-1, j)];
```

(c) (Left Fold of a Binary Operation)

```
1  value_t result = 1;
2
3  // v is a __constant__ array of length N
4  for (i = 0; i < N; i++)
5          result = result + v[i] + result * v[i];
6
7  // bonus: can you solve this without a custom
8  // declare reduction directive?
```

(d) (Repetitive Smoothing of a Vector)

```
1  // v is a pre-initialized array of length N
2  // s is the smoothed version of v, preinitialized with v
3  // M is the number of iterations
4
5  for (i = 0; i < M; i++) {
6     for (j = 2; j < N-2; j++) {
7         s[j] = 0;
8         for (k = -2; k < 3; k++)
9             s[j] += 0.2*v[j+k];
10     }
11     for (j = 0; j < N; j++)
12         v[j] = s[j];
13 }
```

2. Have a look at the following code snippet of a sorting algorithm. The algorithm exhibits a quadratic dependency in terms of the length of X and can be parallelized easily.

```
1  void sequential_sort(std::vector<unsigned int>& X) {
2
3      unsigned int i, j, count, N = X.size();
4      std::vector<unsigned int > tmp(N);
5
6      for (i = 0; i < N; i++) {
7          count = 0;
8          for (j = 0; j < N; j++)
9              if (X[j] < X[i] || X[j] == X[i] && j < i)
10                 count++;
11         tmp[count] = X[i];
12     }
13
14     std::copy(tmp.begin(), tmp.end(), X.begin());
15 }
```

(i) Explain how this sorting algorithm works.
(ii) Analyze the data dependencies of each loop. Which loop is ideally suited for a parallelization with OpenMP pragmas? Consider the problem of shared variables.
(iii) Implement a parallel version of sequential_sort according to your former considerations. Discuss speedup and efficiency.

3. The Euler–Riemann zeta function $\zeta(s) = \sum_{n=1}^{\infty} n^{-s}$ is often used in natural science, notably in applied statistics and during the regularization of quantum field theoretical descriptions of the Casimir effect. An alternative formula for the computation of $\zeta(s)$ on real domains is given by the following expression:

$$\zeta(s) = 2^s \cdot \lim_{k\to\infty} \sum_{i=1}^{k} \sum_{j=1}^{k} \frac{(-1)^{i+1}}{(i+j)^s} \quad .$$

Thus, we can approximate $\zeta(s)$ up to the degree k if we omit the leading lim-operation. The following code snippet implements this idea:

```
1  double Riemann_Zeta(double s, int k) {
2
3      double result = 0.0;
4
5      for (int i = 1; i < k; i++)
6          for (int j = 1; j < k; j++)
7              result += (2*(i&1)-1)/pow(i+j, s);
8
9      return result*pow(2, s);
10 }
```

The asymptotic time complexity of a single call to Riemann_Zeta is obviously in $\mathcal{O}(k^2)$. Let us now investigate the approximation quality depending on the parameter k. To achieve that, we calculate the result of Riemann_Zeta(x, k) for all $k \in \{0, \ldots, N-1\}$ and write it to a vector X of length N:

```
1       for (unsigned int k = 0; k < N; k++)
2           X[k] = Riemann_Zeta(2, k);  // = pi^2/6
```

(i) Parallelize this loop using OpenMP. Are there any data dependencies or shared variables?
(ii) Elaborate on the load balancing of the individual threads.
(iii) Discuss different scheduling modes and the corresponding runtimes.

4. Have a look at the following code snippet. The functions `relax_A` and `relax_B` implement two algorithms for the elastic assignment of distance values between a pair of vectors $Q \in \mathbb{R}^M$ and $S \in \mathbb{R}^N$.

```
1   #define INFTY (std::numeric_limits<double>::infinity())
2   #define AT(i,j) ((i)*(N+1)+(j))
3
4   void init_matrix(double * matrix, size_t M, size_t N) {
5       matrix[AT(0, 0)] = 0.0;
6       for (size_t j = 1; j < N+1; j++)
7           matrix[AT(0, j)] = INFTY;
8       for (size_t i = 1; i < M+1; i++)
9           matrix[AT(i, 0)] = INFTY;
10  }
11
12  double relax_A(double * Q, double * S, size_t M, size_t N) {
13      std::vector<double> matrix((M+1)*(N+1));
14      init_matrix(matrix.data(), M, N);
15
16      for (size_t i = 1; i < M+1; i++)
17          for (size_t j = 1; j < N+1; j++) {
18              double bsf = matrix[AT(i-1, j-1)];
19              if (i > 1)
20                  bsf = std::min(bsf, matrix[AT(i-2, j-1)]);
21              if (j > 1)
22                  bsf = std::min(bsf, matrix[AT(i-1, j-2)]);
23              matrix[AT(i,j)] = bsf + (Q[i-1]-S[j-1])*
24                                      (Q[i-1]-S[j-1]);
25          }
26      return matrix[AT(M, N)];
27  }
28
29  double relax_B(double * Q, double * S, size_t M,  size_t N) {
30      std::vector<double> matrix((M+1)*(N+1));
31      init_matrix(matrix.data(), M, N);
32
33      for (size_t i = 1; i < M+1; i++)
34          for (size_t j = 1; j < N+1; j++)
35              matrix[AT(i,j)] = (Q[i-1]-S[j-1])*
36                                (Q[i-1]-S[j-1])+
37                                std::min(matrix[AT(i-1, j-1)],
38                                std::min(matrix[AT(i-1, j+0)],
39                                         matrix[AT(i+0, j-1)]));
40      return matrix[AT(M, N)];
41  }
```

(i) Visualize the data dependencies of both methods with the help of directed graphs. Draw an exemplary matrix for $M = N = 6$ and sketch the dependencies between the cells with arrows.

(ii) Can you find an appropriate parallelization scheme for both provided methods? Which cells can be relaxed (updated) simultaneously?

(iii) Implement an efficiently parallelized version of `relax_A` using proper pragmas. Also parallelize the initialization method `init_matrix`.

(iv) Implement a parallel version of `relax_B` using OpenMP. You will have to alter the indexing scheme.

5. Which of the following operations can be used for parallel reduction in OpenMP?

(i) The greatest common divisor of two or more integers $gcd(a_0, a_1, \ldots)$.

(ii) The matrix product of two or more general matrices of shape 2×2.

(iii) The complex product of two or more complex numbers.

(iv) The quaternion product of two or more quaternions.

(v) The incremental evaluation of pairwise means of two real numbers

$$\circ : \mathbb{R} \times \mathbb{R} \to \mathbb{R},\ (a, b) \mapsto \tfrac{a+b}{2} \quad .$$

6. Assume we have generated a huge number of m noisy vectors $y^{(i)} \in \mathbb{R}^c$ from given vectors $x^{(i)} \in \mathbb{R}^d$ using the generative model $y^{(i)} = W \cdot x^{(i)} + \epsilon^{(i)}$:

$$y_j^{(i)} = \sum_{k=0}^{d-1} W_{jk} \cdot x_k^{(i)} + \epsilon_j^{(i)} \quad \text{for all } i \in \{0, \ldots, m-1\} \tag{6.22}$$

where $\epsilon_j^{(i)}$ are the entries of a noisy vector sampled from a Gaussian distribution with zero mean and small variance.

(i) Analyze the parallelization potential of the forward evaluation pass for a fixed index i in the equation above.

(ii) Assume that we do not have access to the weight matrix $W \in \mathbb{R}^{c \times d}$ and want to recover its entries just from the knowledge of the vectors $x^{(i)}$ and $y^{(i)}$. Hence, we define the loss function

$$\begin{aligned} L(W) &= \frac{1}{m} \sum_{i=0}^{m-1} \| y^{(i)} - W \cdot x^{(i)} \|_2^2 \\ &= \frac{1}{m} \sum_{i=0}^{m-1} \sum_{j=0}^{c-1} \left(y_j^{(i)} - \sum_{k=0}^{d-1} W_{jk} \cdot x_k^{(i)} \right)^2 \quad . \end{aligned} \tag{6.23}$$

Compute the gradient of $L(W)$ with respect to the entries W_{jk}. Develop an iterative scheme to compute the optimal weight matrix W that minimizes the loss function.

(iii) Discuss the parallelization and vectorization potential of your update procedure in detail.

(iv) Finally, implement your approach. Report speedup and parallelization efficiency.

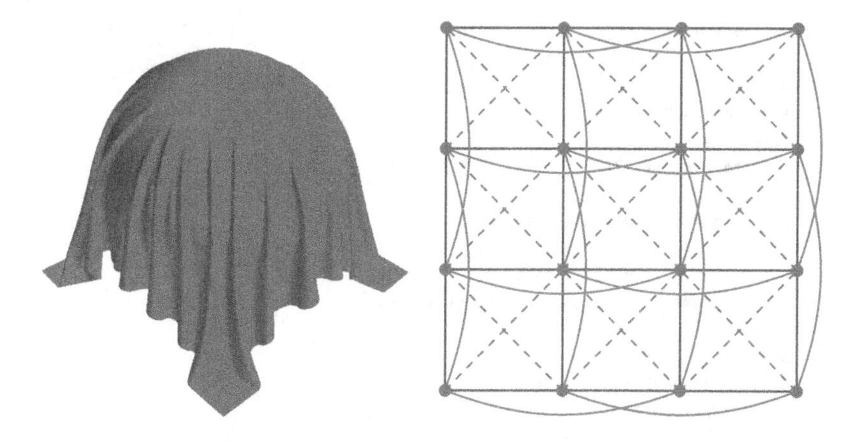

FIGURE 6.4

Constrained-based cloth simulation. The left panel depicts a quadratic rag represented by a uniform $N \times N$-grid interacting with a solid sphere (covered by the rag). The right panel visualizes the dependencies between the nodes of the rag. The vertical/horizontal edges between neighboring vertices ensure the structural integrity of the rag (solid lines), the diagonal edges enforce robustness against shearing (dashed lines) and the horizontal/vertical edges between the next but one neighbors implement resistance against bending of the material (curved lines). See https://www.youtube.com/watch?v=m9gW_29pFgE for an animation.

7. In contrast to common n-body simulations in physics with $\mathcal{O}(n^2)$ many interaction terms between the n particles, we want to investigate a constrained-based cloth simulation with only a linear dependency in the number of mass points. For the sake of simplicity, assume a quadratic shape of cloth sampled by $n = N^2$ mass points on a uniform grid. The topology of the rag is governed by three distinct constraints which enforce structural integrity and robustness against shearing as well as bending of the material (see Fig. 6.4). The major idea behind the to be parallelized cloth simulation algorithm is given by the following workflow:

a. Calculate the future positions for each mass point independently using an explicit integration algorithm for ordinary differential equations with constant step size `eps`. This approach assumes the nodes to be free of any interaction similar to gas particles.

```
1  auto update_positions = [&](T& x, T& y ,T& z,
2                              T& u, T& v, T& w){
3
4          w = (x*x+y*y+z*z > 1 && z > -1) ? w-eps : 0;
5          x += eps*u;
6          y += eps*v;
7          z += eps*w;
8  };
```

b. Afterwards, for each combination of nodes (according to the proposed topology) determine the violation of constraints. If a constraint is violated, readjust the positions accordingly by shifting the particles slightly. Note that `bias` is a small number (e.g. 0.05). Repeat this procedure 8 to 32 times:

```
auto relax_constraint = [&](size_t l, size_t m,
                            T constraint){

        T delta_x = X[l]-X[m];
        T delta_y = Y[l]-Y[m];
        T delta_z = Z[l]-Z[m];

        T length = sqrt(delta_x*delta_x+
                        delta_y*delta_y+
                        delta_z*delta_z);
        T displacement = (length-constraint)*bias;

        delta_x /=length;
        delta_y /=length;
        delta_z /=length;

        tmp_X[l] -= delta_x*displacement;
        tmp_X[m] += delta_x*displacement;
        tmp_Y[l] -= delta_y*displacement;
        tmp_Y[m] += delta_y*displacement;
        tmp_Z[l] -= delta_z*displacement;
        tmp_Z[m] += delta_z*displacement;
};
```

c. Finally, check if the rag intersects with the geometry (sphere and ground surface). If yes, readjust the positions accordingly.

```
auto adjust_positions = [&](T& x, T& y ,T& z) {

        T rho = x*x+y*y+z*z;
        if (rho < 1) {
                rho = sqrt(rho);
                x /= rho;
                y /= rho;
                z /= rho;
        }
        z = std::max<T>(z, -1);
};
```

Your task is to investigate the potential for parallelization of the proposed algorithm.

(i) Read this paper on cloth simulation using position-based dynamics [8]. Explain each of the three aforementioned methods `update_positions`, `relax_constraint` and `adjust_positions`. What is happening there in detail?

(ii) Discuss the potential for parallelization and vectorization considering the techniques OpenMP and AVX. Design a strategy for the computation of a massive rag simulation on a cluster of multi-core CPUs providing support for Intel's advanced vector extensions (AVX).

(iii) Consider a small scale variant of your proposed implementation where $N \approx 100$. Use one or more of the above-mentioned techniques to parallelize the given sequential code. Discuss speedup and efficiency.

REFERENCES

[1] Tim Blechmann, Boost C++ libraries: lock-free data structures, http://www.boost.org/doc/libs/1_64_0/doc/html/lockfree.html (visited on 01/05/2017).

[2] OpenMP Architecture Review Board, OpenMP 4.5 complete specifications, http://www.openmp.org/wp-content/uploads/openmp-4.5.pdf, 2015 (visited on 05/10/2017).

[3] OpenMP Architecture Review Board, OpenMP 4.5 summary card – C/C++, http://www.openmp.org/wp-content/uploads/OpenMP-4.5-1115-CPP-web.pdf, 2015 (visited on 05/10/2017).

[4] T. Cover, P. Hart, Nearest neighbor pattern classification, IEEE Transactions on Information Theory (ISSN 0018-9448) 13 (1) (1967) 21–27, http://dx.doi.org/10.1109/TIT.1967.1053964.

[5] Michael Klemm, SIMD vectorization with OpenMP, https://doc.itc.rwth-aachen.de/download/attachments/28344675/SIMD+Vectorization+with+OpenMP.PDF (visited on 04/20/2017).

[6] Yann LeCun, Corinna Cortes, Christopher J.C. Burges, The MNIST database of handwritten digits, http://yann.lecun.com/exdb/mnist/ (visited on 01/12/2016).

[7] R.B. Marimont, M.B. Shapiro, Nearest neighbour searches and the curse of dimensionality, IMA Journal of Applied Mathematics 24 (1) (1979) 59, http://dx.doi.org/10.1093/imamat/24.1.59.

[8] Stan Melax, AVX based cloth simulation, https://software.intel.com/sites/default/files/m/1/5/2/5/6/33189-AVXCloth.pdf (visited on 03/20/2017).

[9] Ruud van der Pas, OpenMP tasking explained, http://openmp.org/wp-content/uploads/sc13.tasking.ruud.pdf (visited on 05/20/2017).

[10] Xinmin Tian, Bronis R. de Supinski, Explicit vector programming with OpenMP 4.0 SIMD extensions, in: HPC Today, 2014, http://www.hpctoday.com/hpc-labs/explicit-vector-programming-with-openmp-4-0-simd-extensions/.

CHAPTER

COMPUTE UNIFIED DEVICE ARCHITECTURE

7

Abstract

A Graphics Processing Unit (GPU) is an integral piece of hardware in many electronic devices. It is used to expose graphical interfaces to the user and can be found in almost every workstation or mobile device. Starting from the mid-nineties of the last century, the increasing popularity of 3D-accelerated games (such as the Quake and Unreal series) caused a rapid growth of computational capabilities of modern GPUs allowing for the rendering of more and more complex scenes. In the late nineties, manufacturers extended the GPU's core functionality, the efficient rendering of 3D scenes, to additionally support the processing of geometric operations, notably the pioneering NVIDIA GeForce 256 [5] and its dedicated Transform and Lighting (T&L) unit. This trend continued in the new millennium with the introduction of pixel and vertex shader languages which allowed for the manipulation of the prior to this hard-wired rendering pipeline. Consequently, computer scientists used the increasing computational power of GPUs to implement more general algorithms by expressing them in the aforementioned shader languages. This was the birth of the so-called *general-purpose computing on GPUs (GPGPU)*. A major drawback of GPGPU programming was its lack of abstraction: the processed data had to be encoded in textures and manipulated with the instruction set of the shader language which substantially limited the design of complex programs.[1]

In summer 2007, NVIDIA released the *Compute Unified Device Architecture (CUDA)* [3], that allows for the convenient programming of complex applications using a C- or FORTRAN-based language. As a result, general algorithms can be expressed in easily readable code and, at the same time, benefit from the up to two orders-of-magnitude faster execution time compared to single-threaded CPU implementations. Although there exist unified approaches for the programming of GPUs from other vendors (e.g., OpenCL [9] and OpenACC [17]), CUDA is nowadays the predominant parallelization framework on GPUs. This chapter will teach you the basics of CUDA-C++ covering the programming model, the efficient use of the underlying memory hierarchy, and mechanisms for the synchronization of massively parallel tasks with many thousands of threads.

Keywords

CUDA, Massively parallel computing, GPU, Kernel, Thread block, Device synchronization, Warp, Thrust, Compute-to-global-memory-access, Tesla, Streaming multiprocessor, SIMD, Principal component analysis, Loop unrolling, Coalesced memory access, Eigenvalue decomposition, Shared memory

[1]An exhaustive overview of the history of GPGPU can be found in [13].

Parallel Programming. DOI: 10.1016/B978-0-12-849890-3.00007-1

CONTENTS

7.1 INTRODUCTION TO CUDA (HELLO WORLD)

In this chapter you will learn how to write massively parallel programs on NVIDIA GPUs with the help of the CUDA programming model. Since CUDA is basically an extension to C and C++, we can easily port embarrassingly parallel programs without the overhead of learning a completely new language. Let us start with a simple *Hello World* program that prints a greeting message to the command line. The source code in Listing 7.1 is separated into two parts. It consists of a `main` function that is executed on the CPU and a kernel method `hello_kernel` that will be running on the GPU.

```
1  #include <stdio.h>                       // printf
2
3  __global__ void hello_kernel() {
4
5      // calculate global thread identifier, note blockIdx.x=0 here
6      const int thid = blockDim.x*blockIdx.x + threadIdx.x;
7
8      // print a greeting message
9      printf("Hello from thread %d!\n", thid);
10 }
11
12 int main (int argc, char * argv[]) {
13
14     // set the ID of the CUDA device
15     cudaSetDevice(0);
```

```
16
17      // invoke kernel using 4 threads executed in 1 thread block
18      hello_kernel<<<1, 4>>>();
19
20      // synchronize the GPU preventing premature termination
21      cudaDeviceSynchronize();
22  }
```

Listing 7.1: CUDA Hello World.

Let us have a look at the `main` function. The first command `cudaSetDevice` selects the used CUDA-capable GPU. Usually, you will only have one GPU attached to your workstation and thus the identifier can be safely set to `0`. We will talk about the possibility to simultaneously utilize several GPUs later in this chapter. The second call `hello_kernel<<<1, 4>>>()` will invoke one block of four CUDA threads running in parallel on the GPU. Each thread is supposed to print a greeting message. For the moment, we ignore the implementation details of the kernel and proceed with the `cudaDeviceSynchronize()` call. Kernels on the GPU (device) and code on the CPU (host) are executed asynchronously. The final synchronization forces the host to wait until the kernel on the device has finished its computation. This prevents an immediate termination of the whole program. Otherwise, the host-sided `main` function will instantly return without printing the desired message when using no explicit or implicit synchronization mechanism.

An obvious novelty in the signature of `hello_kernel` is the `__global__` qualifier right in front of the return value `void` indicating that this method is callable from the host but executed on the device. The CUDA programming language provides further qualifiers that can be used to specify the scope of execution and inlining behavior. The most important are the following:

- `__global__`: callable from host,[2] executed on the device,
- `__host__`: callable from host, executed on the host,
- `__device__`: callable from device, executed on the device.

Note that `__host__` is implicitly defined for every host function and thus could be added in front of every traditional function which is executed on the CPU. The qualifiers `__host__` and `__device__` can be combined in order to force the compiler to generate binary code for both the CPU and GPU. Additionally, the `__noinline__` and `__forceinline__` qualifiers can be used to control inlining behavior. More details can be found in the CUDA programming guide [4].

As mentioned before, we invoke the kernel on the GPU using one CUDA thread block consisting of four CUDA threads. Further details on the specification of blocks and threads are discussed in the next section. Up to now it is enough to know that we spawn one block with block identifier `blockIdx.x=0` consisting of `blockDim.x=4` threads, each of them enumerated with a local thread identifier `threadIdx.x` ranging from 0 to 3 (inclusive). As a result, the expression in Line 6 evaluates to `thid = 0*4 + threadIdx.x = threadIdx.x`. Alternatively, we could have generated the same indices when spawning two blocks each executing two threads, or four blocks each executing one thread. As we will see later in Section 7.4, the particular distribution of threads can have a massive impact on the

[2]Note that kernels can also be called recursively by the same kernel. However, the concept of Dynamic Parallelism will not be discussed in this book due to its infrequent occurrence in the literature.

overall performance of our program. Finally, the global thread identifier `thid` is printed via the `printf` command. The example code can be compiled with a call to

```
nvcc hello_world.cu -O2 -o hello_world
```

and executed like any other binary built with a host compiler obtaining the following output:

```
Hello from thread 0!
Hello from thread 1!
Hello from thread 2!
Hello from thread 3!
```

Note that the optimization flag `-O2` only affects the host part of the code and does not influence the performance of device functions. Let us briefly discuss the output in the end. You might wonder why the messages are printed in order in contrast to typical multithreaded applications. This is because the print statements are always serialized in batches of 32 consecutive threads within a thread block, a so-called **warp** of threads. If the kernel invocation is modified to use 512 threads within one block, we will observe shuffled batches each consisting of 32 contiguous thread identifiers.

7.2 HARDWARE ARCHITECTURE OF CUDA-ENABLED GPUS

Admittedly, the Hello World example is not very useful since the printing is serialized and we have no data to process. But before we start crunching numbers, we have to learn about the hardware layout of a typical CUDA-capable GPU and how the CUDA programming model maps onto its individual components. Although this section is a bit theoretical, we will benefit from this knowledge by exploiting hardware-specific details that are not part of the CUDA programming language later on. Rephrasing it with the words of Jean-Jacques Rousseau: *Patience is bitter, but its fruit is sweet.*

INTERCONNECTION BETWEEN HOST AND DEVICE

Let us start with the obvious. CUDA-capable GPUs are attached to a server or workstation via dedicated slots on the mainboard. At the time of writing, the majority of graphics cards are connected via the PCIe v3 bus providing up to 16 GB/s bandwidth (see Fig. 7.1). At first sight this sounds reasonably fast but you will soon observe that this is often the main bottleneck of CUDA applications. Addressing this, NVIDIA has introduced NVLink, a novel bus providing up to 80 GB/s peak bandwidth [6], allowing for the efficient communication between the host and the device or between multiple devices attached to the same compute node.

The memory of the host (RAM) and the video memory of the device (VRAM) are physically separated. As a result, we have to allocate memory on both platforms independently and afterwards manage memory transfers between them. It means that data which shall be processed on the GPU has to be explicitly transferred from the host to the device. The computed results residing in the VRAM of the GPU have to be copied back to the host in order to write them to disk. This implies that data allocated on the host cannot be directly accessed from the device and vice versa. The corresponding CUDA commands will be discussed in detail in Section 7.3.

Note that NVIDIA provides powerful libraries bundled with CUDA, e.g. Thrust [7], which features device vectors that can be manipulated from the host. Nevertheless, you should be aware that these fancy abstraction layers may obscure suboptimal parts of your code. As an example, altering all entries

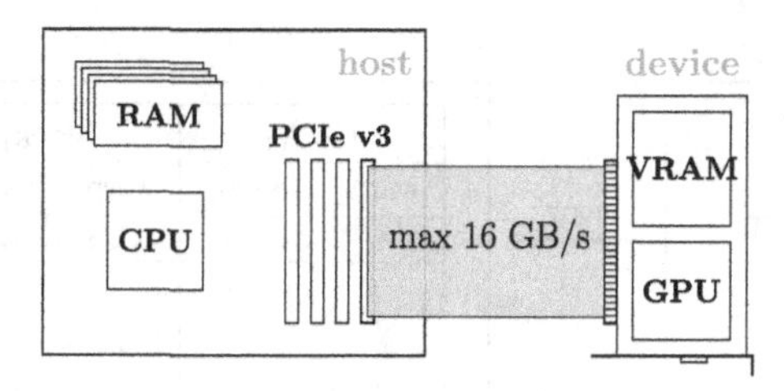

FIGURE 7.1

Schematic overview of the processors (CPU and GPU) and memory components for host and device. Memory transfers are executed over the PCIe v3 bus providing up to 16 GB/s bandwidth. Conceptually, the host and the device can be treated as independent compute platforms which communicate over an interconnection network.

of a Thrust device vector in a host-sided for-loop results in the excessive spamming of tiny memory transfers. A similar though less harsh statement can be made about NVIDIA's unified memory layer that treats the address spaces of the RAM and VRAM as one. Consequently, we will pursue the traditional distinction between both memory spaces. This transparent approach will help you to pinpoint performance bottlenecks of your applications without having to guess the internal details of unified memory addressing.

VIDEO MEMORY AND PEAK BANDWIDTH

Conceptually, the graphics card can be treated as an independent compute platform with its own compute unit and memory. In contrast to the memory attached to a typical mainboard (at the time of writing, DDR4 DRAM) the video memory of a graphics cards is significantly faster. As an example, the Pascal-based Titan X features 12 GB of GDDR5X DRAM modules providing up to 480 GB/s bandwidth compared to the memory bandwidth of current Xeon workstations of less than 100 GB/s. Accelerator cards from the professional Tesla series such as the Tesla P100 or Tesla V100 feature even faster HBM2 stacked memory with up to 720 GB/s or 900 GB/s bandwidth, respectively.

Despite the fact that we can access the global memory with almost one TB/s, we still have to care about the underlying memory hierarchy. As an example, assume you want to increment single-precision floating point (FP32) values residing in VRAM (global memory). First, each of the 32-bit values is loaded into a register. Second, the register is incremented by one which accounts for a single floating point operation (Flop). Finally, the incremented value is written back to global memory. Fig. 7.2 visualizes the described scheme. This corresponds to a *compute-to-global-memory-access* (CGMA) ratio of 1 Flop/8 B (four bytes for reading and four bytes for writing after incrementation) resulting in an overall compute performance of 125 GFlop/s for a memory bandwidth of 1 TB/s. This is merely 1% of the 11 TFlop/s FP32 peak performance of a Tesla P100 card. Therefore proper memory access patterns and the efficient use of fast caches will be one of our main concerns.

ORGANIZATION OF COMPUTATIONAL RESOURCES

The enormous compute capability of modern GPUs can be explained by the sheer number of processing units which usually exceeds a few thousand cores. In contrast to multi-core CPUs with a few tens

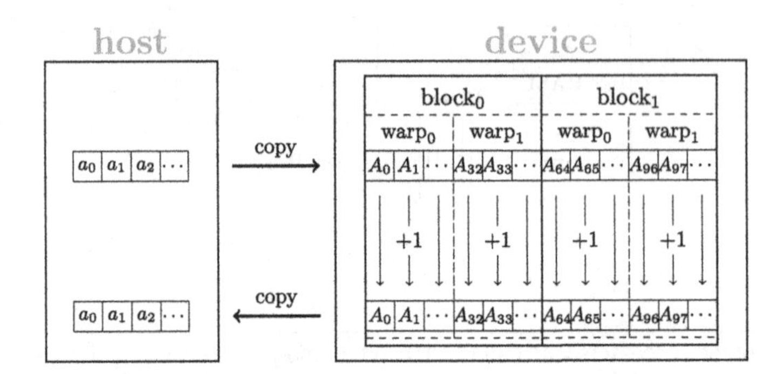

FIGURE 7.2

Memory transfers and control flow of a kernel that increments the entries of an array $a = (a_0, a_1, \ldots)$ stored on the host. First, the host array a is copied to the device array A of same size. Second, each warp of a thread block simultaneously increments 32 consecutive entries of the array A. Third, the incremented values residing in the device array A are copied back to the host array a.

Graphics Processing Unit

GPC					GPC					GPC				
SM	SM	SM	SM	SM	SM	SM	SM	SM	SM	SM	SM	SM	SM	SM
SM	SM	SM	SM	SM	SM	SM	SM	SM	SM	SM	SM	SM	SM	SM
L2 cache														
GPC					GPC					GPC				
SM	SM	SM	SM	SM	SM	SM	SM	SM	SM	SM	SM	SM	SM	SM
SM	SM	SM	SM	SM	SM	SM	SM	SM	SM	SM	SM	SM	SM	SM

FIGURE 7.3

Schematic layout of the GP102 GPU built in the Tesla P40 card. The GPU is partitioned into six Graphics Processing Clusters (GPCs) each providing ten Streaming Multiprocessors (SMs). Each of the 60 SMs share the same L2 cache and global memory. Consumer GPUs such as the Pascal based GeForce GTX 1080 or Titan X have similar layouts with a smaller number of GPCs or SMs (see Table 7.1).

monolithic processing units featuring complex instruction sets and control flows, GPUs can be treated as a huge array of lightweight processing units with limited instruction set and constrained control flow. The hardware layout of a modern GPU is organized in a hierarchical tree.

Mapping Thread Blocks Onto SMs

The first level consists of a small number (see Table 7.1) of Graphics Processing Clusters (GPCs) each of them containing roughly ten (see Table 7.2) Streaming Multiprocessors (SMs) as shown in

Table 7.1 Technical specification of high-end single GPU video cards across generations. Dual GPUs like the Titan Z, Tesla K80, or Tesla M60 are omitted since they are basically stacked versions of their single GPU counterparts. Tesla cards from the Kepler and Maxwell generation are equipped with slower error-correcting (ECC) RAM. The Pascal-based Tesla P100 and Volta-based Tesla V100 provide significantly faster HBM2 stacked memory. Note that the number of cores and VRAM size are slightly increasing over time.

Video card	Generation	GPCs	SMs	FP32/GPU	VRAM @ Bandwidth
Titan (Black)	Kepler	8	15	2880	6 GB @ 336 GB/s
Tesla K40	Kepler	8	15	2880	12 GB @ 288 GB/s
Titan X	Maxwell	3	24	3072	12 GB @ 336 GB/s
Tesla M40	Maxwell	3	24	3072	24 GB @ 288 GB/s
GTX 1080	Pascal	4	40	2560	8 GB @ 320 GB/s
Titan X	Pascal	6	56	3584	12 GB @ 480 GB/s
Tesla P40	Pascal	6	60	3840	24 GB @ 346 GB/s
Tesla P100	Pascal	6	56	3584	16 GB @ 720 GB/s
Tesla V100	Volta	6	80	5120	16 GB @ 900 GB/s

Table 7.2 Technical specification of GPU architectures across four generations. The number of SMs per GPC increases while the amount of cores per SM decreases over time. Thus a core on a Pascal/Volta SM can use more registers than on SMs from previous generations. Further, note the relatively small FP64/FP32 ratio of the Maxwell architecture.

Generation	SM/GPC	FP32/SM	FP64/SM	FP64/FP32	Registers/SM
Kepler	2	192	64	1/3	65,536 × 32bit
Maxwell	8	128	4	1/32	65,536 × 32bit
Pascal	10	64	32	1/2	65,536 × 32bit
Volta	14	64	32	1/2	65,536 × 32bit

Fig. 7.3. The GPCs are a rather hardware-specific detail that have no correspondence in the CUDA programming model. However, the SMs in the second level loosely map onto the aforementioned CUDA thread blocks. One or more thread blocks can be simultaneously executed on a single SM. The specific number is dependent on the ratio of required and available hardware resources as shown in Section 7.4. Note that the runtime environment does not provide any information about a specific order of execution, i.e., we can neither influence nor deduce the planning scheme of the block scheduler. As a result, the parts of our program that are executed on different blocks should be truly data independent.

Within the CUDA programming model the block identifiers can be defined using 1D, 2D, and 3D grids. This allows for a convenient indexing when accessing data on multidimensional domains. As an example, when processing distinct parts of an image one could assign a block with coordinates `(blockIdx.x, blockIdx.y)` to each tile. The grid dimension can be defined as an integer for 1D grids or as a `dim3` struct for the general case. They can be accessed within a kernel via `gridDim.x`, `gridDim.y`, and `gridDim.z`. Our Hello World example from the previous section utilized a 1D grid with either one, two, or four blocks. We demonstrate the invocation of a 3D grid consisting of 3D blocks in Listing 7.2.

```
1  #include <stdio.h>                    // printf
2
3  __global__ void kernel() {
```

```
4
5      // print grid and block dimensions and identifiers
6      printf("Hello from thread (%d %d %d) "
7             "in a block of dimension (%d %d %d) "
8             "with block identifier (%d %d %d) "
9             "spawned in a grid of shape (%d %d %d)\n",
10            threadIdx.x, threadIdx.y, threadIdx.z,
11            blockDim.x,  blockDim.y,  blockDim.z,
12            blockIdx.x,  blockIdx.y,  blockIdx.z,
13            gridDim.x,   gridDim.y,   gridDim.z);
14 }
15
16 int main (int argc, char * argv[]) {
17
18     // set the ID of the CUDA device
19     cudaSetDevice(0);
20
21     // define a grid of 1*2*3 = 6 blocks
22     // each containing  4*5*6 = 120 threads
23     // i.e. altogether 6! = 720 threads
24     dim3  grid_dim(1, 2, 3);
25     dim3 block_dim(4, 5, 6);
26
27     // invoke the kernel
28     kernel<<<grid_dim, block_dim>>>();
29
30     // synchronize the GPU preventing premature termination
31     cudaDeviceSynchronize();
32 }
```

Listing 7.2: Kernel invocation of a 3D grid consisting of 3D blocks.

Before we proceed with the internal components of an SM we have to make an important remark: different CUDA thread blocks of the same kernel cannot be synchronized during the execution of a kernel since there exists no device-wide barrier up to CUDA version 8 which could be called from within a kernel. Note that CUDA 9 introduces the concept of cooperative groups that provide device-sided multi-grid synchronization calls. However, CUDA 9 is not released at the time of writing this book and thus we stick to the traditional approach: inter-block dependencies have to be realized by stacking several kernels on the host enforcing a barrier between individual kernel invocations. This architectural decision made by NVIDIA has several motivations. First, the hardware layouts of low-end and high-end GPUs mainly differ in the number of provided SMs. Thus an elaborate programming model should scale transparently over an arbitrary number of blocks. Second, synchronization is expensive especially when enforcing a global barrier where all SMs have to wait for the last block to finish. This observation can be summarized in a statement that *synchronization does not scale.*[3] As a consequence, we are forced to refactor or even redesign code that relies excessively on global barriers or mutexes.

[3] The semi-automatic parallelization framework OpenACC follows the same philosophy [14].

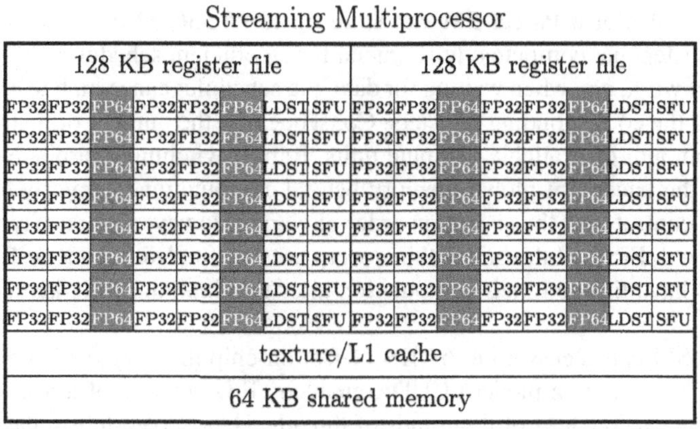

Streaming Multiprocessor

128 KB register file								128 KB register file							
FP32	FP32	FP64	FP32	FP32	FP64	LDST	SFU	FP32	FP32	FP64	FP32	FP32	FP64	LDST	SFU
FP32	FP32	FP64	FP32	FP32	FP64	LDST	SFU	FP32	FP32	FP64	FP32	FP32	FP64	LDST	SFU
FP32	FP32	FP64	FP32	FP32	FP64	LDST	SFU	FP32	FP32	FP64	FP32	FP32	FP64	LDST	SFU
FP32	FP32	FP64	FP32	FP32	FP64	LDST	SFU	FP32	FP32	FP64	FP32	FP32	FP64	LDST	SFU
FP32	FP32	FP64	FP32	FP32	FP64	LDST	SFU	FP32	FP32	FP64	FP32	FP32	FP64	LDST	SFU
FP32	FP32	FP64	FP32	FP32	FP64	LDST	SFU	FP32	FP32	FP64	FP32	FP32	FP64	LDST	SFU
FP32	FP32	FP64	FP32	FP32	FP64	LDST	SFU	FP32	FP32	FP64	FP32	FP32	FP64	LDST	SFU
FP32	FP32	FP64	FP32	FP32	FP64	LDST	SFU	FP32	FP32	FP64	FP32	FP32	FP64	LDST	SFU
texture/L1 cache															
64 KB shared memory															

FIGURE 7.4

Schematic layout of a Streaming Multiprocessor (SM) used in the Pascal generation. An SM provides two blocks each consisting of 32 single-precision compute units (FP32), 16 double-precision compute units (FP64), 8 load and store units (LDST), and 8 special function units (SFU). Each of the two blocks can access 32,768 32-bit registers resulting in an overall size of 256 KB for the storage of variables. L1 cache and 64 KB shared memory are shared among all units in the SM.

Mapping Threads Into Warps

Each SM is composed of multiple compute and instruction units as shown in Fig. 7.4. The Pascal generation features 64 single-precision (FP32), 32 double-precision (FP64) compute units, 16 load and store units (LDST), and 16 special function units (SFU) per SM. The number and ratio of FP32 and FP64 units depend on the GPU generation. Therefore, FP64 performance cannot be directly inferred from the number of FP32 units. As an example, the Maxwell generation provided a FP64 to FP32 ratio of merely 1/32 in contrast to the reasonably higher fractions of the Kepler (1/3), Pascal (1/2), and Volta (1/2) generations.

Up to CUDA 8, a warp consisting of 32 contiguous CUDA threads is processed simultaneously on an SM in lock-step manner. Thus all 32 compute units have to perform the same operation at the same time similar to the Single Instruction Multiple Data (SIMD) paradigm. In contrast to traditional SIMD architectures, the compute units may access non-contiguous memory or mask instructions allowing for the branching of control flow. The latter is called **branch divergence** and should be avoided since the hardware executes the branches one by another. This slightly more flexible computation model was coined by NVIDIA as Single Instruction Multiple Thread (SIMT). Strictly speaking one can state that SIMD is a subset of the SIMT model. In this context, a warp can be considered as SIMD vector unit with 32 SIMD lanes. Thus we can easily express SIMD algorithms in terms of the CUDA programming language. Note that CUDA 9 shifts the traditional paradigm of warp-centered programming to cooperative groups which are a generalization of the warp concept. As a consequence, warps in CUDA 9 might not be executed in lock-step anymore having severe implications for implicit synchronization. At the time of writing (Summer 2017), we cannot predict the impact of cooperative groups on future code development. However, be aware of this major paradigm shift since it might become a game changer.

The two register files of a Pascal SM can store up to 65,536 32-bit variables. This rather high number of 1024 registers per compute unit is crucial for an efficient scheduling of lightweight threads. If a warp runs out of work, e.g, when waiting for data, the scheduler can switch to another warp without dumping or loading the corresponding registers. Consequently, the number of maintained threads may easily exceed the amount of available compute units. Rapid switching (warp streaming) can be used to effectively hide the latency of global memory behind computation. Similarly to the scheduling of blocks, we have no control over the execution order of warps. However, you can rely on two properties up to CUDA version 8. First, all threads within a warp are executed simultaneously. Second, all warps and thus all blocks have finished their computation after termination of the kernel. Only the latter is true for CUDA 9.

The units of an SM can access up to 64 KB of fast on-chip memory for inter-thread communication and caching. Concluding, a modern CUDA-enabled GPU consists of a few thousand cores (see Table 7.1) which can execute tens of thousands of threads. Hence, parallelism has to be organized on a fine-grained scale compared to coarse-grained parallelization schemes on multi-core architectures. Besides the aspect of massively parallel computation, we have to take care of the additional constraints imposed by the SIMT computation model in order to efficiently utilize modern GPUs.

7.3 MEMORY ACCESS PATTERNS (EIGENFACES)

After having revisited the hardware architecture of a typical CUDA-capable GPU, we can now start to write some more useful code. Our task is the CUDA implementation of **Principal Component Analysis (PCA)**, a frequently used dimensional reduction technique in the field of machine learning. The applications of PCA are manifold and range from lossy audio/video compression in signal processing over the computation of intrinsic coordinate systems of rigid bodies in physics to the determination of latent variables in data mining. Throughout this section we process the *CelebA* dataset [16] consisting of 202,599 aligned celebrity faces (see Fig. 7.5). In particular, we compute numerous quantities such as the mean celebrity face, the centered data matrix, the corresponding covariance matrix, and an eigenvector basis which can be used for lossy image compression.

From a programming point of view you will learn how to transfer data between host and device, the proper addressing of global memory using coalesced access patterns and the manual caching of redundant data in shared memory. After having finished this section, you will be able to write basic CUDA programs concurrently processing gigabytes of data. Furthermore, we will discuss when a CUDA parallelization of your program is beneficial in terms of runtime but also cases where a multithreaded CPU implementation might be the better choice.

COMPUTATION OF THE MEAN CELEBRITY FACE

The 202,599 images of the CelebA dataset are stored as RGB-valued matrices of shape 178×218. For the sake of simplicity we have collapsed the three color channels to grayscale by computing the pixel-wise mean of the red, green, and blue intensities. Furthermore, the images have been downsampled by a factor of roughly four in a preprocessing phase. The resulting images still need more than one gigabyte of memory since $(45 \times 55) \times$ `sizeof(float)` $\times$ 202,599 > 1.8 GB. Nevertheless, the algorithms and source code that we discuss perfectly work on the original resolution too (e.g. when

FIGURE 7.5

Eight placeholder images similar to those of the CelebA dataset. Note that the original photos are not depicted due to unclear copyright. In our experiments we use a variant of the dataset where facial features are aligned and RGB channels are collapsed to a single grayscale channel by ordinary averaging.

utilizing a GPU with 32 GB of VRAM). Moreover, an extension to color images is straightforward by treating the pixels in each color channel independently.

In the following, we interpret each image as a flattened vector $v^{(i)} \in \mathbb{R}^n$ where the $n = 45 \cdot 55 = 2475$ pixels are indexed in row-major order. Loosely speaking, we forget the shape of the matrix and consecutively map each pixel onto a slot of the vector. Afterwards, the $m = 202{,}599$ vectors $v^{(i)}$ are stored in the rows of a data matrix $D = \left(v^{(0)}, v^{(1)}, \ldots, v^{(m-1)}\right) \in \mathbb{R}^{m \times n}$ such that $D_{ij} = v^{(i)}_j$ denotes the j-th dimension (pixel) of the i-th vector (image). We subsequently determine the mean image μ by computing the normalized pixel-wise sum of intensity values over all images:

$$\mu_j = \frac{1}{m} \sum_{i=0}^{m-1} v^{(i)}_j \quad \text{for all } j \in \{0, \ldots, n-1\} \quad . \tag{7.1}$$

The resulting mean vector μ is shown in Fig. 7.6. Note that this one-dimensional representation of higher-dimensional data generalizes to any dataset consisting of m fixed length vectors in $\mathbb{R}^n$. However, the slots of the vectors $v^{(i)}$ should be loosely correlated in order to gain meaningful results. Thus you cannot expect this approach to work well on unaligned data without using an approximate alignment of facial features or employing translation-invariant representations. Unaligned data could be tackled with convolutional neural networks [20] or image registration techniques [10].

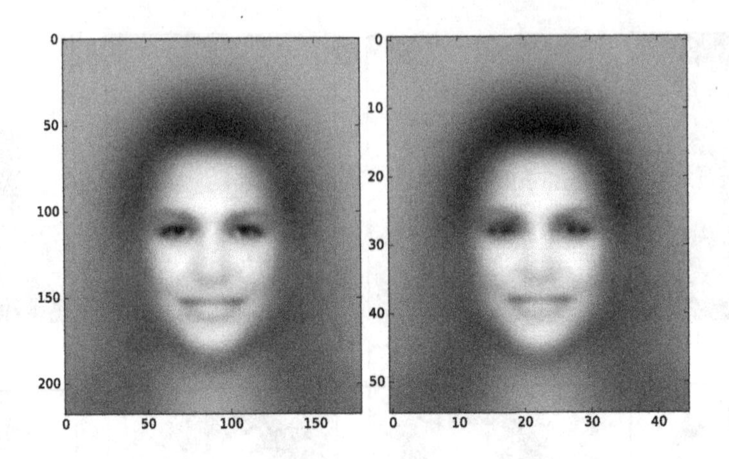

FIGURE 7.6

The mean celebrity face of the CelebA dataset in the original resolution (left panel: 178 × 218 pixels) and its downsampled variant (right panel: 45 × 55 pixels).

Let us start coding. Initially, we include some useful headers which are distributed with this book (see Listing 7.3). The `hpc_helper.hpp` file contains convenient macros for measuring execution times (`TIMERSTART(label)` and `TIMERSTOP(label)`) and for reporting errors (`CUERR`). Error checking is important since CUDA tends to fail silently after incorrect memory accesses or unsuccessful memory allocations while the host code continues its work. Thus you will observe the frequent use of the `CUERR` macro in our code. The remaining two headers provide functions for the loading of binary files and the writing of images in Microsoft bitmap format. Subsequently, we define a template kernel for the mean computation which can be specialized with custom data types for the indexing (usually `uint32_t` or `uint64_t`) and representation of floating point values (usually `float` or `double`). The device pointers `Data` and `Mean` are used to address the data matrix and mean vector. The integers `num_entries` and `num_features` correspond to the number of images (m) and number of features/dimensions (n).

```
1  #include "../include/hpc_helpers.hpp" // timers, error macros
2  #include "../include/bitmap_IO.hpp"   // write images
3  #include "../include/binary_IO.hpp"   // load data
4
5  template <
6      typename index_t,                 // data type for indices
7      typename value_t> __global__      // data type for values
8  void compute_mean_kernel(
9      value_t * Data,                   // device pointer to data
10     value_t * Mean,                   // device pointer to mean
11     index_t num_entries,              // number of images (m)
12     index_t num_features);            // number of pixels (n)
```

Listing 7.3: Header of the mean computation program.

Let us proceed with the main function in Listing 7.4. From Lines 15 to 19, we select the CUDA device and subsequently define three constants specifying the number of images (`imgs`) and their corresponding shape (`rows` and `cols`). Next, we allocate the memory for the data matrix and the mean vector on both the CPU and the GPU using the dedicated commands `cudaMallocHost` and `cudaMalloc`, respectively. Both functions take as first argument the address of a pointer storing the memory location after allocation. Consequently, the argument is a pointer of a pointer in order to allow `cudaMallocHost` and `cudaMalloc` to alter the address from `nullptr` to the corresponding value. The second argument is the number of accessible bytes from that position. The return value denotes whether the allocation was successful and will be handled by the `CUERR` macro. While host-sided memory could also be allocated with alternative commands, e.g., by a call to `malloc` or `new`, we are limited to `cudaMalloc` when explicitly reserving memory on a specific device. Note that throughout this chapter we use capitalized variable names for device-sided pointers and lower case letters for the host. Another popular convention, though not used in this book, appends `_h` (host) and `_d` (device) suffixes to variable names to visually distinguish between address spaces.

```
13  int main (int argc, char * argv[]) {
14
15      // set the identifier of the used CUDA device
16      cudaSetDevice(0);
17
18      // 202599 grayscale images each of shape 55 x 45
19      constexpr uint64_t imgs = 202599, rows = 55, cols = 45;
20
21      // pointer for data matrix and mean vector
22      float * data = nullptr, * mean = nullptr;
23      cudaMallocHost(&data, sizeof(float)*imgs*rows*cols);     CUERR
24      cudaMallocHost(&mean, sizeof(float)*rows*cols);          CUERR
25
26      // allocate storage on GPU
27      float * Data = nullptr, * Mean = nullptr;
28      cudaMalloc(&Data, sizeof(float)*imgs*rows*cols);         CUERR
29      cudaMalloc(&Mean, sizeof(float)*rows*cols);              CUERR
```

Listing 7.4: Main function: memory allocation.

We proceed with the loading of the matrix D from disk to the `data` array on the host using the `load_binary()` method from the provided `binary_IO.hpp` header file. The `TIMERSTART` and `TIMERSTOP` macros determine the execution time. At this point you would manually fill the array using row-major order addressing when processing your own data.

```
31      // load data matrix from disk
32      TIMERSTART(read_data_from_disk)
33      std::string file_name =
34          "./data/celebA_gray_lowres.202599_55_45_32.bin";
35      load_binary(data, imgs*rows*cols, file_name);
36      TIMERSTOP(read_data_from_disk)
```

Listing 7.5: Main function: loading data from disk.

As mentioned before, data that shall be processed on the GPU has to be explicitly transferred from the host to the device. The copying can be achieved with a call to `cudaMemcpy` as shown in Line 39 of Listing 7.6. The command follows the semantics of the traditional `memcpy` call, i.e., the first argument corresponds to the target address and the second to the source pointer. The third argument denotes the number of transferred bytes and the fourth argument specifies the involved platforms. The rather unhandy constants `cudaMemcpyHostToDevice` and `cudaMemcpyDeviceToHost` are used to distinguish between copying from host to device and from device to host. Note that these expressions could be significantly shortened by defining custom variables or macros in order to reduce redundant typing of boilerplate. As an example, the `hpc_helpers.hpp` header file contains the following two lines:

```
#define H2D (cudaMemcpyHostToDevice)
#define D2H (cudaMemcpyDeviceToHost)
```

The call to `cudaMemset` in Line 41 overwrites the device vector `Mean` with zeros. This is a safety mechanism to avoid the following pitfall that we have frequently observed during programming practicals. Assume you have a working CUDA program but neglect return value checking despite the repetitive warnings of your tutor. Moreover, assume you have introduced a bug in the next version which instantly causes your kernel to fail silently. It is highly probable (almost guaranteed) that the subsequent `cudaMemcpy` of the result vector (in our case `Mean`) from the device to the host will transfer the old (correct) data from the previous run still residing in global memory. What could be worse than a defective program that passes unit tests until reboot? Concluding, resetting of results and error checking is mandatory.

After having copied the matrix D to the device we invoke the kernel in Line 46. Referring to Eq. (7.1), parallelization over j (the pixel indices) is advisable since each of the $n = 55 \cdot 45 = 2475$ sums can be evaluated independently.[4] When utilizing 32 CUDA threads per thread block we have to invoke at least $\lfloor n/32 \rfloor$ many blocks. If the remainder $n \,\%\, 32$ is different from zero, i.e., n is not a multiple of the block size, we have to spawn an additional block to process the few remaining pixels. The described scheme can be summarized in a closed expression for safe integer division:

$$\texttt{SDIV}(x, y) = \left\lfloor \frac{x + y - 1}{y} \right\rfloor \geq \frac{x}{y} \quad \text{for all } x, y \in \mathbb{R}^+ \quad . \tag{7.2}$$

The `SDIV` macro is defined in the `hpc_helper.hpp` header file. After termination of the kernel the result is copied back to the host (see Line 52) using `cudaMemcpy`.

```
37      // copy data to device and reset Mean
38      TIMERSTART(data_H2D)
39      cudaMemcpy(Data, data, sizeof(float)*imgs*rows*cols,
40                 cudaMemcpyHostToDevice);                  CUERR
41      cudaMemset(Mean, 0, sizeof(float)*rows*cols);        CUERR
42      TIMERSTOP(data_H2D)
43
44      // compute mean
45      TIMERSTART(compute_mean_kernel)
46      compute_mean_kernel<<<SDIV(rows*cols, 32), 32>>>
47                        (Data, Mean, imgs, rows*cols);     CUERR
48      TIMERSTOP(compute_mean_kernel)
49
```

[4] A parallelization over the index i is conceivable but requires an additional reduction step.

```
50      // transfer mean back to host
51      TIMERSTART(mean_D2H)
52      cudaMemcpy(mean, Mean, sizeof(float)*rows*cols,
53                 cudaMemcpyDeviceToHost);                    CUERR
54      TIMERSTOP(mean_D2H)
```

Listing 7.6: Main function: memory transfers and kernel invocation.

The remaining part of the `main` function in Listing 7.7 is straightforward. The mean image is written to disk using `dump_bitmap` from the `bitmap_IO.hpp` header file. Moreover, allocated memory is explicitly freed using `cudaFreeHost` for the host and `cudaFree` for the device in order to prevent memory leaks.

```
55      // write mean image to disk
56      TIMERSTART(write_mean_image_to_disk)
57      dump_bitmap(mean, rows, cols, "./imgs/celebA_mean.bmp");
58      TIMERSTOP(write_mean_image_to_disk)
59
60      // get rid of the memory
61      cudaFreeHost(data);                                     CUERR
62      cudaFreeHost(mean);                                     CUERR
63      cudaFree(Data);                                         CUERR
64      cudaFree(Mean);                                         CUERR
65  }
```

Listing 7.7: Main function: Writing the result and freeing memory.

Finally, we discuss the `compute_mean_kernel` implementation in Listing 7.8. First, the global thread identifier is calculated in Line 76 analogically to the Hello World kernel from the introductory section. Second, the range of the thread identifier is checked in Line 79 to avoid potential memory access violations when `num_features` is not a multiple of `blockDim.x`. Third, each thread sums up a column of the `Data` matrix in a dedicated register `accum`. Optionally, you can tweak this in Line 86 by giving the compiler a hint to unroll the for-loop in batches of size 32. Finally, we normalize the calculated sum and write it to the `Mean` array.

```
66  template <
67      typename index_t,
68      typename value_t> __global__
69  void compute_mean_kernel(
70      value_t * Data,
71      value_t * Mean,
72      index_t num_entries,
73      index_t num_features) {
74
75      // compute global thread identifier
76      const auto thid = blockDim.x*blockIdx.x + threadIdx.x;
77
78      // prevent memory access violations
79      if (thid < num_features) {
80
81          // accumulate in a fast register,
82          // not in slow global memory
83          value_t accum = 0;
```

```
84
85          // try unrolling the loop with
86          // # pragma unroll 32
87          // for some additional performance
88          for (index_t entry = 0; entry < num_entries; entry++)
89              accum += Data[entry*num_features+thid];
90
91          // write the register once to global memory
92          Mean[thid] = accum/num_entries;
93      }
94  }
```

Listing 7.8: CUDA kernel implementing the mean face computation.

The code can be compiled with the following command:

```
nvcc -O2 -std=c++11 -arch=sm_61 -o mean_computation
```

We need to set the `-std=c++11` flag which activates support for the C++11 standard since our code uses the `auto` keyword. Moreover, `-arch=sm_61` specifies the compute capability of the Pascal generation. Please refer to the CUDA programming guide [4] for an extensive overview of configurable compute capabilities. When executing the program on a Pascal-based Titan X we observe the following output of execution times:

```
TIMING: 13294.3 ms (read_data_from_disk)
TIMING: 170.136 ms (data_H2D)
TIMING: 8.82074 ms (compute_mean_kernel)
TIMING: 0.03174 ms (mean_D2H)
TIMING: 155.921 ms (write_mean_image_to_disk)
```

It takes roughly 13 seconds to read 1.86 GB of image data from spinning disk ($\approx$ 150 MB/s). The memory transfer from host to device over the PCIe bus is accomplished in roughly 170 ms ($\approx$ 11 GB/s). During the execution of the CUDA kernel we access each pixel of the dataset exactly once and apply one addition. Thus the 9 ms correspond to an effective global memory bandwidth of approximately 208 GB/s and a compute performance of about 52 GFlop/s. The time for transferring the average image ($\approx$ 9.5 KB) back to the host is negligible.

Despite the impressive performance of the kernel we utilize merely half of the 480 GB/s theoretical peak bandwidth of the global memory and only 0.5% of the 11 TFlop/s peak performance. Moreover, the amount of $n = 2475$ spawned threads is significantly smaller than the number of 3584 available cores (see Table 7.1). A better approach would spawn tens of thousands of threads. The situation gets even worse if we include the reasonably slower memory transfers over PCIe or the time needed for reading the data from disk. As a result, the described algorithm considered as standalone application is not well suited for the GPU since the 52 GFlop/s compute performance can easily be beaten by state-of-the-art multi-core CPUs. Nevertheless, the kernel can be used as subroutine in high-level algorithms which stack several CUDA kernels. Concluding, raw compute performance is meaningless when having too few data to process.

Finally, let us make an important statement about synchronization. You may have noticed the missing `cudaDeviceSynchronize()` command after calling the kernel in Line 46. The following memory transfer using `cudaMemcpy` implicitly synchronizes the device rendering an explicit synchronization redundant. However, asynchronous memory transfers and kernel launches are possible and will be discussed in Section 8.2. This implicit synchronization behavior is also enforced when allocating/freeing memory using `cudaMalloc(Host)/cudaFree(Host)`, setting memory with `cudaMemset` or a switch between the L1/shared memory configurations using `cudaDeviceSetCacheConfig`. A complete list can be found in the CUDA programming guide [4].

COMPUTING THE CENTERED DATA MATRIX

PCA is usually performed on centered data, i.e., we have to subtract the mean vector μ from each of the vectors $v^{(i)}$. The components of the centered vectors $\overline{v}^{(i)}$ are obtained as follows:

$$\overline{v}_j^{(i)} = v_j^{(i)} - \mu_j \quad \text{for all } i \in \{0, \ldots, m-1\} \\ \text{and all } j \in \{0, \ldots, n-1\} \quad . \tag{7.3}$$

As a result, the columns of the centered data matrix $\overline{D}_{ij} = \overline{v}_j^{(i)}$ sum up to zero. Conceptually, there are two levels of parallelism. On the one hand, we could parallelize the mean correction over the pixel indices j analogically to the previous subsection. On the other hand, each image $v^{(i)}$ can also be treated independently. We will see that the first approach is about eight times faster than the latter. This can be explained by the different memory access patterns of both approaches. However, before we dive into details let us write some code. The first kernel parallelizes the centering over the pixel indices and serializes the loop over the image indices (see Listing 7.9).

```
1   template <
2       typename index_t,
3       typename value_t> __global__
4   void correction_kernel(
5       value_t * Data,
6       value_t * Mean,
7       index_t num_entries,
8       index_t num_features) {
9
10      const auto thid = blockDim.x*blockIdx.x + threadIdx.x;
11
12      if (thid < num_features) {
13
14          const value_t value = Mean[thid];
15
16          for (index_t entry = 0; entry < num_entries; entry++)
17              Data[entry*num_features+thid] -= value;
18
19      }
20  }
```

Listing 7.9: CUDA kernel performing mean correction (pixel indices j).

The code is similar to the mean computation kernel. First, the global thread identifier `thid` is determined in Line 10. Second, we check the range of the thread indices to prevent memory access violations (Line 12). Third, the j-th component of the Mean vector is written to the register value in Line 14. Fourth, we subtract the corresponding value from each vector in the loop body. Finally, the kernel is launched with the following call:

```
correction_kernel<<<SDIV(rows*cols, 32), 32>>>
                    (Data, Mean, imgs, rows*cols);    CUERR
```

An orthogonal approach spawns one thread for each of the images and serializes the loop over the pixel indices. The corresponding source code is provided in Listing 7.10. Here, we compute again the global thread identifier and subsequently check its range. Finally, the loop is executed over pixel indices.

```
1  template <
2      typename index_t,
3      typename value_t> __global__
4  void correction_kernel_ortho(
5      value_t * Data,
6      value_t * Mean,
7      index_t num_entries,
8      index_t num_features) {
9
10     const auto thid = blockDim.x*blockIdx.x + threadIdx.x;
11
12     if (thid < num_entries) {
13
14         for (index_t feat = 0; feat < num_features; feat++)
15             Data[thid*num_features+feat] -= Mean[feat];
16     }
17 }
```

Listing 7.10: CUDA kernel performing mean correction (image indices i).

The corresponding kernel call differs mainly in the number of spawned blocks and threads which now correspond to the number of images.

```
correction_kernel_ortho<<<SDIV(imgs, 32), 32>>>
                    (Data, Mean, imgs, rows*cols);    CUERR
```

When executing both kernels on a Pascal based Titan X we measure roughly 60 ms for the first kernel and approximately 500 ms for the orthogonal approach. On first sight, this observation seems to be counterintuitive since the number of images exceed the number of pixels by far. Nevertheless, the second kernel cannot benefit from the increased level of parallelism due to its suboptimal memory access pattern. Let us have a look at the for-loop body of the orthogonal approach:

```
for (index_t feat = 0; feat < num_features; feat++)
    Data[thid*num_features+feat] = some_value;
```

The entries of Data are not accessed consecutively although the variable feat enumerates the inner loop and thus seems to address contiguous memory. Recall that 32 threads of a warp are executed simultaneously. Therefore the index thid is altered faster than the variable feat. As a result, we access 32 entries residing in the same column during each iteration of the loop. This causes excessive invalidation of cache lines. By contrast, the access pattern of the first kernel

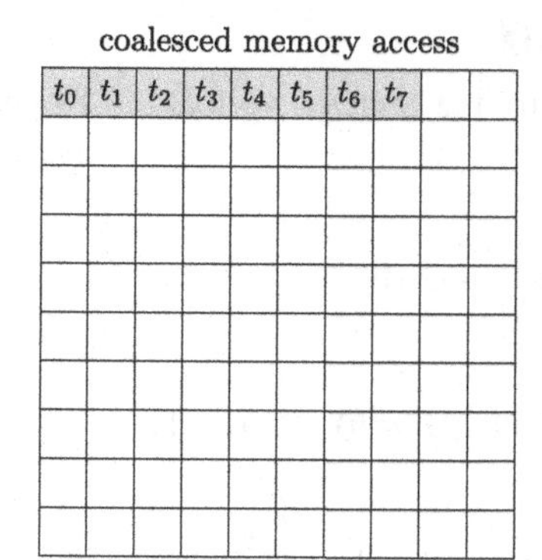

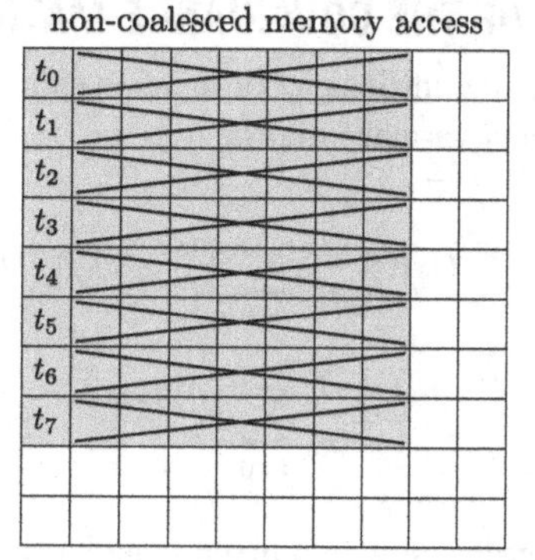

FIGURE 7.7

An example for coalesced and non-coalesced memory access patterns on a 10 × 10 matrix. The left panel shows how eight consecutive threads can simultaneously utilize a whole cache line (gray box). When accessing columns of a matrix as shown in the right panel, each thread issues a load of a complete cache line. Moreover, the remaining seven entries of each cache line are instantly invalidated when manipulating the first entry which enforces a new load during the next iteration. The size of the gray shaded areas corresponds to the expected execution times.

```
for (index_t entry = 0; entry < num_entries; entry++)
    Data[entry*num_features+thid] = some_value;
```

addresses consecutive memory during each iteration.

Let us revisit what we have seen. Whenever consecutive threads access contiguous memory, more precisely when a sequence of threads $(t_k, t_{k+1}, \ldots, t_{k+31})$ simultaneously reads or writes contiguous memory positions $(p_l, p_{l+1}, \ldots, p_{l+31})$ for fixed values of k and l, we call this pattern **coalesced** and otherwise **non-coalesced** (see Fig. 7.7). Coalesced access patterns are highly advisable in order to saturate the bandwidth to global memory. Non-coalesced reads or writes are often as slow as random accesses and should be avoided at any cost. Several techniques can be applied to effectively prevent random access, e.g. the reordering of indices, transposition of input data (which is an exercise), or the execution of highly irregular access patterns on faster memory types. The latter will be demonstrated in Section 7.4. Finally, let us make a concluding remark related to coalesced addressing in multidimensional blocks. Usually, the local thread identifier `threadIdx.x` is altered faster than `threadIdx.y` and `threadIdx.z`. Therefore variables that depend on `threadIdx.x` (in our case `thid`) should always manipulate the least significant bits of the indexing scheme. As an example, the coalesced scheme

```
Data[threadIdx.y*matrix_width+threadIdx.x] = some_value;
```

is typically significantly faster than its non-coalesced counterpart

```
Data[threadIdx.x*matrix_width+threadIdx.y] = some_value;
```

Keep in mind that this subtle distinction may slow down your algorithm by roughly one order-of-magnitude. A similar behavior can be observed on traditional CPUs where row-major addressing, which corresponds to coalesced memory access, reasonably outperforms column-major indexing (see Section 3).

COMPUTATION OF THE COVARIANCE MATRIX

PCA determines a new intrinsic coordinate system for the centered vectors $\overline{v}^{(i)}$ by performing a diagonalization of their covariance matrix

$$C_{jj'} = \frac{1}{m}\sum_{i=0}^{m-1}(v_j^{(i)} - \mu_j)\cdot(v_{j'}^{(i)} - \mu_{j'})$$
$$= \frac{1}{m}\sum_{i=0}^{m-1}\overline{v}_j^{(i)}\cdot\overline{v}_{j'}^{(i)} \quad \text{for all } j, j' \in \{0, \ldots, n-1\} \quad . \tag{7.4}$$

An entry $C_{jj'}$ describes how two pixels j and j' are correlated since it reassembles the scalar product of the corresponding intensity vectors along the image axis.

$$C_{jj'} = \frac{1}{m}\langle\, (\overline{v}_j^{(0)}, \overline{v}_j^{(1)}, \ldots, \overline{v}_j^{(m-1)}) \mid (\overline{v}_{j'}^{(0)}, \overline{v}_{j'}^{(1)}, \ldots, \overline{v}_{j'}^{(m-1)})\,\rangle \tag{7.5}$$

Here, $\langle\cdot|\cdot\rangle$ denotes the standard scalar product in Euclidean space. As an example, if j denotes the position of the left eye and j' the position of the right eye, we expect $C_{jj'}$ to be a reasonably high value. In the general case, we determine all pairwise correlations of the n features represented in the vector space $\mathbb{R}^n$. The covariance matrix C exhibits a number of desirable mathematical properties:

- (**Symmetry**) An entry $C_{jj'}$ is invariant under the swapping of the indices j and j' since multiplication of two real numbers is commutative.
- (**Normality**) C is a normal matrix since $C^T \cdot C = C \cdot C^T$. Thus C can be diagonalized using an eigenvalue decomposition.
- (**Positive Spectrum**) The scalar product in Eq. (7.5) is a positive-definite bilinear form. Consequently, C exhibits exclusively non-negative eigenvalues.

The real-valued eigenvectors $\{u^{(0)}, u^{(1)}, \ldots, u^{(n-1)}\}$ of C are mutually orthogonal and span an n-dimensional vector space over $\mathbb{R}$. You might wonder why we are interested in this particular basis, as we could have stuck to the canonical basis or any other set of linearly independent vectors. Assume we want to describe the images from the high-dimensional vector space $\mathbb{R}^n$ with just one coordinate. Mathematically speaking, we are interested in the optimal basis vector $b \in \mathbb{R}^n \setminus \{0\}$ which on average captures most of the information stored in the centered data matrix $\overline{D}$. Let $\langle\overline{v}^{(i)}|\hat{b}\rangle$ be the projection of the centered vector $\overline{v}^{(i)}$ onto the normed basis vector $\hat{b} = b/\sqrt{\langle b|b\rangle}$ in terms of the scalar product. Then we need to determine the minimizer u that provides the optimal least square approximation of all vectors $\overline{v}^{(i)}$:

$$u = \underset{b\in\mathbb{R}^n\setminus\{0\}}{\operatorname{argmin}} \frac{1}{m}\sum_{i=0}^{m-1}(\overline{v}^{(i)} - \langle\overline{v}^{(i)}|\hat{b}\rangle\cdot\hat{b})^2$$
$$= \underset{b\in\mathbb{R}^n\setminus\{0\}}{\operatorname{argmin}} \frac{1}{m}\sum_{i=0}^{m-1}\left(\overline{v}^{(i)} - \frac{\langle\overline{v}^{(i)}|b\rangle\cdot b}{\langle b|b\rangle}\right)^2 \tag{7.6}$$

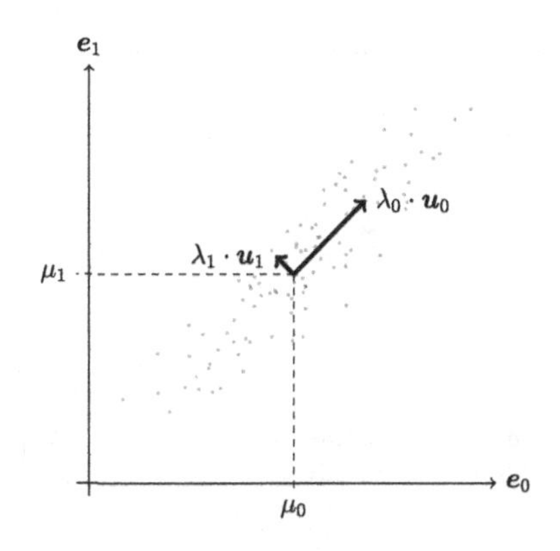

FIGURE 7.8

Exemplary eigenvalue decomposition of the covariance matrix computed from 100 points in $\mathbb{R}^2$. The point cloud is sampled from a multivariate normal Gaussian which has been stretched by a factor of 4 along the x-axis, rotated by 45 degrees, and subsequently shifted by $\mu = (\mu_0, \mu_1)$. Thus the two eigenvectors $u_0 = (1/\sqrt{2}, 1/\sqrt{2})$ and $u_1 = (-1/\sqrt{2}, 1/\sqrt{2})$ describe the columns of the rotation matrix. The eigenvalues $\lambda_0 = 4$ and $\lambda_1 = 1$ are equivalent to the standard deviations in the rotated intrinsic coordinate system.

Let us expand the square in the last equation and substitute the scalar products by explicit sums over j and j'. We obtain a constant term which can be neglected during optimization and a second term depending on the covariance matrix C. Note that the second (mixing) term and the third term from the binomial expansion coincide up to multiplicative factors.

$$\begin{aligned} u &= \underset{b \in \mathbb{R}^n \setminus \{0\}}{\operatorname{argmin}} \sum_{i=0}^{m-1} \left(\frac{1}{m} \sum_{j=0}^{n-1} (\overline{v}_j^{(i)})^2 - \frac{1}{m} \sum_{j=0}^{n-1} \sum_{j'=0}^{n-1} \frac{b_j \cdot \overline{v}_j^{(i)} \cdot \overline{v}_{j'}^{(i)} \cdot b_{j'}}{\langle b | b \rangle} \right) \\ &= \underset{b \in \mathbb{R}^n \setminus \{0\}}{\operatorname{argmax}} \sum_{j=0}^{n-1} \sum_{j'=0}^{n-1} \frac{b_j C_{jj'} b_{j'}}{\langle b | b \rangle} = \underset{b \in \mathbb{R}^n \setminus \{0\}}{\operatorname{argmax}} \frac{\langle b | C \cdot b \rangle}{\langle b | b \rangle} \quad . \end{aligned} \tag{7.7}$$

The last fraction is called *Rayleigh quotient* in the literature. We can maximize this expression for a diagonalizable matrix C by simply choosing the eigenvector of C with highest eigenvalue. This can be easily validated by setting $C \cdot b = \lambda \cdot b$. Analogically, one can show that the optimal basis for the approximation of our data with $0 < k \leq n$ vectors is given by the subset of k eigenvectors with the greatest eigenvalues. An extensive proof can be found in [8]. Fig. 7.8 visualizes the described procedure.

In the following, we will compute a suitable basis for the (lossy) approximation of celebrity faces, so-called eigenfaces. The eigenvalue decomposition will be accomplished using the `cusolverDnSgesvd` method from the cuSOLVER library which is bundled with CUDA. Hence, the implementation of the covariance matrix computation is left to us. We assume that the mean adjusted images are stored in the centered data matrix $\overline{D}_{ij}$ and that we have already allocated memory for the covariance matrix $C_{jj'}$ on the device. A naive implementation of the covariance kernel would look as follows:

```
template <
    typename index_t,
    typename value_t> __global__
void covariance_kernel(
    value_t * Data,           // centered data matrix
    value_t * Cov,            // covariance matrix
    index_t num_entries,      // number of images (m)
    index_t num_features) {   // number of pixels (n)

    // determine row and column indices of Cov (j and j' <-> J)
    const auto J = blockDim.x*blockIdx.x + threadIdx.x;
    const auto j = blockDim.y*blockIdx.y + threadIdx.y;

    // check range of indices
    if (j < num_features && J < num_features) {

        // store scalar product in a register
        value_t accum = 0;

        // accumulate contribution over images (entry <-> i)
        for (index_t entry = 0; entry < num_entries; entry++)
            accum += Data[entry*num_features+j] *  // (non-coal.)
                     Data[entry*num_features+J];   // (coalesced)

        // write down normalized projection
        Cov[j*num_features+J] = accum/num_entries; // (coalesced)
    }
}
```

Listing 7.11: CUDA kernel performing a naive covariance matrix computation.

The source code is similar to the mean computation kernel where we have summed up entries of the data matrix along the image axis. In this case, the accumulation is performed for all n^2 products of pixel values. The kernel launch is straightforward:

```
// feel free to experiment with different block sizes
dim3 blocks(SDIV(rows*cols, 8), SDIV(rows*cols, 8));
dim3 threads(8, 8); // 64 threads (2 warps) per block
covariance_kernel<<<blocks, threads>>>
                    (Data, Cov, imgs, rows*cols);   CUERR
```

The kernel takes roughly 36 seconds to compute the whole covariance matrix for $m = 202{,}599$ images each consisting of $n = 2475$ pixels. The execution time can be cut in half if we exploit the symmetry of C. Here, we compute only entries below the diagonal.

```
template <
    typename index_t,
    typename value_t> __global__
void symmetric_covariance_kernel(
    value_t * Data,
    value_t * Cov,
    index_t num_entries,
```

```
 8        index_t num_features) {
 9
10        // indices as before
11        const auto J = blockDim.x*blockIdx.x + threadIdx.x;
12        const auto j = blockDim.y*blockIdx.y + threadIdx.y;
13
14        // execute only entries below the diagonal since C = C^T
15        if (j < num_features && J <= j) {
16
17            value_t accum = 0;
18
19            for (index_t entry = 0; entry < num_entries; entry++)
20                accum += Data[entry*num_features+j] *  // (non-coal.)
21                         Data[entry*num_features+J];   // (coalesced)
22
23            // exploit symmetry
24            Cov[j*num_features+J] =                    // (coalesced)
25            Cov[J*num_features+j] = accum/num_entries; // (non-coal.)
26        }
27    }
```

Listing 7.12: CUDA kernel performing symmetric covariance matrix computation.

Both the naive and symmetric kernel can be further accelerated by a factor of at least ten if we remove redundant accesses to global memory. Assume you want to compute the first row of the covariance matrix

$$C_{0j'} = \frac{1}{m} \sum_{i=0}^{m-1} \overline{v}_0^{(i)} \cdot \overline{v}_{j'}^{(i)} \quad \text{for all } j' \in \{0, \ldots, n-1\} \quad . \tag{7.8}$$

The naive kernel loads all m entries of $\overline{v}_0^{(i)}$ for each choice of j' from global memory. This results in $m \cdot n$ global memory accesses. The symmetric approach reduces this by a factor of two but still exhibits an $\mathcal{O}(m \cdot n)$ dependency. This can be avoided if we store the m entries belonging to pixel zero in a separate cache that can be accessed significantly faster than global memory. When neglecting the time for accessing this type of memory we can reduce the loads of all values belonging to pixel zero (left factor) to $\mathcal{O}(m)$ operations during the computation of a single row.

Conceptually, we aim for an implementation which loads the whole $m \times n$ matrix $\overline{D}_{ij}$ only once from global memory and subsequently performs the $n^2 \cdot m$ many additions needed for the computation of C. As a result, we can reasonably increase the computing-to-global-memory-access (CGMA) ratio. This approach could be realized on a hypothetical GPU with roughly 2 GB of fast on-chip memory. Unfortunately, this requirement sounds rather utopian due to the sheer number of needed transistors and the associated cost of such a device. In comparison, a Pascal-based Titan X provides 64 KB of *shared memory* for each of the 56 SMs (see Table 7.1 and Fig. 7.4) resulting in an overall size of roughly 3.5 MB for the whole device. Shared memory has to be explicitly utilized by the programmer and can be accessed by all threads within the same block. It is usually used to store redundant data or to perform highly irregular memory access patterns which would be inefficient when executed on global memory. Note that in the literature some authors tend to refer to shared memory as *scratchpad* due to the described usage as tiny storage for organizational purposes. In the following, we will utilize shared memory to explicitly cache small tiles of the data matrix in order to drastically reduce the execution time of the covariance kernel. Although an SM may physically provide a greater amount of shared memory, e.g., 64 KB for the Pascal generation, we can utilize 48 KB at most within a single thread block. As an example, if we specify a kernel to launch with 32 KB of shared memory on a Pascal GPU then the block scheduler can map two thread blocks onto a single SM.

Assume we want to compute all entries $C_{jj'}$ residing in a quadratic tile of shape $w \times w$. The contributions during the summation over the m images depend only on the pixel indices j and j' that are within the range of that tile. Consequently, we would have to store w columns of the data matrix each of length m for both the indices j and j'. Unfortunately, this does not fit in 48 KB of shared memory since the length of a column $m = 202{,}599$ is extraordinarily large. However, we can exploit the associativity of addition by subsequently summing up w images in a neighborhood of w pixels. Let c be the index enumerating $C = SDIV(m, w) = \frac{m+w-1}{w}$ many chunks of image indices then we can rewrite the computation as

$$C_{jj'} = \frac{1}{m} \sum_{c=0}^{C-1} \sum_{i=c \cdot w}^{(c+1) \cdot w} \sigma(i) \cdot \overline{v}_j^{(i)} \cdot \overline{v}_{j'}^{(i)} \quad \text{for all } j, j' \text{ in a } w \times w \text{ tile} \quad , \tag{7.9}$$

where $\sigma(i)$ equals one if $i < m$ and zero otherwise. Note that the auxiliary term $\sigma(i)$ is only needed if m is not a multiple of w. For each contribution (iteration) of the outer sum (loop) we store all values stemming from w images within a neighborhood of w pixels around j and j' in two separate arrays in scratchpad memory. In the inner sum (loop) we can now load the factors for all index combinations of j and j' within a tile from fast memory. Each of the $SDIV(n, w) \times SDIV(n, w)$ many tiles of the $n \times n$ covariance matrix C will be processed by a single thread block spawned in a 2D grid. The indices j and j' within a tile correspond to local thread identifiers `threadIdx.x` and `threadIdx.y`. Furthermore, we can again exploit the symmetry of C by considering only these tiles that contain no index combination (j, j') below the diagonal. Fig. 7.9 illustrates the described approach.

Let us start coding. First, we define some auxiliary variables for the convenient indexing of the tiles. The window w corresponds to the `chunk_size` parameter in the template. A thread block will consist of `chunk_size`×`chunk_size` many threads organized in a 2D grid. Furthermore, we immediately abort thread blocks which compute tiles above the diagonal in Line 25. This could be optimized by spawning only blocks below the diagonal. However, the resulting index calculation is non-trivial and would unreasonably complicate this example.

```
1  template <
2      typename index_t,
3      typename value_t,
4      uint32_t chunk_size=8 > __global__
5  void shared_covariance_kernel(
6      value_t * Data,
7      value_t * Cov,
8      index_t num_entries,
9      index_t num_features) {
10
11
12     // first index in a window of width chunk_size
13     const index_t base_x = blockIdx.x*chunk_size;
```

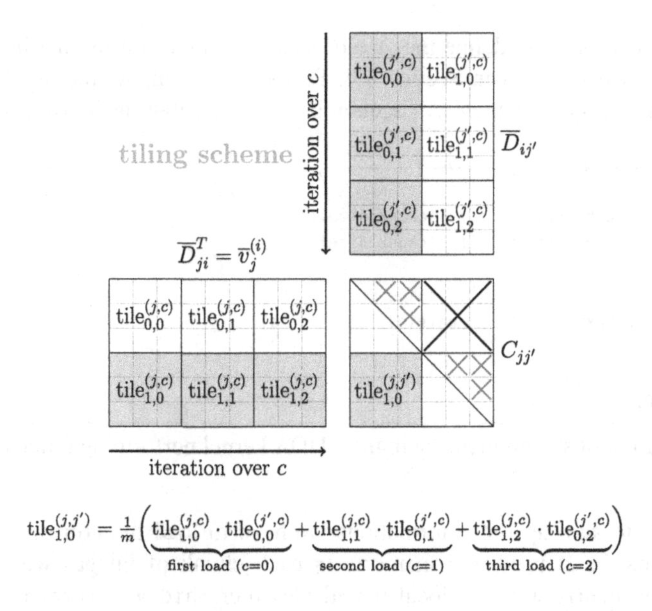

FIGURE 7.9

Tiling scheme for the efficient computation of a covariance matrix C of shape $n \times n = 6 \times 6$ from a centered data matrix $\overline{D}$ of shape $m \times n = 9 \times 6$. Both C and $\overline{D}$ are subdivided into tiles of shape $w \times w = 3 \times 3$. The contributions within a tile of C are consecutively summed up in $m/w = 3$ iterations. During each iteration the tiles have to be loaded into shared memory before we can process the matrix product of the corresponding tiles. The matrix product is performed by $w \times w$ many threads within a thread block which can simultaneously access the values stored in shared memory. As a result, we can decrease the accesses to global memory by a factor of w. Tiles which are located above the diagonal can be neglected due to the symmetry of C. Moreover, the contributions of tiles on the diagonal only have to be evaluated partially.

```
14        const index_t base_y = blockIdx.y*chunk_size;
15
16        // local thread identifiers
17        const index_t thid_y = threadIdx.y;
18        const index_t thid_x = threadIdx.x;
19
20        // global thread identifiers
21        const index_t x = base_x + thid_x;
22        const index_t y = base_y + thid_y;
23
24        // optional early exit for tiles above the diagonal
25        if (base_x > base_y) return;
```

Listing 7.13: Initial part of the CUDA kernel performing efficient covariance matrix computation.

Second, we allocate shared memory for the two tiles using the `__shared__` qualifier in Lines 27 and 28. The semantic is similar to the definition of static arrays in multiple dimensions. The dimensions have to be specified with constant integers, so you cannot use variables defined inside the kernel or its signature. Thus we have to pass `chunk_size` as template parameter which is known at compile time. Alternatively, one can define the size of shared memory during the invocation of a kernel which will be discussed in detail in Section 7.4. Further, we compute the number of chunks to be processed and define a register which will accumulate the contributions during summation along the image axes.

```
26      // allocate shared memory
27      __shared__ value_t cache_x[chunk_size][chunk_size];
28      __shared__ value_t cache_y[chunk_size][chunk_size];
29
30      // compute the number of chunks to be computed
31      const index_t num_chunks = SDIV(num_entries, chunk_size);
32
33      // accumulated value of scalar product
34      value_t accum = 0;
```

Listing 7.14: Allocation of shared memory in the CUDA kernel performing efficient covariance matrix computation.

Before we can start to sum up the contributions we have to load the corresponding portions from the centered data matrix $\overline{D}$ to shared memory. In each chunk of images we enumerate the rows (index i) of the data matrix with the local thread identifier `thid_y = threadIdx.y`. This local index has to be adjusted by an offset `chunk*chunk_size` as shown in Line 39 to determine the global row identifier `row`. The column indices (j and j') are represented by `thid_x = threadIdx.x` shifted by the offsets `base_x` and `base_y` in order to select the correct tile (see Lines 40 and 41). Subsequently, we check if the global identifiers are within a valid range (see Lines 44–46). The auxiliary variables `valid_row`, `valid_col_x` and `valid_col_y` are used for masking in a later step in case the number of images m or the number of pixels n is not a multiple of the window size w.

```
35      // for each chunk
36      for (index_t chunk = 0; chunk < num_chunks; chunk++) {
37
38              // assign thread IDs to rows and columns
39              const index_t row   = thid_y + chunk*chunk_size;
40              const index_t col_x = thid_x + base_x;
41              const index_t col_y = thid_x + base_y;
42
43              // check if valid row or column indices
44              const bool valid_row   = row   < num_entries;
45              const bool valid_col_x = col_x < num_features;
46              const bool valid_col_y = col_y < num_features;
```

Listing 7.15: Start of the main loop in the CUDA kernel performing efficient covariance matrix computation.

After having defined the necessary variables for indexing we can now load the tiles from global memory to fast shared memory. This is accomplished with a simple assignment as show in Lines 53–55. The aforementioned auxiliary variables are used to prevent malicious memory accesses. Entries that exceed the range of m and n are filled with zeros which does not influence the final result. Note that we use local thread identifiers on the left-hand side where we access shared memory in contrast to the right-hand side where we read global memory using global identifiers. Afterwards, we have to ensure that all threads in a block have finished loading with a call to `__syncthreads()` in Line 60. In our case we have spawned a 2D block of shape 8×8, i.e., both warps (2×32 threads) are potentially executed in parallel and thus have to be explicitly synchronized. This is crucial to prevent that one warp proceeds its execution while the other is still loading data. Thread blocks consisting of only one warp are always synchronized rendering a call to `__syncthreads()` redundant. Note that this barrier only affects the threads within a thread block. Device-wide barriers across all threads of the GPU have to be enforced by terminating the whole kernel or using atomic based synchronization schemes (see Section 8.1).

```
47
48          // fill shared memory with tiles where thid_y
49          // enumerates image identifiers (entries) and
50          // thid_x denotes feature coordinates (pixels).
51          // cache_x corresponds to x and cache_y to y
52          // where Cov[x,y] is the pairwise covariance
53          cache_x[thid_y][thid_x] = valid_row*valid_col_x ?
54                                 Data[row*num_features+col_x] : 0;
55          cache_y[thid_y][thid_x] = valid_row*valid_col_y ?
56                                 Data[row*num_features+col_y] : 0;
57
58          // this is needed to ensure that all threads
59          // have finished writing to shared memory
60          __syncthreads();
```

Listing 7.16: Global to shared memory copies in the CUDA kernel performing efficient covariance matrix computation.

At this point we have stored both tiles in shared memory. The partial scalar products are now computed using a simple loop over the pixel dimension (see Line 66). Again, we can exploit symmetry and abandon entries above the diagonal in Lines 63 and 75. This affects only tiles on the diagonal of the covariance matrix (see gray crosses in Fig. 7.9). The second call to `__syncthreads()` in Line 71 enforces a block-wide barrier to prevent premature execution of the next iteration. Finally, the normalized result is written to the covariance matrix in Line 76 for both entries below and their symmetric counterpart above the diagonal.

```
61
62          // optional early exit
63          if (x <= y)
64              // here we actually evaluate the scalar product
65              for (index_t k = 0; k < chunk_size; k++)
66                  accum += cache_y[k][thid_y]*
67                           cache_x[k][thid_x];
68
69          // this is needed to ensure that shared memory can
70          // safely be overwritten again in the next iteration
71          __syncthreads();
72      } // end for-loop over chunk entries
73
74      // since Cov[x,y] = Cov[y,x] we only compute one entry
```

```
75        if (y < num_features && x <= y)
76            Cov[y*num_features+x] =
77            Cov[x*num_features+y] = accum/num_entries;
78    }
```

Listing 7.17: Main computation in the CUDA kernel performing efficient covariance matrix computation.

The described kernel computes the whole covariance matrix in only 980 ms on a Pascal-based Titan X in comparison to 18 s for the symmetrized naive kernel which exclusively operates on global memory. This impressive speedup of roughly 18 is caused by the caching of redundant entries in shared memory. In theory, we can reuse w columns of the data matrix after having read them once from global memory. Consequently, the speedup should monotonically grow if we increase the window parameter. However, the amount of used shared memory per block also influences the execution time. Section 7.4 discusses this topic in detail.

COMPUTATION OF EIGENFACES

After having determined the covariance matrix we can compute the set of eigenvectors $\{u^{(0)}, \ldots, u^{(n-1)}\}$ and the spectrum of eigenvalues $\{\lambda_0, \ldots, \lambda_{n-1}\}$. This is accomplished with the `cusolverDnSgesvd` routine for singular value decomposition (SVD) provided by the cuSOLVER library which is bundled with CUDA. Note that SVD and eigenvalue decomposition can be used interchangeably in the case of positive definite and symmetric matrices. SVD factorizes a general $m \times n$ matrix $\overline{D}$ into a product of three matrices $U \in \mathbb{R}^{m\times m}$, $\Sigma \in \mathbb{R}^{m\times n}$ and $V^T \in \mathbb{R}^{n\times n}$:

$$\overline{D} = U \cdot \Sigma \cdot V^T \quad , \tag{7.10}$$

where both U and V^T are orthogonal linear maps, i.e., $U \cdot U^T = \mathbb{I}_{m\times m}$ and $V^T \cdot V = \mathbb{I}_{n\times n}$. The matrix of singular values Σ is zero everywhere except on the diagonal. It is easy to show that eigenvalue decomposition of $C = \overline{D}^T \cdot \overline{D}$ can be obtained via SVD:

$$\begin{aligned} \overline{D}^T \cdot \overline{D} &= (U \cdot \Sigma \cdot V^T)^T \cdot (U \cdot \Sigma \cdot V^T) \\ &= V \cdot \big(\Sigma^T \cdot \underbrace{(U^T \cdot U)}_{=\mathbb{I}_{m\times m}} \cdot \Sigma\big) \cdot V^T = V \cdot \Sigma^2 \cdot V^T \quad . \end{aligned} \tag{7.11}$$

However, instead of computing a costly SVD of the huge centered data matrix $\overline{D}$, we apply it directly to the much smaller covariance matrix C. Recall that C is symmetric ($C = C^T$) and thus we observe that $U = V$ since

$$C = U \cdot \Sigma \cdot V^T = V \cdot \Sigma \cdot U^T = C^T \quad \Rightarrow \quad C = U \cdot \Sigma \cdot U^T = V \cdot \Sigma \cdot V^T \quad , \tag{7.12}$$

which is the eigenvalue decomposition of the covariance matrix. Note that this is not a rigid mathematical proof but shall suffice for a programming book. We can further exploit the fact that SVD returns the eigenvectors sorted by the magnitude of their eigenvalues which in our case coincide with the singular values stored in Σ. Thus the first eigenvector captures most of the variance stored in the centered data matrix $\overline{D}$ and so forth. Fig. 7.10 depicts the first 16 eigenvectors of the CelebA dataset.

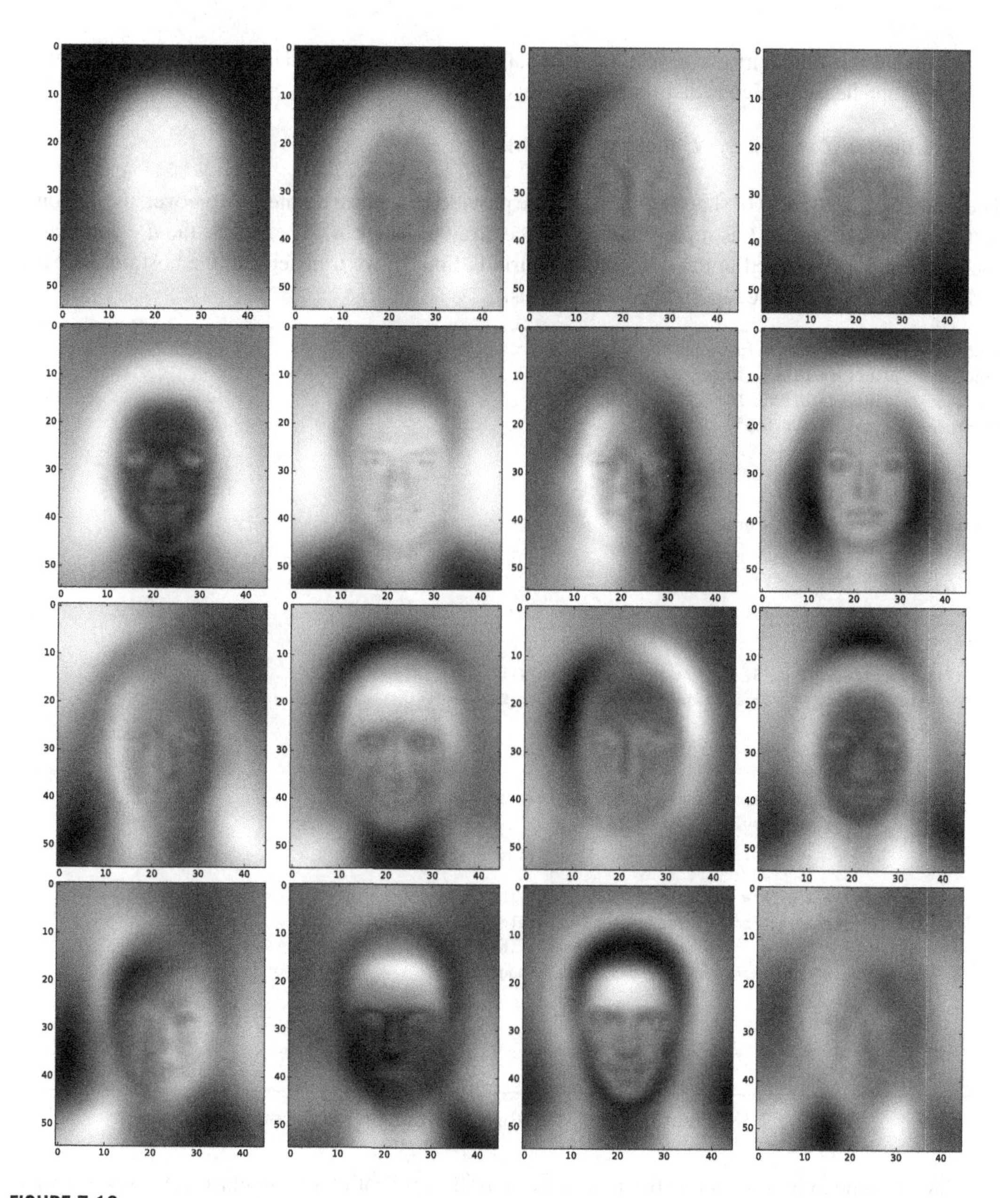

FIGURE 7.10

The top-16 eigenvectors $(u^{(0)}, \ldots, u^{(15)})$ each of shape 55×45 computed from the covariance matrix C of the centered images $\overline{v}^{(i)}$ stored in the CelebA dataset. The images are sorted by the magnitude of their corresponding eigenvalue in descending order (row-major indexing). The eigenvectors are ideally suited for the lossy compression of celebrity photos due to their strong similarity with human faces. Moreover, one can observe that the first eigenfaces encode low frequency structures of a typical face while those with higher indices reassemble facial features in the high frequency domain. Note that eigenvectors can be scaled with any non-zero scalar while preserving their mathematical properties. Thus we occasionally observe negative images.

For the sake of simplicity, we have hidden the call to `cusolverDnSgesvd` in a device function defined in the `svd.hpp` header which is distributed with this book.

```
svd_device(Cov, U, S, V, height, width);
```

The matrices `Cov`, `U`, and `V` of shape $n \times n$ have to be provided as device pointers. Moreover, the diagonal matrix of singular values `S` is represented as device array of length n. Afterwards, the n eigenvectors each of length n are stored in the rows of the matrix `U`. Finally, we transfer `U` to the host matrix `eigs` and subsequently write the top-16 candidates to disk.

```
1   #include "../include/hpc_helpers.hpp"
2   #include "../include/binary_IO.hpp"
3   #include "../include/bitmap_IO.hpp"
4   #include "../include/svd.hpp"
5
6   // definition of previously discussed kernels here
7
8   int main (int argc, char * argv[]) {
9
10      // 1. covariance matrix computation up to here
11      // 2. define and allocate U, S, V as device pointers here
12
13      // 3. call to SVD decomposition using cuSOLVER
14      if(svd_device(Cov, U, S, V, rows*cols, rows*cols))
15          std::cout << "Error: svd not successful." << std::endl;
16
17      // 4. define host array eigs of shape n x n
18      // 5. transfer eigenface back to host array eigs
19      cudaMemcpy(eigs, U, sizeof(float)*rows*cols*rows*cols,
20                 cudaMemcpyDeviceToHost);                    CUERR
21
22      // 6. write down top-k eigenvectors as images
23      for (uint32_t k = 0; k < 16; k++) {
24          std::string image_name = "imgs/eigenfaces/celebA_eig"
25                                 + std::to_string(k)+".bmp";
26          dump_bitmap(eigs+k*rows*cols, rows, cols, image_name);
27      }
28
29      // 7. free memory here with cudaFree(Host) as usual
30  }
```

Listing 7.18: Main function skeleton performing SVD decomposition.

Having come to the end, let us briefly discuss a useful application of eigenfaces. Assume we want to approximate an image of a face $v \in \mathbb{R}^n$ with merely $k << n$ values instead of storing all n pixels. This can be achieved by projecting the mean adjusted face $\overline{v} = v - \mu$ onto the top-k eigenvectors. The new feature representation of the image v consists of k features $(f_0, \ldots, f_{k-1})$:

$$\left(f_0, \ldots, f_i, \ldots, f_{k-1}\right) = \left(\langle \overline{v} | u^{(0)} \rangle, \ldots, \langle \overline{v} | u^{(i)} \rangle, \ldots, \langle \overline{v} | u^{(k-1)} \rangle\right) \quad , \tag{7.13}$$

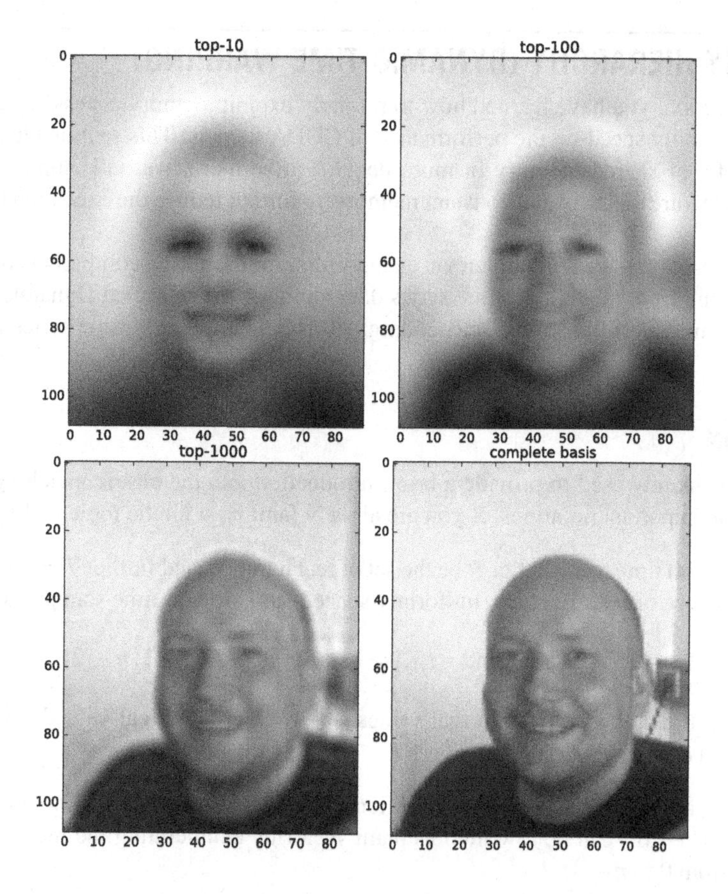

FIGURE 7.11

Lossy compression of a face by expanding the image in the basis of top-k eigenfaces. When using all eigenvectors we can perfectly reconstruct the original image (lower right corner). The images have been sampled at a resolution of $n = 109 \times 89 = 9701$ pixels. Thus the image assembled with 1000 basis vectors (lower left corner) corresponds to a compression ratio of roughly 1:10. The female bias in the CelebA dataset is clearly observable when using only a few eigenvectors since the mean image dominates the reconstruction. The ripple effect for $k = 1000$ is caused by the sub-optimal alignment of the face.

where $\langle \cdot | \cdot \rangle$ denotes the standard scalar product in Euclidean space. The reconstruction of the original image v is straightforward – we simply expand the mean adjusted image $\overline{v}$ in the basis of eigenvectors:

$$v = \mu + \overline{v} \approx \mu + \sum_{i=0}^{k-1} f_i \cdot u^{(i)} = \mu + \sum_{i=0}^{k-1} \langle \overline{v} | u^{(i)} \rangle \cdot u^{(i)} \quad . \tag{7.14}$$

Note that in the case $k = n$ we can reconstruct v perfectly since we have expanded it in a complete basis of n linearly independent vectors $(u^{(0)}, \ldots, u^{(n-1)})$. For $k < n$ we obtain a lossy approximation of the original image (see Fig. 7.11).

7.4 MEMORY HIERARCHY (DYNAMIC TIME WARPING)

In the previous section, you have learned how to properly exploit memory access patterns and shared memory to significantly speed up the performance of CUDA kernels. This section elucidates the benefits and trade-offs of shared memory in more detail. Furthermore, we will employ other types of memory, namely texture memory and constant memory, to further reduce the execution times of CUDA applications.

Throughout this section we will implement an algorithm for the elastic comparison of time-resolved sequences stemming from the field of time series data mining – the so-called Dynamic Time Warping (DTW) similarity measure. Before we start coding, let us define the term time series and discuss the DTW algorithm.

INTRODUCTION

This subsection is mainly used to provide a brief introduction into the elastic matching of time series and to fixate some important notations. If you are already familiar with the topic feel free to skip it.

Definition 1 (Uniform time series)**.** Let $\mathbb{R}$ be the set of real numbers and further $T = (t_0, t_1, \ldots, t_j, \ldots, t_{n-1})$ a finite sequence of n real-valued, uniformly spaced, and ordered time stamps, i.e.,

$$t_{j+1} - t_j = \text{const.} \quad \text{and} \quad t_{j+1} \geq t_j \quad \text{for all } j \in \{0, \ldots, n-2\} \quad , \tag{7.15}$$

then a mapping of the time stamps onto real values $t_j \mapsto S_{t_j}$ is called real-valued and **uniform time series** over the real domain.

The particular values of the time stamps are often neglected. Slightly abusing notation, the mapping of n time stamps $t_j \mapsto S_{t_j}$ can be rewritten as plain vector by enumerating the measured values with indices ranging from 0 to $n - 1$:

$$S = (S_0, S_1, \ldots, S_j, \ldots, S_{n-1}) \in \mathbb{R}^n \quad . \tag{7.16}$$

This definition can be naturally extended to higher-dimensional values $S_j \in \mathbb{R}^d$ or non-uniformly spaced or continuous time domains. However, for the sake of simplicity, we limit ourselves to this simple scenario.

Time series are ubiquitous. They can be found in almost any field of research, e.g., natural science, finance or medical applications. This includes the measurement of physical quantities over time such as voltage, pressure, and speed, exchange rates or stock prices as well as electrocardiograms. Fig. 7.12 shows two further daily life examples. Moreover, pseudo time series exclusively exhibiting discrete values can be found in many other domains, e.g., DNA or protein sequences and strings in natural language processing. The discussed concepts and techniques are also applicable to those domains with only minor modification of the source code. Actually, the to be discussed wavefront relaxation scheme of DTW was first applied during parallelization of a sequence alignment algorithm in bioinformatics.

Assume you have stored m time series $S^{(i)}$ of fixed length n in a $m \times n$ data matrix $D_{ij} = S_j^{(i)}$ where i denotes the identifier of a particular time series $S^{(i)}$ and j enumerates its n time stamps. Further, you have measured another time series Q of length n and subsequently want to determine the most

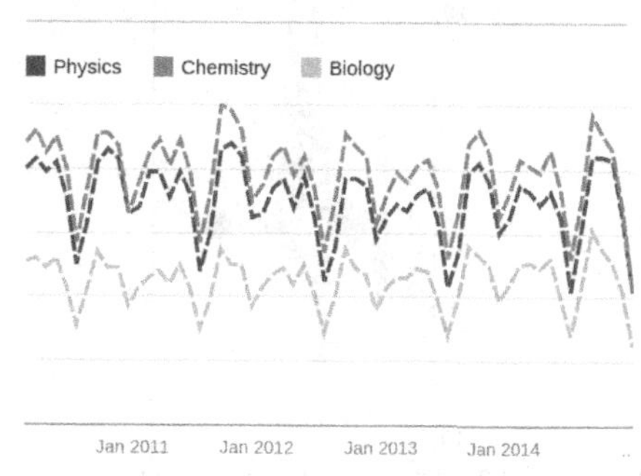

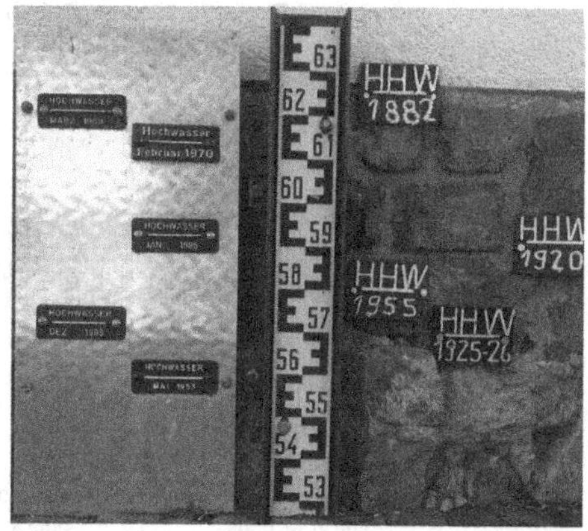

FIGURE 7.12

Two examples for the occurrence of time series in daily life. The left panel depicts the search interest for the three scientific fields "physics," "chemistry," and "biology" as stated by Google Trends. Similar graphs can be obtained for "math" and "computer science." Obviously, the three graphs are strongly correlated and could be used as an indicator function for lecture periods at universities. The right panel illustrates historic signs for the registration of high tides of the river Rhine in Bingen (Germany). Each sign corresponds to one entry in a time series of water levels.

similar entry $S^{(i)}$ from your data matrix by querying Q against your stored data. A canonical notion of similarity $\text{dist}(Q, S^{(i)})$ between Q and a data base entry $S^{(i)}$ could be established by computing the L_p norm of their difference vector:

$$L_p(Q, S^{(i)}) = \|Q - S^{(i)}\|_p = \sqrt[p]{\sum_{j=0}^{n-1} |Q_j - S_j^{(i)}|^p} \quad \text{where} \quad p \geq 1 \ . \tag{7.17}$$

Common values for the external parameter are $p = 2$ for so-called *Euclidean Distance* or $p = 1$ for *Manhattan Distance* which is typically more robust against noise and outliers. Note that we can safely omit the expensive computation of the root due to its monotony property.

Despite the frequent use of L_p based distance measures in time series data mining, the lock-step computation of differences $Q - S^{(i)}$ limits their applicability to (approximately) aligned data. More precisely, every index j of the query Q has to be mapped onto exactly the same index of the subject sequence $S^{(i)}$. This implies, that we have to know the correspondence of time stamps beforehand. Unfortunately, that is not necessarily the case when processing signals from motion sensors or audio recordings since humans tend to perform motions at varying speeds or pronounce syllables of spoken words differently. DTW determines the unknown correspondence of indices by locally shrinking and stretching the time axis.

As an example, consider two voice recordings of the word "exact." The online service of Oxford Learner's Dictionaries [18] specifies the same pronunciation for both American and British English. However, the provided recordings differ slightly in the length of syllables. The American speaker

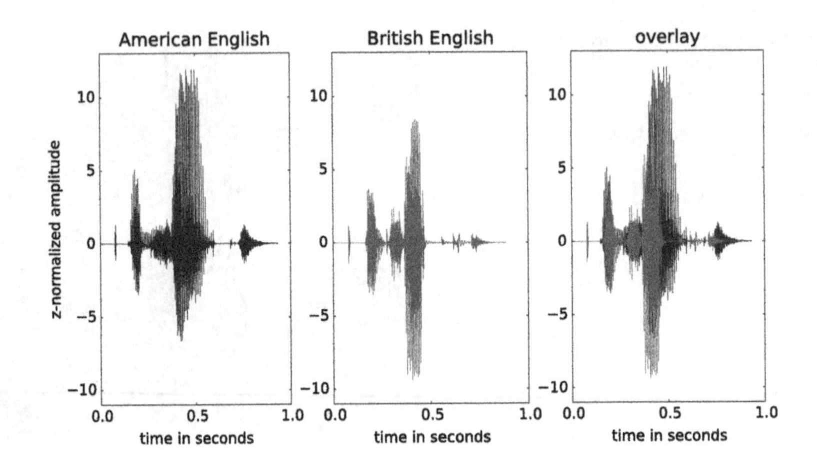

FIGURE 7.13

An example for two voice records of the word "exact" in American (left panel) and British (middle) English pronunciation taken from the online service of Oxford Learner's Dictionaries. The time series were sampled at 40,100 Hz for roughly one second. Both time series are aligned at the initial spike. The overlay of both signals emphasizes the difference in local phases.

spends more time on the "a" than his British counterpart. Fig. 7.13 visualizes both voice records and a direct comparison of both time series. Obviously, both recordings have similar shape but occasionally tend to be out of phase. Thus a rigid one-to-one mapping of indices is apparently not advisable. The simplest approach to fix the phase shifts would be the local adjustment of the time-axis in such a manner that both speakers spend the same time on each syllable. In 1994, Berndt and Clifford introduced such a time-warping mechanism to the data mining community [1] which was developed before by Sakoe and Chiba in 1978 [19] in the field of speech recognition. Fig. 7.14 illustrates the idea of local shrinking and stretching of the time-axis for the two given voice recordings.

In the following, we develop a formal description for the dynamic deformation of the time-axis. However, we do not want to allow arbitrary mappings between the index sets of the query Q and the subject $S^{(i)}$ and therefore impose the following restriction. Consider the Cartesian product of index domains $\mathcal{J} = (0, \ldots, j, \ldots, n-1)$ and $\mathcal{J}' = (0, \ldots, j', \ldots, n-1)$ of all possible combinations of time stamps stemming from the query and subject sequence

$$\mathcal{J} \times \mathcal{J}' = \{(j, j') \mid j \in \mathcal{J} = \text{dom}(Q) \text{ and } j' \in \mathcal{J}' = \text{dom}(S^{(i)})\} \quad , \tag{7.18}$$

then we are interested in a path that begins at the node $(j = 0, j' = 0)$ and ends at $(j = n-1, j' = n-1)$. This implies that the first/last entries of both time series are matched onto each other resulting in a global alignment of time series. During the further matching of indices, we impose that beginning from $(0, 0)$ we have to increment either the indices j, or j', or both by one until we finally reach $(n-1, n-1)$. Loosely speaking, we locally decide either to forward the time of the query, the subject, or both. Hence, each index in the query and the subject is at least matched once. Moreover, this monotonic mapping of indices prohibits loops resulting in a directed acyclic graph (DAG) with nodes $V = \{(j, j')\}$

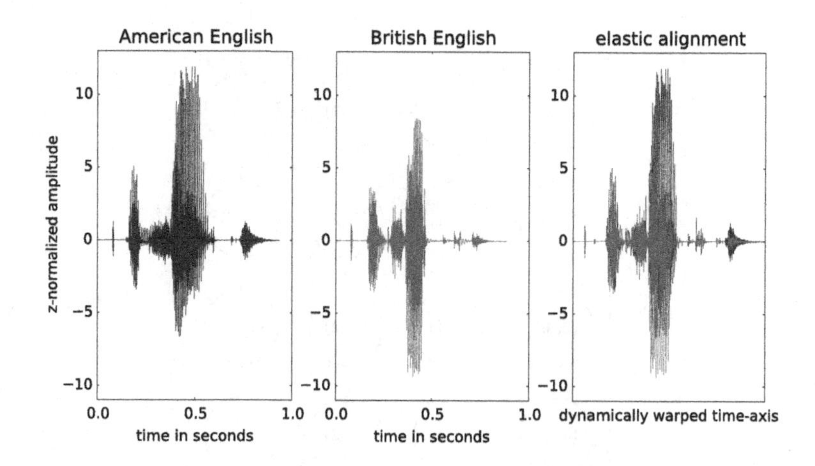

FIGURE 7.14

The same example for two voice records of the word "exact" in American (left panel) and British (middle) English pronunciation taken from the online service of Oxford Learner's Dictionaries. In the right panel, we have locally dilated and contracted the time-axis to find an optimal alignment in terms of least square error. As a result, both voice recordings have an intuitive alignment.

and directed edges from the set $E \subseteq V \times V$ where

$$E = \Big\{((j, j'), (j+1, j')),\ ((j, j'), (j, j'+1)),\ ((j, j'), (j+1, j'+1))\Big\} \quad . \tag{7.19}$$

A warping path γ is a sequence of nodes $\big((0, 0), \ldots, (n-1, n-1)\big)$ from the set of all paths Γ which are compatible with the described edge scheme. Let us summarize that with a definition:

Definition 2. Let $\mathcal{J} := \mathrm{dom}(Q)$ and $\mathcal{J}' := \mathrm{dom}(S^{(i)})$ be the index sets of the time series Q and $S^{(i)}$, then the sequence of tuples $\gamma := \big((j_l, j'_l) \in \mathcal{J} \times \mathcal{J}'\big)_l$ is a monotone, continuous, and bounded warping path if and only if

$$\begin{aligned} &\min\,(j_{l+1} - j_l,\ j'_{l+1} - j'_l) \geq 0 \quad \wedge \\ &\max(j_{l+1} - j_l,\ j'_{l+1} - j'_l) = 1 \quad \text{for all } l \in \{0, \ldots, |\gamma| - 2\}\,, \end{aligned} \tag{7.20}$$

where $(j_0, j'_0) = (0, 0)$ and $(j_{|\gamma|-1}, j'_{|\gamma|-1}) = (|Q| - 1, |S^{(i)}| - 1)$.

Finally, we define a weighting function that assigns the squared distance between the corresponding values belonging to the incoming (right) node of each edge

$$w : \mathcal{J} \times \mathcal{J}' \to \mathbb{R}_0^+\,,\ (j, j') \mapsto w(j, j') = (Q_j - S^{(i)}_{j'})^2 \quad . \tag{7.21}$$

Usually, the weighting function is designed to express similarity. In our case a weight is zero if the difference of combined time series entries vanishes. By contrast, alignment algorithms from the field of Bioinformatics such as Needleman–Wunsch (NW) or Smith–Waterman (SW) assign positive/high

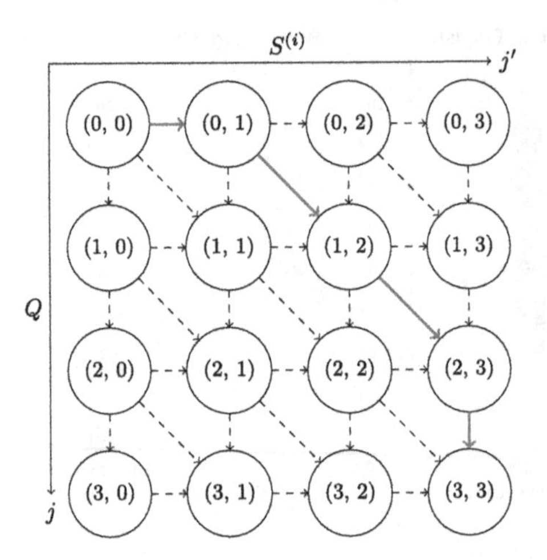

FIGURE 7.15

An example for the DAG representation of the optimization problem for the DTW distance measure between two time series of length four. The nodes are connected by horizontal, vertical and diagonal edges of step size one (continuity). Each edge increments at least one index (monotony). Due to the bounding property of DTW the optimal warping path (bold path) starts in the upper left and ends in the lower right cell.

values for matches and negative/small values for mismatches or insertion and deletions. Thus DTW aims to minimize while NW and SW maximize weights. Concluding, the objective of DTW is to find the optimal warping path $\hat{\gamma}$ out of all valid paths Γ that minimizes the accumulated weights along the matched nodes.

Definition 3 (DTW). Let Γ be the set of all monotone, continuous, and bounded warping paths. The optimal warping path $\hat{\gamma}$ and its associated measure $\hat{d}$ with respect to a given weighting function $w : \mathcal{J} \times \mathcal{J}' \to \mathbb{R}_0^+$ are defined as:

$$\hat{\gamma} := \underset{\gamma \in \Gamma}{\operatorname{argmin}} \sum_{(j,j') \in \gamma} w(j, j') \quad \text{and} \quad \hat{d} := \min_{\gamma \in \Gamma} \sum_{(j,j') \in \gamma} w(j, j') \quad . \tag{7.22}$$

Fig. 7.15 depicts a sketch of the described graph representation. This optimization problem is equivalent to the calculation of a shortest path within the aforementioned DAG. Each cell (j, j') is represented by a node and has maximal three incoming edges with associated weights $w(j, j')$. The **Single-Source Shortest Path** problem on DAGs can be solved in linear time $\mathcal{O}(|V| + |E|)$ by the relaxation of nodes in topological order [2]. Fortunately, we do not need to compute a topological sorting beforehand since it is implicitly given by enumerating the cells in lexicographic order. We can consecutively relax the cells in row-major order without violating node dependencies (see Fig. 7.15). In practice the computation is achieved by **dynamic programming** using a matrix of size $(|Q| + 1) \times (|S^{(i)}| + 1) = (n + 1) \times (n + 1)$. The corresponding relaxation scheme can be written recursively:

$$M[j,j'] = w(j-1,j'-1) + \min \begin{cases} M[j-1,j'] \\ M[j,j'-1], \\ M[j-1,j'-1] \end{cases} \quad \text{where} \quad \begin{array}{l} M[0,0]=0 \\ M[0,j']=\infty \; \forall j' \geq 1 \\ M[j\,,0]=\infty \; \forall j \geq 1 \end{array}$$

As a result, the final DTW measure $M\big[|Q|, |S^{(i)}|\big] = M[n,n]$ can be computed in $\mathcal{O}\big(|Q| \cdot |S^{(i)}|\big) = \mathcal{O}\big(n^2\big)$ time.[5] The optimal warping path $\hat{\gamma}$ is determined by back-tracing the predecessor information stored in a separate matrix. The L_p-norm is a special case where γ consists only of nodes on the main diagonal. Hence, the Euclidean-flavored DTW measure fulfills the inequality $\text{DTW}\big(Q, S^{(i)}\big) \leq \text{ED}\big(Q, S^{(i)}\big)$ for all time series $Q, S^{(i)}$ of the same length.

A LINEAR MEMORY ALGORITHM FOR SEQUENTIAL DTW

Having revisited the theoretical fundamentals we can start to write a first sequential program that compares a query Q with m time series $S^{(i)}$ stored in a data matrix D. In our experiments we will process $m = 2^{20} = 1{,}048{,}576$ time series each of length $n = 128$ from the popular Cylinder–Bell–Funnel (CBF) dataset. CBF is a synthetically created set of time series consisting of three characteristic classes. The generating functions for the classes are given as follows [12]:

$$\begin{aligned} C(t) &:= (6+\eta) \cdot \chi_{[a,b]}(t) + \epsilon(t) &, \\ B(t) &:= (6+\eta) \cdot \chi_{[a,b]}(t) \cdot \tfrac{t-a}{b-a} + \epsilon(t) &, \\ F(t) &:= (6+\eta) \cdot \chi_{[a,b]}(t) \cdot \tfrac{b-t}{b-a} + \epsilon(t) &, \end{aligned} \tag{7.23}$$

where $t \in \{0, \ldots, 127\}$, $\chi_{[a,b]}$ the indicator function on $[a,b]$, η as well as $\epsilon(t)$ drawn from a standard normal distribution $\mathcal{N}(\mu = 0, \sigma = 1)$, a an integer uniformly drawn from the interval $[16,32]$, and $b-a$ an integer uniformly drawn from $[32,96]$. As a result, CBF exhibits a reasonable variability in the amplitudes (η) and time-dependent noise on the measurement-axis ($\epsilon(t)$) as well as variable length and position of the shape's support on the time-axis ($\chi[a,b]$). Fig. 7.16 depicts a family of time series for each of the three classes. For the sake of simplicity, we provide a `generate_cbf` method defined in the `cbf_generator.hpp` header file which allows for the convenient sampling of time series from the CBF dataset.

Now we implement a sequential version of DTW for arbitrary numerical data types that support addition, subtraction, and the minimum operation (see Listing 7.19). The query and subject are passed as pointers of type `value_t` and their length is encoded by the argument `num_features` of type `index_t`. In Line 11 we allocate the aforementioned penalty matrix M of shape $(n+1) \times (n+1)$. The initialization is accomplished in Lines 14–16 by setting every entry of the first row and first column to infinity, except the upper left corner $M[0,0]$ which is set to zero. This mimics the missing incoming edges at the border of the graph. The two for-loops traverse the nodes $(j, j') =$ (`row`, `col`) of the DAG in lexicographic order. The registers `diag`, `abve` and `left` store the accumulated cost of the diagonal

[5]Note, the shortest path algorithm is indeed linear in the number of nodes and edges of the graph. However, this number is proportional to $|Q| \cdot |S^{(i)}|$ which results in quadratic runtime in terms of lengths.

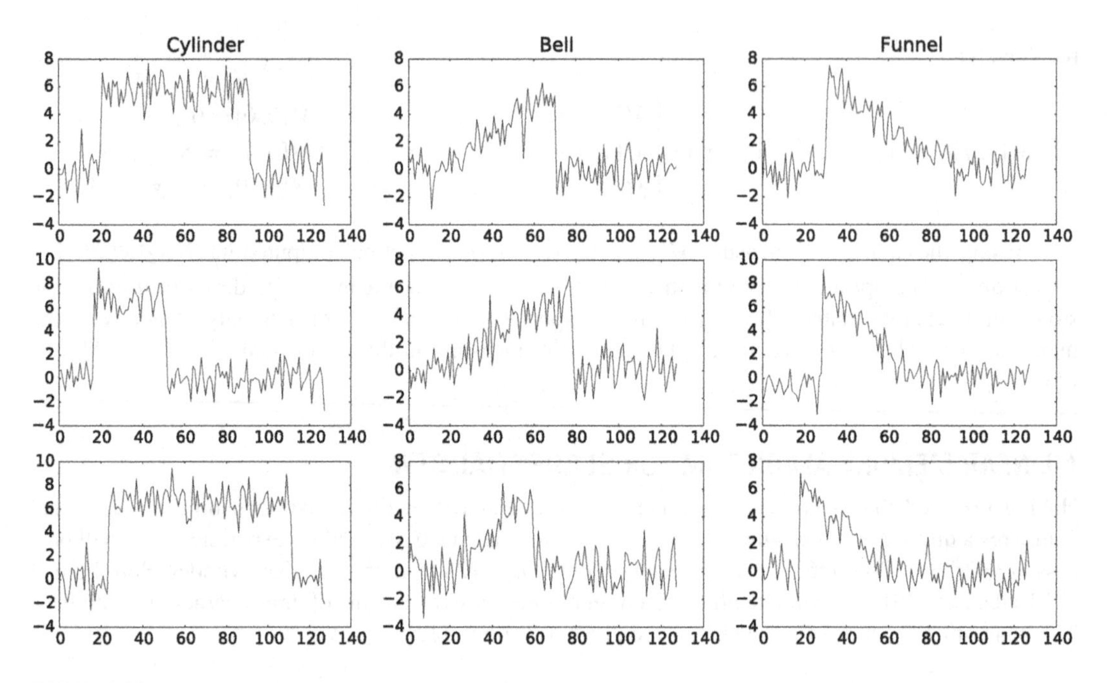

FIGURE 7.16

An example of three time series for each class of the CBF datasets. Note, the non-negligible variability in amplitudes, temporal noise and support.

cell $(j-1, j'-1)$, above cell $(j-1, j')$ and left cell $(j, j'-1)$ of M. Afterwards, their minimum is added to the weight $w(j-1, j'-1) =$ `residue * residue` of the corresponding node and subsequently stored in $M[j, j']$ (see Line 33). Recall that this greedy relaxation scheme produces the globally optimal solution on weighted DAGs since it respects the topological sorting [2]. Finally, the lower right corner of the penalty matrix $M[n, n]$ coincides with the measure of the shortest and thus optimal warping path $\hat{\gamma}$.

```
1  template <
2      typename index_t,
3      typename value_t> __host__
4  value_t plain_dtw(
5      value_t * query,            // pointer to query
6      value_t * subject,          // pointer to subject
7      index_t num_features) {     // number of time ticks (n)
8
9      // allocate the penalty matrix M of shape (n+1) x (n+1)
10     const index_t lane = num_features+1;
11     value_t * penalty = new value_t[lane*lane];
12
13     // initialize the matrix M
14     for (index_t index = 1; index < lane; index++)
15         penalty[index] = penalty[index*lane] = INFINITY;
```

```
16      penalty[0] = 0;
17
18      // traverse graph in row-major order
19      for (index_t row = 1; row < lane; row++) {
20          const value_t q_value = query[row-1];
21
22          for (index_t col = 1; col < lane; col++) {
23
24              // determine contribution from incoming edges
25              const value_t diag = penalty[(row-1)*lane+col-1];
26              const value_t abve = penalty[(row-1)*lane+col+0];
27              const value_t left = penalty[(row+0)*lane+col-1];
28
29              // compute residue between query and subject
30              const value_t residue = q_value-subject[col-1];
31
32              // relax node by greedily picking minimum edge
33              penalty[row*lane+col] = residue*residue +
34                                      min(diag, min(abve, left));
35          }
36      }
37      // report the lower right cell and free memory
38      const value_t result = penalty[lane*lane-1];
39      delete [] penalty;
40
41      return result;
42  }
```

Listing 7.19: Naive sequential implementation of the DTW similarity measure.

The plain DTW function can now be called in parallel using OpenMP or any other threading library. Note that the DTW call remains sequential.

```
1   #include <omp.h>                        // use -Xcompiler="-fopenmp"
2
3   template <
4       typename index_t,
5       typename value_t> __host__
6   void host_dtw(
7       value_t * query,                    // pointer to query
8       value_t * subject,                  // pointer to data matrix
9       value_t * dist,                     // pointer to distance array
10      index_t num_entries,                // number of entries (m)
11      index_t num_features) {             // number of time ticks (n)
12
13      # pragma omp parallel for
14      for (index_t entry = 0; entry < num_entries; entry++) {
15          const index_t off = entry*num_features;
16          dist[entry] = plain_dtw(query, subject+off, num_features);
17      }
18  }
```

Listing 7.20: Naive multithreaded implementation of the DTW similarity measure.

The major drawback of this implementation is the quadratic memory consumption of the cost matrix M. Assume we want to concurrently query a `float`-valued times series of length $n = 128$ against $m = 2^{20}$ time series then we have to allocate `sizeof(float)` $\cdot (n+1)^2 \cdot m \approx 65$ GB of memory. Fortunately, we can modify the quadratic memory algorithm to only use linear memory occupying `sizeof(float)` $\cdot 2 \cdot (n+1) \cdot m \approx 1$ GB which easily fits into the RAM of a GPU. This can be achieved as follows. The relaxation of a cell $M[i, j]$ only depends on two cells (diagonal and above) from the previous row and on one cell (left) in the same row. Thus it is sufficient to store merely two rows each of length $n+1$ and subsequently interleave them using a cyclic indexing scheme.

```
1  template <
2      typename index_t,
3      typename value_t> __host__
4  value_t dtw(
5      value_t * query,            // pointer to query
6      value_t * subject,          // pointer to subject
7      index_t num_features) {     // number of time ticks (n)
8
9      // allocate two rows of the penalty matrix of shape 2 x (n+1)
10     const index_t lane = num_features+1;
11     value_t * penalty = new value_t[2*lane];
12
13     // initialization is slightly different to the quadratic case
14     for (index_t index = 0; index < lane; index++)
15         penalty[index+1] = INFINITY;
16     penalty[0] = 0;
17
18     // traverse the graph in topologically sorted order
19     for (index_t row = 1; row < lane; row++) {
20
21         // here we have to compute cyclic indices (0,1,0,1,0,...)
22         const value_t q_value = query[row-1];
23         const index_t target_row = row & 1;
24         const index_t source_row = !target_row;
25
26         // this is crucial to reset the zero from row zero to inf
27         if (row == 2)
28             penalty[target_row*lane] = INFINITY;
29
30         // now everything as usual
31         for (index_t col = 1; col < lane; col++) {
32
33             // cyclic indices for the score matrix
34             const value_t diag = penalty[source_row*lane+col-1];
35             const value_t abve = penalty[source_row*lane+col+0];
36             const value_t left = penalty[target_row*lane+col-1];
37
38             // traditional indices for the time series
39             const value_t residue = q_value-subject[col-1];
40
41             // relax the cell
42             penalty[target_row*lane+col] = residue*residue +
```

```
43                                              min(diag,
44                                              min(abve, left));
45          }
46      }
47
48      // here we have to compute the index of the last row
49      const index_t last_row = num_features & 1;
50      const value_t result = penalty[last_row*lane+num_features];
51      delete [] penalty;
52
53      return result;
54  }
```

Listing 7.21: Linear memory implementation of the DTW similarity measure.

At this point, we can call the `host_dtw` method in the `main` function. The corresponding code snippet is shown in Listing 7.22. First, we allocate host-sided memory for the storage of the time series `data` and the distance vector `dist` in Lines 20–23. Second, we generate $m = 2^{20}$ database entries each of length $n = 128$ in Line 32 using the `generate_cbf` method defined in the header file `cbf_generator.hpp`. For the sake of simplicity, we have chosen $Q = S^{(0)}$ in this example. Hence, we pass the `data` pointer as first argument which is reserved for the query and, as expected, again as second argument which corresponds to the database. The array `labels` stores the class labels being either 0 (Cylinder), 1 (Bell), or 2 (Funnel). They are used later for a simple consistency check. Third, we compute all DTW distances using a call to `host_dtw` in Line 32. Finally, we print some distance scores and clean up the memory.

```
1   #include "../include/cbf_generator.hpp" // generate_cbf
2   #include "../include/hpc_helpers.hpp"   // timers
3
4   // define the used data types
5   typedef uint64_t index_t;
6   typedef uint8_t  label_t;
7   typedef float    value_t;
8
9   int main () {
10
11      // the shape of the data matrix
12      constexpr index_t num_features = 128;
13      constexpr index_t num_entries  = 1UL << 20;
14
15      // small letters for host arrays
16      value_t * data   = nullptr, * dist = nullptr;
17      label_t * labels = nullptr;  // C,B,F encoded as 0,1,2
18
19      // malloc memory
20      cudaMallocHost(&data,   sizeof(value_t)*num_features
21                                             *num_entries);  CUERR
22      cudaMallocHost(&dist,   sizeof(value_t)*num_entries);  CUERR
23      cudaMallocHost(&labels, sizeof(label_t)*num_entries);  CUERR
24
25      // create CBF data set on host
26      TIMERSTART(generate_data)
```

```
27      generate_cbf(data, labels, num_entries, num_features);
28      TIMERSTOP(generate_data)
29
30      // here we use the first time series of data as query
31      TIMERSTART(DTW_openmp)
32      host_dtw(data, data, dist, num_entries, num_features);
33      TIMERSTOP(DTW_openmp)
34
35      // let us print some distance values
36      for (index_t index = 0; index < 9; index++)
37          std::cout << index_t(labels[index]) << " "
38                    << dist[index] << std::endl;
39
40      // get rid of the memory
41      cudaFreeHost(labels);                                CUERR
42      cudaFreeHost(data);                                  CUERR
43      cudaFreeHost(dist);                                  CUERR
44  }
```

Listing 7.22: Host implementation of one-versus-all DTW computation.

When executing this program on a dual socket Intel Xeon E5-2683 v4 (2.1 GHz) CPU we need roughly 92 seconds to sequentially compute the 2^{20} DTW scores using only the master thread. The multithreaded variant which harnesses all 2×16 physical cores scales almost linearly resulting in approximately 2.9 seconds execution time. The program's output shows the first nine computed distance scores (right column) with their corresponding class labels (left column). As expected, the distance between the query $Q = S^{(0)}$ and the first database entry $S^{(0)}$ vanishes. Moreover, we observe that the DTW distance to other Cylinder instances (0) is significantly smaller than to Bell (1) and Funnel (2) entries.

```
TIMING: 901.758 ms (generate_data)
TIMING: 2894.01 ms (DTW_openmp)
0 0
1 175.457
2 319.968
0 6.02721
1 158.446
2 281.643
0 18.9647
1 157.522
2 179.027
```

A common performance metric for dynamic programming-based alignment algorithms is the number of cell updates per second (CUPS). In our case, we have relaxed 128^2 cells in each of the 2^{20} score matrices which results in 190 MCUPS for the sequential version using a single core and 6 GCUPS for the multithreaded variant harnessing all cores of two Xeon CPUs. In the following, we will develop a CUDA parallelization that performs at 34 GCUPS on a single Pascal-based Titan X GPU.

A NAIVE CUDA PORT OF LINEAR MEMORY DTW

Assume we have already copied the host-sided database `data` from the host to an appropriate device array `Data` and further allocated memory for the device-sided distance vector `Dist`. An obvious parallelization strategy would be the computation of one DTW score per CUDA thread. Consequently, we have to allocate $2 \cdot (n + 1)$ cells of the score matrix in each thread when employing the aforementioned linear memory scheme. Conceptually, we already know four kinds of user-modifiable memory:

- **global memory:** We could pass an auxiliary array `Cache` to the kernel which resides in the global memory of the GPU in order to store the score matrices.
- **local memory:** The same could be achieved by defining static arrays, so-called local memory, inside the kernel body.
- **shared memory:** Alternatively, we could reserve scratchpad memory to benefit from faster access times. Unfortunately, we have to deal with its small size.
- **registers:** The reasonable amount of registers per SM could be used to store the score matrix. However, it is not obvious how to enumerate them in a for-loop.

In this subsection, you will learn how to employ the first two strategies. Furthermore, we will demonstrate, that in most cases accesses to local memory coincide with interleaved accesses to global memory, a fact that is surprisingly unknown among CUDA programmers. In detail, we will investigate another example for coalesced and non-coalesced memory access patterns which are now hidden behind primitives of the CUDA programming language.

Let us start coding. The signature of the kernel in Listing 7.23 is straightforward: we pass the device pointers to the query of length n, the subject database of shape $m \times n$, the distance vectors of length m, and the auxiliary memory `Cache` of shape $m \times 2 \cdot (n + 1)$. Afterwards, we compute the global thread identifier `thid` which enumerates the time series in the database and determine the index offset `base` addressing the first time tick of each subject sequence. The constant `lane` is defined for convenience to simplify the indexing scheme.

```
1  template <
2      typename index_t,
3      typename value_t> __global__
4  void DTW_naive_kernel(
5      value_t * Query,          // pointer to the query
6      value_t * Subject,        // pointer to the database
7      value_t * Dist,           // pointer to the distance
8      value_t * Cache,          // auxiliary memory for matrices
9      index_t num_entries,      // number of time series (m)
10     index_t num_features) {   // number of time ticks (n)
11
12     // compute global thread indentifier, lane length and offset
13     const index_t thid = blockDim.x*blockIdx.x+threadIdx.x;
14     const index_t lane = num_features+1;
15     const index_t base = thid*num_features;
```

Listing 7.23: Identification of thread id in the naive DTW kernel utilizing external memory.

At this point, we have to check if the thread identifier `thid` is within the range of allowed indices. The remaining code is almost identical to the host-sided linear memory implementation in Listing 7.21. The only differences are the missing memory deallocation and the distance assignment at the end of the kernel.

```
16      // prevent malicious memory accesses
17      if (thid < num_entries) {
18
19          // set penalty to the correct position in memory
20          value_t * penalty = Cache + thid*2*lane;
21
22          // init penalty matrix
23          penalty[0] = 0;
24          for (index_t index = 0; index < lane; index++)
25              penalty[index+1] = INFINITY;
26
27          // relax the graph in topologically sorted order
28          for (index_t row = 1; row < lane; row++) {
29
30              const value_t q_value = Query[row-1];
31              const index_t target_row = row & 1;
32              const index_t source_row = !target_row;
33
34              if (row == 2)
35                  penalty[target_row*lane] = INFINITY;
36
37              for (index_t col = 1; col < lane; col++) {
38
39                  const index_t src_off = source_row*lane;
40                  const index_t trg_off = target_row*lane;
41
42                  const value_t diag = penalty[src_off+col-1];
43                  const value_t abve = penalty[src_off+col-0];
44                  const value_t left = penalty[trg_off+col-1];
45
46                  const value_t s_value = Subject[base+col-1];
47                  const value_t residue = q_value - s_value;
48
49                  penalty[target_row*lane+col] = residue * residue
50                                               + min(diag,
51                                                 min(abve, left));
52              }
53          }
54
55          // write down the result
56          const index_t last_row = num_features & 1;
57          Dist[thid] = penalty[last_row*lane+num_features];
58      }
59  }
```

Listing 7.24: Main computation in the naive DTW kernel utilizing external memory.

The kernel can be called from within the main function as follows:

```
const uint64_t threads = 32;
DTW_naive_kernel<<<SDIV(num_entries, threads), threads>>>
    (Data, Data, Dist, Cache, num_entries, num_features);
```

The runtime of this kernel is roughly 30 seconds on a Pascal-based Titan X. This is three times faster than the sequential computation from the previous subsection. Nevertheless, it performs ten times worse than the multithreaded variant harnessing all 2×16 cores of a dual socket Xeon CPU. Actually, this is caused by a suboptimal memory access pattern.

An alternative approach would be the use of static arrays defined inside the kernel, so-called local memory. The signature of the kernel is similar to the previous one with the exception that we do not need to pass the array `Cache`. Auxiliary memory is now defined inside the kernel on a per thread basis as shown in Line 20 of Listing 7.25. Each CUDA thread can access its own instance of the score matrix `penalty`. Similarly to the definition of shared memory inside a kernel, the size of the array must be known at compile time. This can be realized by passing the number of time ticks as an additional template parameter `const_num_features` (see Line 4).

```
1  template <
2      typename index_t,
3      typename value_t,
4      index_t const_num_features> __global__
5  void DTW_static_kernel(
6      value_t * Query,           // pointer to the query
7      value_t * Subject,         // pointer to the database
8      value_t * Dist,            // pointer to the distance
9      index_t num_entries,       // number of time series (m)
10     index_t num_features) {    // number of time ticks (n)
11
12     // compute global thread indentifier, lane length and offset
13     const index_t thid = blockDim.x*blockIdx.x+threadIdx.x;
14     const index_t lane = num_features+1;
15     const index_t base = thid*num_features;
16
17     if (thid < num_entries) {
18
19         // define penalty matrix as thread-local memory
20         value_t penalty[2*(const_num_features+1)];
21
22         // fill in here the initialization and the relaxation
23         // steps from the previous kernel (Lines 22--53)
24
25         // write down the result
26         const index_t last_row = num_features & 1;
27         Dist[thid] = penalty[last_row*lane+num_features];
28     }
29 }
```

Listing 7.25: Naive DTW kernel utilizing thread-local memory.

The corresponding kernel call additionally specifies the data types and `num_features` as compile time constant in single angle brackets:

```
const uint64_t threads = 32;
DTW_static_kernel<uint64_t, float, num_features>
        <<<SDIV(num_entries, threads), threads>>>
        (Data, Data, Dist, num_entries, num_features);
```

Surprisingly, this kernel terminates after merely 2.7 seconds being on a par with the multithreaded CPU implementation. But how can we explain this massive difference in runtime of more than one order-of-magnitude? To answer this question, we have to understand how CUDA maintains local memory. Some programmers falsely claim that arrays in local memory are always stored in fast registers and thus memory accesses must be significantly faster. However, an all-encompassing answer is more complicated. CUDA may assign arrays defined inside a kernel to registers if their size is reasonably small but tends to put long arrays (like in our case) in ordinary global memory. The only way to validate an assignment to registers is a careful inspection of PTX ISA instructions – the CUDA equivalent of assembly code. However, an exhaustive introduction to PTX ISA is way beyond the scope of this book and thus we assume for simplicity that local memory resides in global memory. Consequently, the massive difference in execution time is caused by something else.

Local memory when stored in global memory has a different layout than our naive aliasing of auxiliary memory using the array `Cache`. In the naive case, the two rows of our score matrices are stored in chunks of size $c := 2 \cdot (n+1)$. Consider the case where we simultaneously access the upper left entry of the score matrix `penalty[0]` during initialization. The 32 threads of a warp concurrently write to memory positions which are c slots apart. Hence, we cannot benefit from caching when accessing the entries of the score matrices. By contrast, local memory exhibits a distinct layout which interleaves the entries, i.e., all upper left entries `penalty[0]` are stored in contiguous positions of the memory followed by all entries `penalty[1]` and so forth. This coalesced access pattern results in significantly faster execution times since we can benefit from optimal cache utilization of global memory. Hence, a better term to describe the characteristics of local memory would be *thread-local interleaved global memory*. We can easily validate this by changing the addressing of the auxiliary array `Cache`. This can be achieved by defining an index transformation `iota` which reorders the memory accesses analogically to local memory (see Line 19 in Listing 7.26).

```
1  template <
2      typename index_t,
3      typename value_t> __global__
4  void DTW_interleaved_kernel(
5      value_t * Query,          // pointer to the query
6      value_t * Subject,        // pointer to the database
7      value_t * Dist,           // pointer to the distance
8      value_t * Cache,          // auxiliary memory for matrices
9      index_t num_entries,      // number of time series (m)
10     index_t num_features) {   // number of time ticks (n)
11
12     // compute global thread indentifier, lane length and offset
13     const index_t thid = blockDim.x*blockIdx.x+threadIdx.x;
14     const index_t lane = num_features+1;
15     const index_t base = thid*num_features;
16
17     // define lambda for local index transposition
18     // resulting in a coalesced memory access pattern
```

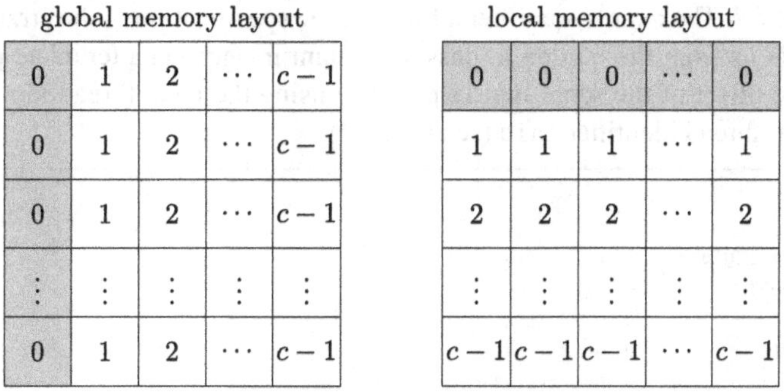

global memory layout

0	1	2	⋯	$c-1$
0	1	2	⋯	$c-1$
0	1	2	⋯	$c-1$
⋮	⋮	⋮	⋮	⋮
0	1	2	⋯	$c-1$

local memory layout

0	0	0	⋯	0
1	1	1	⋯	1
2	2	2	⋯	2
⋮	⋮	⋮	⋮	⋮
$c-1$	$c-1$	$c-1$	⋯	$c-1$

FIGURE 7.17

Memory layouts of the score matrices for the naive chunk based approach using the auxiliary array `Cache` residing in global memory (left) and the interleaved variant employed by local memory (right). The latter guarantees coalesced access during relaxation of the $c = 2 \cdot (n + 1)$ cells (two lanes). The gray shaded cells are simultaneously accessed by all threads of a warp.

```
19      auto iota = [&] (const index_t& index)
20                      {return index*num_entries+thid;};
21
22      if (thid < num_entries) {
23
24          // set penalty to Cache without offset
25          value_t * penalty = Cache;
26
27          // fill in initialization and relaxation from
28          // DTW_naive_kernel (Lines 22-57) and substitute
29          // penalty[x] with penalty[iota(x)]
30
31      }
32  }
```

Listing 7.26: Naive DTW kernel utilizing interleaved global memory.

This kernel performs at the same speed as the static kernel which employs local memory. Concluding, our take-home message is that local memory coincides for long arrays with device memory using interleaved addressing. Hence, it has the same high latency and low memory bandwidth as global memory. (See Fig. 7.17.)

WAVEFRONT RELAXATION IN SHARED MEMORY

Shared memory can be employed to drastically speed up the computation of memory-bound algorithms as we have already seen in Section 7.7. Shared memory is block-local memory in contrast to local memory which is thread-local. Thus we have to slightly modify the indexing scheme since all threads within a CUDA thread block share the same memory. Listing 7.27 shows the implementation of the corresponding kernel. The allocation of the memory in Line 17 differs slightly from our previous approach in Section 7.7. The `extern` qualifier allows for the specification of the array size during kernel launch which reliefs us from the burden to pass a constant integer via a template parameter. Furthermore, we adjust the offset of the score matrix `penalty` using the local thread identifier `threadIdx.x` instead of the global thread identifier `thid` (see Line 22).

```
1   template <
2       typename index_t,
3       typename value_t> __global__
4   void DTW_shared_kernel(
5       value_t * Query,           // pointer to the query
6       value_t * Subject,         // pointer to the database
7       value_t * Dist,            // pointer to the distance
8       index_t num_entries,       // number of time series (m)
9       index_t num_features) {    // number of time ticks (n)
10
11      // compute global thread indentifier, lane length and offset
12      const index_t thid = blockDim.x*blockIdx.x+threadIdx.x;
13      const index_t lane = num_features+1;
14      const index_t base = thid*num_features;
15
16      // define array in shared memory with externally defined size
17      extern __shared__ value_t Cache[];
18
19      if (thid < num_entries) {
20
21          // set penalty to the correct position in memory
22          value_t * penalty = Cache + threadIdx.x*(2*lane);
23
24          // init penalty matrix
25          penalty[0] = 0;
26          for (index_t index = 0; index < lane; index++)
27              penalty[index+1] = INFINITY;
28
29          // relax graph in topologically sorted order
30          for (index_t row = 1; row < lane; row++) {
31
32              const value_t q_value = Query[row-1];
33              const index_t target_row = row & 1;
34              const index_t source_row = !target_row;
35
36              if (row == 2)
37                  penalty[target_row*lane] = INFINITY;
38
39              for (index_t col = 1; col < lane; col++) {
40
41                  const index_t src_off = source_row*lane;
42                  const index_t trg_off = target_row*lane;
43
44                  const value_t diag = penalty[src_off+col-1];
45                  const value_t abve = penalty[src_off+col-0];
46                  const value_t left = penalty[trg_off+col-1];
47
```

```
48                const value_t s_value = Subject[base+col-1];
49                const value_t residue = q_value - s_value;
50
51                penalty[target_row*lane+col] = residue * residue
52                                             + min(diag,
53                                               min(abve, left));
54            }
55        }
56
57        const index_t last_row = num_features & 1;
58        Dist[thid] = penalty[last_row*lane+num_features];
59    }
60 }
```

Listing 7.27: Naive DTW kernel utilizing shared memory.

The corresponding kernel call specifies the size of the reserved shared memory in bytes with a third parameter in triple angle brackets:

```
uint64_t threads = 32;
uint64_t sh_mem  = 2*(num_features+1)*threads*sizeof(float);
DTW_shared_kernel<uint64_t, float>
    <<<SDIV(num_entries, threads), threads, sh_mem>>>
    (Data, Data, Dist, num_entries, num_features);
```

This kernel terminates after approximately 630 ms when executed on a Pascal-based Titan X resulting in a speedup of roughly four over the local memory implementation. Unfortunately, the use of shared memory imposes a major limitation. We cannot use this approach when processing significantly longer time series. The size of sh_mem is $2 \cdot (n+1) \cdot$ blockDim.x $\cdot$ sizeof(float) $= 2 \cdot (128+1) \cdot 32 \cdot 4 \approx$ 32.25 KB in our example, i.e., it fits within the 48 KB limitation of allowed shared memory per block, however, for $n = 192$ we exceed this value.

To overcome this limitation, a distinct parallelization technique can be employed – the so-called wavefront relaxation scheme. The naive approach either using global, local, or shared memory employs a coarse-grained parallelization scheme processing one DTW score per thread. By contrast, wavefront relaxation is based on a fine-grained parallelization strategy where one CUDA thread block computes one DTW score. As a consequence, we have significantly more shared memory left per block allowing for the computation of longer time series. Moreover, we can utilize more than one warp within a thread block. A disadvantage is that we can neither control nor deduce the execution order of threads within a block in order to respect cell dependencies in the graph structure. Wavefront relaxation reorders the cell updates in such a manner that all threads simultaneously work on independent entries of the score matrix. A close look at Fig. 7.15 reveals that we can concurrently relax all nodes which are located in the same minor diagonal (lower left to upper right) of the penalty matrix without breaking dependencies. The $2 \cdot n + 1$ minor diagonals can be updated one by another since an entry only depends on three entries stored in the two previous diagonal lanes. Fig. 7.18 visualizes the described access scheme. Unfortunately, the memory and compute footprint of wavefront relaxation is slightly higher than the traditional approach. We have to store three lanes of length $(n+1)$ instead of two and moreover the number of $(2 \cdot n + 1) \cdot (n+1) = 2 \cdot n^2 + 3 \cdot n + 1$ relaxed cells is almost two times higher than the naive scheme which updates only $(n+1)^2 = n^2 + 2 \cdot n + 1$ cells. Nevertheless, we will observe that the described strategy is highly competitive to the naive approach.

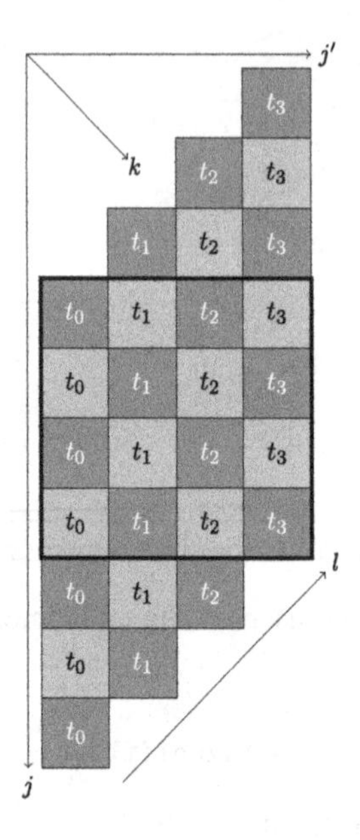

wavefront relaxation

local update

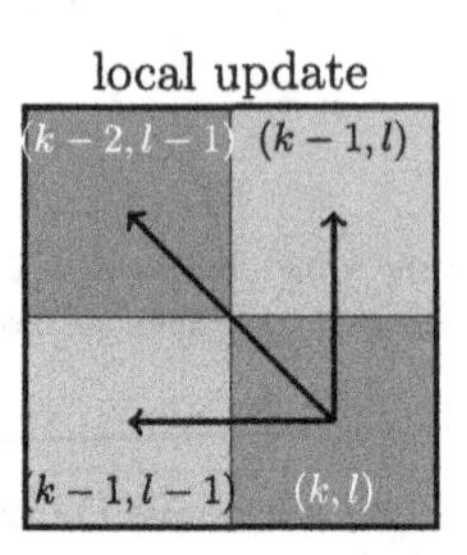

index transformation:

$$(k,l) \mapsto (j,j') = (k-l,l)$$
$$(j,j') \mapsto (k,l) = (j+j',j')$$

FIGURE 7.18

An example for wavefront relaxation of a $(n+1) \times (n+1) = 4 \times 4$ score matrix. Cells on minor diagonals (lower left to upper right) can be updated concurrently without violating dependencies in the graph structure. To achieve that, the traditional index scheme enumerating rows and columns (j, j') has to be rewritten to address the $2 \cdot n + 1$ minor diagonals by the index k. The n entries in each minor diagonal are processed by distinct threads t_l. The described scheme can be realized in linear memory by accessing three consecutive minor diagonals in cyclic order.

Let us start coding. Listing 7.28 shows the kernel signature and the initialization phase of the DTW algorithm. First, we define the block identifier `blid` which enumerates the subject sequences (one DTW score per block) and the local thread identifier `thid` which is used to address the cells on a diagonal in Lines 12–13. Second, the constant `lane` is defined for convenient indexing and the offset `base` is adjusted to use the block identifier instead of the global thread identifier (see Lines 16–17). Third, we reserve three lanes of shared memory using the `extern` and `__shared__` keywords (see Line 20). The size of shared memory will be later specified as kernel launch parameter. Subsequently, the lanes are initialized with infinity except for the first entry (upper left cell) which is set to zero (see Lines 20–31). Finally, we synchronize all threads within a thread block using a block-wide barrier in Line 34. This step is crucial since we want to ensure that all threads have finished initialization before proceeding with the relaxation of the score matrix. Note that the call to `__syncthreads()` is redundant when utilizing 32 or less threads since all threads within a warp are always synchronized up to CUDA 8. Starting from CUDA 9 and its novel cooperative group paradigm, this is not guaranteed anymore. As a consequence, we strongly advise to always synchronize threads explicitly in future code!

```
1  template <
2      typename index_t,
3      typename value_t> __global__
```

```
4  void DTW_wavefront_kernel(
5      value_t * Query,           // pointer to the query
6      value_t * Subject,         // pointer to the database
7      value_t * Dist,            // pointer to the distance
8      index_t num_entries,       // number of time series (m)
9      index_t num_features) {    // number of time ticks (n)
10
11     // compute block and local thread identifier
12     const index_t blid = blockIdx.x;
13     const index_t thid = threadIdx.x;
14
15     // calculate lane length and time series offset
16     const index_t lane = num_features+1;
17     const index_t base = blid*num_features;
18
19     // define score matrix in shared memory
20     extern __shared__ value_t Cache[];
21     value_t * penalty = Cache;
22
23     // initialize score matrix with infinity
24     for (index_t l = thid; l < lane; l += blockDim.x) {
25         penalty[0*lane+l] = INFINITY;
26         penalty[1*lane+l] = INFINITY;
27         penalty[2*lane+l] = INFINITY;
28     }
29
30     // upper left corner set to zero
31     penalty[0*lane+0] = 0;
32
33     // force all threads within a block to synchronize
34     __syncthreads();
```

Listing 7.28: Initialization of the wavefront DTW kernel utilizing shared memory.

At this point, we can start relaxing the score matrix. The $2 \cdot (n + 1) - 1$ diagonal lanes are enumerated with the index k. Next, the cyclic lane indices have to be computed: the current lane $k\%3$ is denoted as `target_row`, the row before $(k - 1)\%3$ as `source_row`, and the row before that row $(k - 2)\%3$ as `before_row`. The use of the ternary conditional statement `( )?( ):( )` is advisable since it reduces the number of modulo operations and can be easily optimized by the compiler. The number of expensive modulo operations could be reduced further by moving the initialization of `target_row` outside of the outer loop. The inner loop in Line 47 enumerates the cells with the index l using a block-cyclic distribution of length `blockDim.x`. The loop body consists of four steps. First, we compute the traditional indices (j, j') in Lines 50–51 in order to access the time series entries in a later stage. Second, we determine if the index combination (k, l) corresponds to a valid position (j, j') in the score matrix (see Line 54). Third, the residual value for the index combination (j, j') is determined if the cell is located inside the score matrix. In the opposite case it is set to infinity. Fourth, cells within a diagonal are updated concurrently using the access pattern shown in Fig. 7.18. The block-wide barrier in Line 73 ensures that all cells within a lane have been updated before proceeding with the next diagonal. Finally, we write down the result stored in the lower right cell of the score matrix (see Line 77).

```
36      // relax diagonals
37      for (index_t k = 2; k < 2*lane-1; k++) {
38
39          // compute cyclic lane indices
40          const index_t target_row = k % 3;
41          const index_t before_row = target_row == 2 ? 0 :
42                                     target_row + 1;
43          const index_t source_row = before_row == 2 ? 0 :
44                                     before_row + 1;
45
46          // each thread updates one cell
47          for (index_t l = thid; l < lane; l += blockDim.x) {
48
49              // compute traditional indices (j, j') from (k, l)
50              const index_t j = k-l;
51              const index_t J = l;
52
53              // determine if indices are outside of score matrix
54              const bool outside = k <= l || J == 0 || j >= lane;
55
56              // compute the residue Q_{j-1} - S^{(i)}_{j'-1}
57              const value_t residue = outside ? INFINITY :
58                                      Query[j-1]-Subject[base+J-1];
59
60              // concurrently relax the cells
61              const index_t bfr_off = before_row*lane;
62              const index_t src_off = source_row*lane;
63              const index_t trg_off = target_row*lane;
64
65              penalty[trg_off+l] = outside ? INFINITY :
66                                   residue*residue
67                                 + min(penalty[bfr_off+l-1],
68                                   min(penalty[src_off+l+0],
69                                       penalty[src_off+l-1]));
70          }
71
72          // force all threads within a block to synchronize
73          __syncthreads();
74      }
75
76      const index_t last_diag = (2*num_features) % 3;
77      Dist[blid] = penalty[last_diag*lane+num_features];
78  }
```

Listing 7.29: Main computation in the wavefront DTW kernel utilizing shared memory.

The kernel call has to be modified to spawn one thread block for each database entry. Furthermore, we have to adjust the shared memory size.

```
uint64_t threads = 32;
uint64_t sh_mem  = 3*(num_features+1)*sizeof(float);
DTW_wavefront_kernel<<<num_entries, threads, sh_mem>>>
    (Data, Data, Dist, num_entries, num_features);
```

The execution time of 940 ms is higher than the 630 ms for the naive kernel due to the increased number of relaxed cells. However, the kernel can now process time series up to a length of $n = 4095$ in comparison to the naive kernel which is limited to a maximum length of $n = 191$.

CONCURRENT SCHEDULING AND BANK CONFLICTS

Before we proceed with further optimization of the wavefront kernel let us briefly analyze the measured execution times. The wavefront kernel is only 1.5 times slower than the naive kernel employing shared memory despite the fact it updates the double amount of cells. Usually, it is hard to pinpoint one explicit reason which causes deviations in estimated runtime since modern GPUs are complex dynamic systems which cannot be described by simple performance models. However, we can make educated guesses based on frequent observations.

As an example, we know that the block scheduler may assign several thread blocks onto the same SM if it provides sufficient resources for execution. The naive kernel needs roughly 33 KB of shared memory to store two lanes of length 129 for each of the 32 processed time series. Thus an SM with less than 66 KB of shared memory cannot simultaneously process two blocks. The Pascal generation features only 64 KB of shared memory per SM in contrast to some Maxwell GPUs (e.g. GM200) providing up to 96 KB. As a consequence, we may encounter a performance loss of roughly two if exceeding half the size of physically provided memory. This theoretical model can be validated as follows. Starting from a kernel that uses no shared memory at all we define an array of shared memory within the kernel via `extern __shared__ dummy[]`. Now we can measure the execution time of the kernel for different shared memory sizes. The performance of the kernel will monotonically decrease if we increase the amount of scratchpad memory. The wavefront kernel occupies less than 2 KB of shared memory allowing for the concurrent execution of up to 24 blocks on the same SM. Due to this increased level of parallelism we expect a decrease in runtime.

Proper memory access patterns are another aspect of shared memory performance. Since the release of the Fermi generation, scratchpad is organized in 32 memory banks which are assigned to its entries in a block-cyclic fashion, i.e., reads and writes to a four-byte word stored at position k are handled by the memory bank $k\%32$. Thus memory accesses are serialized if two or more threads within a warp access the same bank resulting in performance degradation. An exception is the case where all threads in a warp address the same shared memory address, which is handled efficiently via a broadcast. As an example, consider the case where we process eight-byte words stored in shared memory (such as `int64_t` or `double`) using a memory bank width of four. Thread 0 loads the lower four bytes from memory bank zero and the upper portion from memory bank one. Analogically, Thread 1 loads from memory bank two and three and so forth. In the same warp, Thread 16 also uses the memory banks zero and one resulting in a so-called bank conflict. Fortunately, starting from the Kepler generation we can reconfigure the width of the memory banks to be either four or eight via `cudaDeviceSetSharedMemConfig()` on the whole device or on a per-kernel basis via `cudaFuncSetSharedMemConfig()` using the arguments

- `cudaSharedMemBankSizeDefault` (use the device default)
- `cudaSharedMemBankSizeFourByte` (4 B)
- `cudaSharedMemBankSizeEightByte` (8 B)

in order to address this problem. Note that before Kepler it was impossible to write a bank conflict-free algorithm that processes eight-byte words. The naive DTW kernel employing shared memory stores the two lanes of the score matrix with an offset of $c = 2 \cdot (128 + 1) = 258$ entries. As a consequence, we access memory banks with a stride of two since Thread 0 uses memory bank $(0 \cdot c)\%32 = 0$ for the entry $M[0, 0]$, Thread 1 uses memory bank $(1 \cdot c)\%32 = 2$ and so forth. By contrast, the wavefront kernel simultaneously accesses consecutive entries in shared memory avoiding any bank conflict. Concluding, the wavefront kernel has a smaller shared memory footprint which allows for the concurrent execution of several thread blocks on the same SM and further employs a better memory access pattern.

Another possibility to tweak shared memory shall not be unmentioned. CUDA allows for the reduction of manually maintained shared memory in favor of L1 cache which both share the same transistors on the chip. The scratchpad to L1 ratio can be configured globally using either `cudaDeviceSetCacheConfig()` or on a per-kernel level using `cudaFuncSetCacheConfig()`. Possible arguments for the preferred ratio (scratchpad / L1) are given as follows:

- `cudaFuncCachePreferNone` (use the device default)
- `cudaFuncCachePreferShared` (48 KB / 16 KB)
- `cudaFuncCachePreferL1` (16 KB / 48 KB)
- `cudaFuncCachePreferEqual` (32 KB / 32 KB)

Note that CUDA ignores your configuration if the amount of shared memory cannot be realized with the specified ratio. This level of optimization rarely affects overall performance and thus it is not advisable to employ it during early development stages of your program. A similar statement can be made about bank conflicts. In general, it is desirable to design bank conflict-free algorithms, however, it is not guaranteed to result in optimal code. As an example, if we transpose the shared memory layout of the naive kernel (similar to local memory) in order to avoid bank conflicts we end up with a slower implementation.

TEXTURE MEMORY AND CONSTANT MEMORY

We now optimize the wavefront kernel to outperform the naive variant by exploiting advanced caching mechanisms of CUDA-enabled GPUs. Due to their original purpose GPUs are specialized to efficiently process graphics. One important aspect of scene rendering is linear, bilinear, and trilinear (re-)interpolation of textures. In terms of programming, textures can be described as multidimensional arrays which exhibit a spatial locality property. As an example, the texture elements (texels) of a 2D texture T should be accessed efficiently in both dimensions, i.e., reading the entries $T[i, j]$, $T[i, j + 1]$, and $T[i + 1, j]$ should have approximately the same latency since they encode similar visual features in a neighborhood. However, we already know that $T[i, j]$ and $T[i, j + 1]$ most probably reside in the same cache line in contrast to $T[i, j]$ and $T[i + 1, j]$, which are stored far apart. Addressing this, CUDA-enabled GPUs provide efficient hardware-based texture interpolation mechanisms which makes use of intelligent caching strategies. Furthermore, textures can be evaluated at fractional texel positions using hardware-based multilinear interpolation or automatically map intensity values to the normalized interval $[0, 1]$. Although not used in our example, both features can be exploited to save some Flop/s in interpolation tasks. A wavefront implementation of DTW can benefit from texture memory since we access both the query and the subject in random order while traversing the diagonals.

Textures are defined at compile time and can be bound to arrays at runtime. Note that CUDA makes a subtle distinction between arrays and linear memory – throughout this chapter we use the terms linear memory and array interchangeably. Further details can be found in the CUDA programming guide [4]. The maximum number of allowed texels bound to linear memory are 2^{27} in the 1D case, $65{,}000^2$ in the 2D case and 4096^3 in the 3D case. In the following we will bind a 1D texture to the subject database stored in the array `float * Data`. First, we have to declare the texture itself as a static global variable:

```
texture<float, 1, cudaReadModeElementType> tSubject;
```

In general, the template parameters of `texture<DataType, Type, ReadMode>` are specified as follows:

- DataType: a basic 32-bit type such as `float` or `int`
- Type: the dimension being either 1, 2, or 3 (optional: default=1)
- ReadMode: the mode being either `cudaReadModeElementType` which passes the data as is or `cudeReadModeNormalizedFloat` which maps integer types to the interval [0, 1] (optional: default=`cudaReadModeElementType`)

After declaration in the document body we can now bind the texture at runtime in the main function using

```
cudaBindTexture(0, tSubject, Data,
                sizeof(float)*num_entries*num_features);
```

The first argument is a pointer of type `size_t *` storing an offset which we neglect. The remaining arguments specify the declared texture object, the address pointing to linear memory, and its corresponding size in bytes. Note that in our example we store $m = 2^{20}$ time series each of length $n = 128 = 2^7$ in the database resulting in exactly 2^{27} texels. A texture can be unbound with a call to `cudaUnbindTexture()`. The signature of the wavefront kernel is slightly modified: we simply remove the pointer to `Subject` since texture objects are visible across the whole device.

```
1  template <
2      typename index_t,
3      typename value_t> __global__
4  void DTW_wavefront_tex_kernel(
5      value_t * Query,         // pointer to the query
6      value_t * Dist,          // pointer to the distance
7      index_t num_entries,     // number of time series (m)
8      index_t num_features) {  // number of time ticks (n)
9
10     // basically the wavefront kernel with the exception
11     // that we substitute the expression in Line 58
12     // Subject[base+J-1] with tex1Dfetch(tSubject, base+J-1)
13 }
```

Listing 7.30: Wavefront DTW kernel utilizing texture memory.

Finally, we have to alter the body of the wavefront kernel (Line 58) to access entries in the subject database via `tex1Dfetch(tSubject, base+J-1)` instead of using `Subject[base+J-1]`. The kernel call is identical to the non-optimized wavefront kernel except the missing argument `Subject`. A measurement of the runtime reveals that the use of texture memory reduces the runtime from 940 ms to 900 ms.

This 4% improvement in runtime is rather marginal for $n = 128$. However, for higher values of n we gain reasonable improvements as shown in [11]. Moreover, this optimization is basically free since it depends only on minor modifications of the source code: technically one does not even need to remove the pointer to `Subject` since it is defined anyway.

An even more powerful caching strategy can be pursued if we employ so-called constant memory. CUDA-enabled GPUs provide the possibility to store up to 48 KB of heavily cached read-only storage in global memory. Because of its small size we may only store a limited amount of information which can be accessed during the execution of a kernel across the whole device. Typical use cases include the distribution of runtime constants which specify the characteristics of an algorithm such as exponents of the L_p-norm or gap penalties in elastic matching algorithms. Actually, CUDA uses constant memory to broadcast the kernel arguments to all thread blocks. In our experiments, we will cache the query of length $n = 128$ since it is frequently accessed by each of the thread blocks.

Similarly to texture memory, constant memory has to be declared globally at compile time and can be altered at runtime. In our example, we declare the memory directly after the include section of our program:

```
__constant__ float cQuery[12*1024];
```

Unfortunately, we have to know the size at compile and thus we decide to take the maximum value. If we are sure that we do not want to process longer queries, we could have hard-coded the size to $n = 128$. Subsequently we have to copy the query which here corresponds to the first time series in the subject database from the device array `float * Data` to `cQuery` in the main function:

```
cudaMemcpyToSymbol(cQuery, Data,
                   sizeof(value_t)*num_features);
```

The final modification in the original wavefront kernel is straightforward: we simply substitute the access to `Query[j-1]` with `cQuery[j-1]`. Note that similarly to texture memory we do not have to pass `cQuery` as kernel argument since it is visible across the whole device

```
1  template <
2      typename index_t,
3      typename value_t> __global__
4  void DTW_wavefront_const_kernel(
5      value_t * Subject,       // pointer to the subject
6      value_t * Dist,          // pointer to the distance
7      index_t num_entries,     // number of time series (m)
8      index_t num_features) {  // number of time ticks (n)
9
10     // basically the wavefront kernel with the exception
11     // that we substitute the expression in Line 58
12     // Query[j-1] with cQuery[j-1]
13 }
```

Listing 7.31: Wavefront DTW kernel utilizing constant memory.

The kernel call is identical to the non-optimized wavefront kernel except the missing argument `Query`. The corresponding runtime of the kernel is only 510 ms in contrast to the 940 ms of the non-optimized variant resulting in almost double the performance with only minor modifications of the source code.

We can even outperform the naive shared memory approach which performs at 640 ms by almost 20%. Note that we could combine constant memory and texture memory within the same kernel; however, the reduction in runtime is barely noticeable. Concluding, the wavefront kernel using constant memory is significantly faster than all aforementioned kernels based on the naive relaxation scheme and further supports the processing of reasonably longer queries.

7.5 OPTIMIZATION GUIDELINES

This section briefly summarizes the optimization methods and execution times in order to derive general rules that might help you to optimize code. Moreover, we discuss further possible optimization techniques which are not covered by this example. The major take home-message is that *everything is about efficient memory utilization*. GPUs provide excessive amounts of FLOP/s, however, we cannot benefit from that if we ignore the characteristics of the attached memory. Hence, we have to ensure that the compute units have something to compute and do not waste their potential while waiting for data. This can be summarized as follows:

"Computation in vacuum is futile."

Starting from the sequential algorithm on the CPU which performs at approximately 90 seconds we have observed that a simple coarse-grained task parallelization scheme over 32 CPU cores scales almost linearly. However, when porting the algorithm directly to the GPU, we have observed a massive decrease in performance of approximately one order-of-magnitude due to the non-coalesced memory access pattern. However, if we restructure the memory layout to allow coalesced memory accesses using either local memory or a transposed indexing scheme in global memory, we could achieve comparable performance. Furthermore, we could significantly speed up the kernel by using shared memory for the score matrix.

"Ensure your access pattern is coalesced. If not – make it so."

The use of shared memory imposes hard limitations due to its small size. This can be addressed by using fine-grained parallelization strategies which assign a task to a whole thread block instead of using one thread per task. This approach may dramatically reduce the shared memory footprint of algorithms but has the disadvantage that we have to parallelize the task itself. Unfortunately, this requires the complete restructuring of the source code as shown in the wavefront kernel. Furthermore, a fine-grained parallelization scheme might not be obvious or is completely unknown in the worst case. Consequently, one has to spend a reasonable amount of time to spot potential for fine-grained parallelization of core algorithms.

"Fine-grained parallelization is better than coarse-grained parallelization."

We have employed advanced caching strategies to further boost the performance by using either texture or constant memory for frequently accessed data. We have observed that constant memory is worth it if we repeatedly read the same data across the whole device. Texture memory can be employed to benefit from spatial locality of data accesses, however, reasonable improvements in runtime can only be expected if we exploit the hardware-based interpolation capabilities of higher-dimensional textures.

Note that this book covers only a fraction of available memory types. Further types such as surface memory, layered textures, or CUDA arrays are documented in the CUDA programming guide [4]. We strongly advise to read the provided documentation on a regular basis since each GPU generation introduces novel instructions which might dramatically speed up your algorithm.

"Do not rely on theoretical assumptions – experiment with novel techniques."

Finally, let us state that we could further optimize the wavefront kernel by using the enormous amount of registers per SM for the score matrix instead of shared memory. State-of-the-art implementations of elastic matching algorithms actually unroll the relaxation of several minor diagonals in registers resulting in even faster execution times [15]. Analogically, the next chapter shows you how to efficiently share information between threads in warp without the need for shared memory.

7.6 ADDITIONAL EXERCISES

1. Write a CUDA program that computes the sum of two vectors $x, y \in \mathbb{R}^n$ stored in the host arrays `float * x` and `float * y`. Compare the execution times of the corresponding kernel and memory transfers with a sequential program running on a single CPU core. Is your CUDA parallelization beneficial?
2. Let $x = (x_0, x_1, x_2, x_3) \in \mathbb{R}^4$ be an arbitrary vector in four-dimensional space and accordingly

$$v = \left(x^{(0)}, \ldots, x^{(n-1)}\right)$$
$$= \left(x_0^{(0)}, x_1^{(0)}, x_2^{(0)}, x_3^{(0)}, \ldots, x_0^{(n-1)}, x_1^{(n-1)}, x_2^{(n-1)}, x_3^{(n-1)}\right) \in \mathbb{R}^{4 \cdot n}$$

an array of $n \in \mathbb{N}$ vectors. Our goal is to normalize each vector $x^{(k)}$ individually:

$$\hat{\cdot} : \mathbb{R}^4 \to S^3 \subset \mathbb{R}^4 \, , \, x \mapsto \hat{x} := \frac{x}{\sqrt{x_0^2 + x_1^2 + x_2^2 + x_3^2}} \qquad \text{such that} \qquad \|\hat{x}\| = 1 \quad .$$

In practice, the array of vectors v is given by a `float` array of length $4 \cdot n$. An important application for this task is the normalization of Hamiltonian quaternions to have unit modulus in order to represent rotations in three-dimensional Euclidean space. In the following, we will use n CUDA-threads to accomplish this task. Your implementation should use several blocks and work with an arbitrary number of vectors n.

   (i) Write a naive CUDA-kernel that normalizes n vectors using n threads.

   (ii) Do the same but treat the input array `float * V` as an array of structs `float4 * V`. To achieve this, read four values at once using the `float4` data type. Is the runtime affected?

   (iii) Look up the mathematical functions `rsqrt` and `rsqrtf` in the CUDA documentation. Can you use them to speed up your implementation?

   (iv) Propose another optimization, implement it and measure the runtime. Is it beneficial to use a GPU for this dedicated task? Justify your claims.

3. Let $A, B \in \mathbb{R}^{n \times n}$ be two square matrices that can be mapped onto each other by transposition i.e., $A = B^T$, explicitly expressed in coordinates:

$$A_{ij} = B_{ij}^T = B_{ji} \quad \text{for all} \quad i, j \in \{0, \ldots, n-1\} \quad .$$

Assume A and B fit into global memory, then we can compute B from the knowledge of A or vice versa. Our goal is to store the transposed version of A in B using $\mathcal{O}(n^2)$ many CUDA-threads. Your implementation should use several blocks and work with arbitrary matrix length n.

(i) Write a naive CUDA kernel, that transposes A using n^2 threads on a two-dimensional grid.

(ii) Let $\texttt{TILE} \in \{1, 2, 4, 8, 16\}$ be the length of a submatrix in A and $n = \texttt{TILE} \cdot k$ for some $k \in \mathbb{N}$. Write a CUDA kernel that uses $n^2/\texttt{TILE}^2$ many threads such that one thread transposes a submatrix with $\texttt{TILE}^2$ entries. Your implementation should work for all above-stated choices of `TILE`. Which variant performs best? Explain your observation.

(iii) Assume $\texttt{TILE} = 4$, i.e. one tile has 16 entries. Use four `float4` structs to store the 16 entries in a CUDA thread. Afterwards transpose the 4×4 submatrix stored in four `float4` structs using six swaps of `float` values. Finally, write the transposed submatrix to the corresponding part of B using four writes of `float4` structs. Visualize the indexing scheme with a sketch.

Measure the execution times of the three described implementations. Which one performs best? Explain your observation.

4. Revisit the shared memory-based kernel for the efficient computation of the covariance matrix in Listing 7.13.

$$C_{jj'} = \frac{1}{m}\sum_{i=0}^{m-1} \overline{v}_j^{(i)} \cdot \overline{v}_{j'}^{(i)} = \frac{1}{m}\sum_{i=0}^{m-1} \overline{D}_{ji}^T \cdot \overline{D}_{ij'} \quad \text{for all } j, j' \in \{0, \ldots, n-1\} \quad .$$

Adapt the code to be applicable to the multiplication of two arbitrarily shaped matrices A and B such that $C = A \cdot B$. Discuss the memory layout of the involved matrices. Can you guarantee coalesced accesses to both A and B?

5. The Knapsack problem describes an optimization task, where a thief tries to pack up n objects $i \in \{0, \ldots, n-1\}$ of value $V[i]$ and weight $W[i]$ into a container of limited capacity c. For each object i, the thief may decide if he takes it or leaves it behind. The goal is to maximize the accumulated value of the chosen objects without violating the capacity constraint. Mathematically speaking, we are interested in the optimal subset $\mathcal{J}^* \subseteq \mathcal{I}$ amongst all subsets $\mathcal{J} \subseteq \mathcal{I} := \{0, \ldots, n-1\}$ such that

$$\mathcal{J}^* := \underset{\mathcal{J} \subseteq \mathcal{I}}{\operatorname{argmax}} \sum_{j \in \mathcal{J}} V[j] \quad \text{subject to:} \quad \sum_{j \in \mathcal{J}} W[j] \leq c \quad .$$

For exclusively integral weights there exists a pseudo-polynomial algorithm, which solves this NP-complete problem in $\mathcal{O}(n \cdot c)$ time and space with the help of dynamic programming. To achieve this, a matrix M of size $(n+1) \times (c+1)$ is initialized with zeros and incrementally relaxed using the following code:

```
1  unsigned int M[n+1][c+1], // filled with zeros
2               W[n], V[n];  // weights and values
3
4  for (size_t i = 1; i < n+1; i++) {
5      for (size_t j = 0; j < c+1; j++)
6          if (W[i-1] <= j)
7              M[i,j] = max(M[i-1,j)],
8                           M[i-1,j-W[i-1]]+V[i-1]);
9          else
10             M[i,j] = M[i-1,j];
```

The optimal accumulated value is stored in `M[n, c]`. Our goal is to write an efficient CUDA parallelization using multiple thread blocks.

(i) Discuss the dependencies between the individual entries of the matrix `M`. Which cells can be updated independently? Are there global barriers? If yes, how can we enforce global synchronization on CUDA-enabled GPUs?

(ii) Implement your proposed method. Is the use of shared, constant or texture memory beneficial? Benchmark several implementations and create appropriate plots.

(iii) The asymptotic time complexity and the memory consumption of the aforementioned algorithm are in $\mathcal{O}(N \cdot C)$. Using your considerations from (i), provide a linear memory algorithm that only uses $\mathcal{O}(C)$ cells to compute the result. Implement your approach.

REFERENCES

[1] Donald J. Berndt, James Clifford, Using dynamic time warping to find patterns in time series, in: KDD Workshop, 1994, pp. 359–370.

[2] Thomas H. Cormen, et al., Introduction to Algorithms, 3rd edition, The MIT Press, 2009.

[3] NVIDIA Corporation, CUDA, parallel programming and computing platform, https://developer.nvidia.com/cuda-zone (visited on 10/12/2015).

[4] NVIDIA Corporation, CUDA programming guide version 8.0, https://docs.nvidia.com/cuda/cuda-c-programming-guide/, 2016 (visited on 09/25/2016).

[5] NVIDIA Corporation, GeForce 256, the world's first GPU, http://www.nvidia.com/page/geforce256.html (visited on 10/12/2015).

[6] NVIDIA Corporation, NVLink high-speed interconnect: application performance (whitepaper), http://www.nvidia.com/object/nvlink.html, 2014 (visited on 09/25/2016).

[7] NVIDIA Corporation, Thrust parallel algorithms and data structures library, https://developer.nvidia.com/thrust (visited on 10/12/2015).

[8] Richard O. Duda, Peter E. Hart, David G. Stork, Pattern Classification, 2nd edition, Wiley-Interscience, ISBN 0471056693, 2000.

[9] Khronos Group, OpenCL, the open standard for parallel programming of heterogeneous systems, https://www.khronos.org/opencl/ (visited on 10/12/2015).

[10] Translational Imaging Group, NiftyReg: medical image registration using CUDA, http://cmictig.cs.ucl.ac.uk/wiki/index.php/NiftyReg (visited on 10/12/2015).

[11] C. Hundt, B. Schmidt, E. Schömer, CUDA-accelerated alignment of subsequences in streamed time series data, in: 2014 43rd International Conference on Parallel Processing, 2014, pp. 10–19.

[12] Mohammed Waleed Kadous, Learning comprehensible descriptions of multivariate time series, in: Ivan Bratko, Saso Dzeroski (Eds.), Proceedings of the 16th International Conference of Machine Learning (ICML-99), Morgan Kaufmann, 1999, pp. 454–463.

[13] David B. Kirk, Wen-mei W. Hwu, Programming Massively Parallel Processors: A Hands-on Approach, 2nd edition, Morgan Kaufmann Publishers Inc., San Francisco, CA, USA, 2013.

[14] Jeff Larkin, James Bayer, Comparing OpenACC 2.5 and OpenMP 4.5, in: NVIDIA GPU Technology Conference, 2016, http://on-demand.gputechconf.com/gtc/2016/presentation/s6410-jeff-larkin-beyer-comparing-open-acc-openmp.pdf (visited on 10/01/2016).

[15] Yongchao Liu, Bertil Schmidt, CUSHAW2-GPU: empowering faster gapped short-read alignment using GPU computing, IEEE Design & Test 31 (1) (2014) 31–39, http://dx.doi.org/10.1109/MDAT.2013.2284198.

[16] Ziwei Liu, et al., Deep learning face attributes in the wild, in: Proceedings of International Conference on Computer Vision (ICCV), 2015, http://mmlab.ie.cuhk.edu.hk/projects/CelebA.html (visited on 09/25/2016).

[17] OpenACC Organisation, OpenACC, directives for accelerators, http://www.openacc.org/ (visited on 10/12/2015).

[18] Oxford Learner's Dictionaries: entry for the word 'exact', http://www.oxfordlearnersdictionaries.com/definition/english/exact_1 (visited on 10/12/2015).
[19] H. Sakoe, S. Chiba, Dynamic programming algorithm optimization for spoken word recognition, Acoustics, Speech and Signal Processing, IEEE Transactions on 26 (1) (1978) 43–49, http://dx.doi.org/10.1109/TASSP.1978.1163055.
[20] Christian Szegedy, et al., Going deeper with convolutions, in: Proceedings of the IEEE Conference on Computer Vision and Pattern Recognition, 2015, pp. 1–9.

CHAPTER

ADVANCED CUDA PROGRAMMING 8

Abstract

In the recent past, CUDA has become the major framework for the programming of massively parallel accelerators. NVIDIA estimates the number of CUDA installations in the year 2016 to exceed one million. Moreover, with the rise of Deep Learning this number is expected to grow at an exponential rate in the foreseeing future. Hence, extensive CUDA knowledge is a fundamental pursuit for every programmer in the field of High Performance Computing. The previous chapter focused on the basic programming model and the memory hierarchy of modern GPUs. We have seen that proper memory utilization is key to obtain efficient code. While our examples from the previous chapter focused on thread-level implementations, we investigate now warp-level parallelization and the efficient use of atomic functions. Both techniques in combination enable further code optimization. Moreover, we discuss overlapping of communication and computation in single-GPU and multi-GPU scenarios using streams. We conclude the chapter with a brief discussion of CUDA 9 and its novel features.

Keywords

CUDA, GPU, Warp intrinsics, Atomic operations, Z-normalization, Compare and swap loop, Parallel prefix scan, Multiple GPUs, CUDA streams, Asynchronous memory transfer, Dynamic parallelism, CUDA-aware MPI

CONTENTS

Parallel Programming. DOI: 10.1016/B978-0-12-849890-3.00008-3

8.1 WARP INTRINSICS AND ATOMIC OPERATIONS (PARALLEL REDUCTION)

Up to this point, we have used registers to exclusively store state variables such as indices or intermediate values. This section shows you how to use the enormous amount of registers per SM as data storage. Historically, registers have been designed as thread-local memory that can exclusively be manipulated in the scope of a single thread. Since 32 threads within a warp are concurrently executed in lock-step manner, one might want to share information between them. The traditional approach would employ shared memory for inter-thread communication. Starting from the Kepler generation, CUDA introduced so-called warp intrinsics to achieve the same. They offer two advantages: firstly they communicate more efficiently and secondly we can save valuable space in shared memory which can be now used to cache other quantities. Another important technique allowing for the concurrent access of memory without race-conditions is the use of atomic operations. Throughout the rest of this section we will demonstrate both techniques in detail.

SEGMENTED PARALLEL REDUCTION

In the following, we develop a simple algorithm for parallel reduction based on warp intrinsics. Assume you want to process a data matrix $D_{ij} = S_j^{(i)}$ storing m one-dimensional time series of fixed length n. The index i enumerates the time series and j denotes the time ticks. A popular preprocessing technique in the field of time series data mining is *z-normalization* which adjusts the mean and variance of each sequence

$$z(S_j^{(i)}) = \frac{S_j^{(i)} - \mu^{(i)}}{\sigma^{(i)}} \quad , \tag{8.1}$$

where $\mu^{(i)}$ and $\sigma^{(i)}$ are the individual means and standard deviations of each of the m time series. Note that a similar technique is applied during batch normalization of deep neural networks [12]. Afterwards, each time series has vanishing mean and unit variance. Z-normalization is usually applied to remove offsets and variability in amplitudes in order to allow robust classification in a subsequent phase. A traditional fine-grained parallelization approach would process one time series per thread block. The corresponding sums could be evaluated in shared memory using parallel reduction

$$\mu^{(i)} = \frac{1}{n}\sum_{j=0}^{n-1} S_j^{(i)} \quad \text{and} \quad \sigma^{(i)} = \sqrt{\frac{1}{n-1}\sum_{j=0}^{n-1}(S_j^{(i)} - \mu^{(i)})^2} \quad . \tag{8.2}$$

Note that a coarse-grained parallelization scheme where each thread processes one time series is also conceivable. However, you would have to transpose the data matrix to guarantee coalesced memory accesses. Here, we can achieve the same without transposition. For the sake of simplicity, assume that $n = 32$ such that we can process all time ticks within a single warp. Our to be implemented algorithm can be split into four phases:

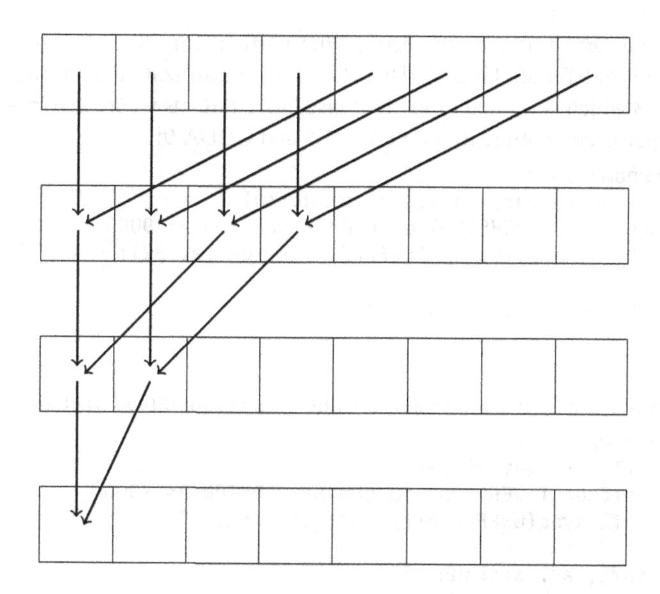

FIGURE 8.1

An example of parallel reduction within a warp of size 8. In each of the $\log_2(8) = 3$ iterations k we combine two values with a stride of $8/2^{k+1}$. The final result is stored in the register of thread zero.

1. Each thread j within a warp loads its corresponding value $S_j^{(i)}$ into a register.
2. We perform an all-reduce on the registers to compute the average $\mu^{(i)}$.
3. Another all-reduce is performed in registers in order to calculate $\sigma^{(i)}$.
4. Finally, we normalize each entry in registers and write them back to global memory.

Steps 2 and 3 are accomplished using so-called **warp intrinsics** which allow for the efficient sharing of registers within a warp. Initially, we will implement the all-reduce primitive with a traditional scheme for parallel reduction and subsequently broadcast the result to all threads within a warp as shown in Fig. 8.1. Later we will use another approach which directly accumulates the result in all threads.

Up to CUDA 8 the strided access can be realized with a shuffle-down instruction `__shfl_down()` which was introduced by the Kepler generation. This CUDA 8 intrinsic operation takes three arguments: the first one corresponds to the used register, the second one denotes the stride, and the third one specifies the warp size up to a length of 32 (optional).

```
// CUDA 8 shuffle down
T __shfl_down_(T var, unsigned int delta, int width=32)
```

Note that starting from CUDA 9 the intrinsic command `__shfl_down()` is deprecated and replaced by `__shfl_down_sync()` since threads in a warp are not implicitly synchronized anymore.

```
// CUDA 9 shuffle down
T __shfl_down_sync(unsigned int mask, // <- the mask is new
                   T var, unsigned int delta, int width=32)
```

The new parameter `unsigned int mask` encodes the participating threads within a warp in binary representation (`0xFFFFFFFF` for all threads). This allows for the utilization of arbitrary subsets of a warp in contrast to CUDA 8 which is limited to `width` consecutive threads where `width` $\in \{1, 2, 4, 8, 16, 32\}$. The following wrapper is compatible to both CUDA 8 and CUDA 9:

```
template <typename T>
T my_warp_shfl_down(T var, unsigned int delta) {
    #if defined(CUDART_VERSION) && CUDART_VERSION >= 9000
    return __shfl_down_sync(0xFFFFFFFF, var, delta, 32);
    #else
    return __shfl_down(var, delta, 32);
    #endif
}
```

Another shuffle intrinsic can be used to read the value of a certain thread `srcLane`:

```
template <typename T>
T my_warp_shfl(T var, int srcLane) {
    #if defined(CUDART_VERSION) && CUDART_VERSION >= 9000
    return __shfl_sync(0xFFFFFFFF, var, srcLane, 32);
    #else
    return __shfl(var, srcLane, 32);
    #endif
}
```

Assume we have already provided a device-sided pointer to the data matrix `Data` which is filled with instances of the CBF data set of length $n = 32$ using `generate_cbf()` as shown in the main function of Listing 7.22. Then we can compute the sum for the mean $\mu^{(i)}$ as follows:

```
1  template <
2      typename index_t,
3      typename value_t> __global__
4  void znorm_kernel(
5      value_t * Subject,        // pointer to the subject
6      index_t num_entries,      // number of time series (m)
7      index_t num_features) {   // number of time ticks (n)
8
9      // get thread and block identifiers
10     const index_t blid = blockIdx.x;
11     const index_t thid = threadIdx.x;
12     const index_t base = blid*num_features;
13
14     // 1. coalesced load of entries
15     value_t v = Subject[base+thid];
16     value_t x = v; // copy for later usage
17
18     // 2a. perform a warp reduction (sum stored in thread zero)
19     for (index_t offset = num_features/2; offset > 0; offset /= 2)
20         x += my_warp_shfl_down(x, offset);
21
22     // 2b. perform the first broadcast
23     value_t mu = my_warp_shfl(x, 0)/num_features;
```

Listing 8.1: Initialization in Z-Normalization kernel.

The for-loop performs the parallel reduction by fetching the entries from above using `my_warp_shfl_down()`. In a second phase we broadcast the final sum back to the upper entries via `my_warp_shfl()` which reads the register of Thread 0. Now we can proceed with the computation of the variance. Here, we employ the same approach with the exception that we compute the sum of squared residues. Finally, we perform the normalization and write the result back to global memory.

```
24        // define the square residues
25        value_t y = (v-mu)*(v-mu);
26
27        // 3a. perform a warp reduction (sum stored in thread zero)
28        for (index_t offset = num_features/2; offset > 0; offset /= 2)
29            y += my_warp_shfl_down(y, offset, num_features);
30
31        // 3b. perform the second broadcast
32        value_t var = my_warp_shfl(y, 0)/(num_features-1);
33
34        // 4. write result back
35        Subject[base+thid] = (v-mu)*cuda_rsqrt(var);
36    }
```

Listing 8.2: Main computation in the Z-Normalization kernel.

The call to `cuda_rsqrt` is mapped to either the single-precision or double-precision variant of `rsqrt` – an efficient instruction for the reverse square root. In general, you can define overloaded functions to allow for the mixing of C++ templates with C-signatures of some CUDA calls.

```
1  __forceinline__ __device__
2  double cuda_rsqrt(const double& value) {
3      return rsqrt(value);
4  }
5
6  __forceinline__ __device__
7  float cuda_rsqrt(const float& value) {
8      return rsqrtf(value);
9  }
```

Listing 8.3: Mapping C-style functions to templated C++.

Note that we can completely get rid of the broadcast phases 2b and 3b by using the CUDA 8/9 instruction `__shfl_xor()/__shfl_xor_sync()` instead of calling the aforementioned `__shfl_down()/__shfl_down_sync()` instructions as shown in [9]. Further optimization techniques can be applied when processing longer time series. An obvious approach is the stacking of several reduction per warp operating on distinct registers. Basically, we have pursued a similar technique by computing both the mean and variance in one warp. However, both reductions depend on each other in our experiment – the general case would interleave several independent reductions in order to benefit from instruction parallelism. Moreover, we could use several warps per block to increase the thread occupancy of the SMs. Unfortunately, we have to use shared memory in this case in order to communicate the partial results between distinct warps.

GLOBAL PARALLEL REDUCTION

Sometimes one has to accumulate a huge array instead of independently reducing many short arrays. In this case, we can pursue a similar approach by initially computing the partial reductions in all thread blocks and afterwards merge the results. To achieve that, we have basically two options. On the one hand, one could store the partial results in an auxiliary array `value_t * Aux` residing in global memory and subsequently call the same kernel recursively on the auxiliary memory until we have accumulated all contributions in a single memory position. As an example, when reducing 32 blocks consisting of 32 threads, we would store the 32 partial results of the blocks in `Aux` and afterwards perform another parallel reduction using one thread block. On the other hand, one could concurrently write the partial contributions to a single memory location. Unfortunately, this introduces a race condition on the memory position storing the final result.

This issue can be resolved if we use atomic operations in order to guarantee correct results. The CUDA programming language provides an extensive set of atomic instructions such as `AtomicAdd()`, `AtomicSub()`, `AtomicMin()`, and `AtomicMax()` [4]. All aforementioned commands accept as first argument a pointer to global or shared memory and as second argument the corresponding value.

In the following, we will compute the sum of all entries in an array `value_t * Input` of size `length` and subsequently store the final result in `value_t * Output` using atomics. The workflow of the algorithm consists of three phases:

1. We load all entries of `Input` into registers.
2. Each block consisting of a single warp performs a warp reduction.
3. Finally, each warp atomically adds its partial result to `Output`.

The corresponding implementation is straightforward (see Listing 8.4).

```
1   template <
2       typename index_t,
3       typename value_t,
4       index_t warp_size=32> __global__
5   void global_reduction_kernel(
6       value_t * Input,          // pointer to the data
7       value_t * Output,         // pointer to the result
8       index_t   length) {       // number of entries
9
10      // get thread and block identifiers
11      const index_t thid = threadIdx.x;
12      const index_t blid = blockIdx.x;
13      const index_t base = blid*warp_size;
14
15      // store entries in registers
16      value_t x = 0;
17      if (base+thid < length)
18          x = Input[base+thid];
19
20      // do the Kepler shuffle
21      for (index_t offset = warp_size/2; offset > 0; offset /= 2)
22          x += my_warp_shfl_down(x, offset, warp_size);
23
```

```
24      // write down result
25      if (thid == 0)
26          atomicAdd(Output, x);
27  }
```

Listing 8.4: Global reduction kernel.

This kernel has the disadvantage that we only utilize 32 threads per block and further have to spawn quite a high number of `SDIV(length, 32)` many blocks. A solution that works with an arbitrary number of blocks and warps per block could be implemented as follows:

1. We spawn a fixed number of blocks with a fixed number of warps.
2. All warps read their values from `Input` into registers. If we run out of warps, we simply read several values in a round robin fashion and add them up until we exhaust `length`.
3. Each warp performs a parallel reduction and subsequently accumulates the final result atomically in `Output`.

The corresponding source code is shown in Listing 8.5.

```
1   template <
2       typename index_t,
3       typename value_t,
4       index_t warp_size=32> __global__
5   void static_reduction_kernel(
6       value_t * Input,          // pointer to the data
7       value_t * Output,         // pointer to the result
8       index_t length) {         // number of entries (n)
9
10      // get global thread identifier
11      const index_t thid = blockDim.x*blockIdx.x+threadIdx.x;
12
13      // here we store the result
14      value_t accum = value_t(0);
15
16      // block-cyclic summation over all spawned blocks
17      for (index_t i = thid; i < length; i += blockDim.x*gridDim.x)
18          accum += Input[i];
19
20      // reduce all values within a warp
21      for (index_t offset = warp_size/2; offset > 0; offset /= 2)
22          accum += my_warp_shfl_down(accum, offset, warp_size);
23
24      // first thread of every warp writes result
25      if (thid % 32  == 0)
26          atomicAdd(Output, accum);
27  }
```

Listing 8.5: Global reduction kernel using more than one warp.

Note that the presented implementations have to be taken with a grain of salt. If we choose the template parameter `value_t` to be an integral data type such as `int32_t` or `uint32_t`, then the result stored in

`Output` will always be the same independently of the execution order of warps. However, when using floating-point data types such as `float` or `double`, we cannot rely on the associativity of addition since $(a+b)+c \neq a+(b+c)$ for all a, b and c in general. Hence, the numerically stable summation of floating-point values requires a little more effort [14].

ARBITRARY ATOMIC OPERATIONS

Although CUDA provides a comprehensive set of atomic instructions [4], you might encounter a situation where you have to atomically update a value with an unsupported operation. Assume we define the following associative[1] binary operation

$$\circ : \mathbb{R} \times \mathbb{R}, (x, y) \mapsto x \circ y := x + y + x \cdot y \quad , \tag{8.3}$$

then you will desperately search for an atomic `o=` assignment in the CUDA documentation. The bad news is that you will not find it. The good news is that you can implement virtually every atomic operation on your own as long as the data type fits into 64 bits of memory. This can be achieved with a *compare-and-swap* (CAS) instruction which atomically performs the following three actions without interruption.

1. Compare a given value `expectation` with a value `source` stored in memory.
2. If both values `expectation` and `source` coincide then set `source` to a given value `target`, otherwise do nothing.
3. Return `target` if the swap in 2. was successful, otherwise return `source`.

The return value in 3. can be rewritten as ternary conditional statement: `(expected == source) ? target : source`. Conceptually, CAS tries to swap two values under the constraint that our assumptions about the source location are fulfilled. But how can you use that to atomically update a value? We have to take into consideration that our expectation could be wrong, i.e., another thread has already altered the source position and thus the swap will (and should consequently) fail. We might have to update our expectation and reapply the CAS operation. Hence, **CAS has always to be performed within a loop** until we finally succeed or alternatively we can cancel the attempt due to a certain constraint violation. In order to demonstrate the latter, let us artificially extend the binary operation in Eq. (8.3) to be only applied if $0 \leq x \circ y < 10$. Concluding, we aim to implement a custom function `atomicUpdateResultBoundedByTen()`.

The CUDA programming language provides CAS primitives for three basic integral data types `value_t` $\in$ {`int, unsigned int, unsigned long long int`}:

```
value_t atomicCAS(value_t* source_address,
                  value_t  expected,
                  value_t  target)
```

Unfortunately, we have to recast any other data type that fits into either 32 or 64 bits at pointer level into one of those three types (even the modern `int32_t`, `uint32_t`, and `uint64_t` equivalents). As

[1] You have already proved that in the exercises of Chapter 1.

an example, you can reinterpret a `float y` on byte-level as `int x` via `int x = (int*)(&y)` and vice versa. At this point, we can implement our `atomicUpdateResultBoundedByTen()` function. For the sake of simplicity, we assume `value_t` to be a basic integer.

```
1  __device__ __forceinline__
2  int atomicUpdateResultBoundedByTen(
3      int* address,
4      int value) {
5
6      // get the source value stored at address
7      int source = *address, expected;
8
9      do {
10         // we expect source
11         expected = source;
12
13         // compute our custom binary operation
14         int target = expected+value+expected*value;
15
16         // check the constraint
17         if (target < 0 || target >= 10)
18             return source;
19
20         // try to swap the values
21         source = atomicCAS(address, expected, target);
22
23     // expected == source on success
24     } while (expected != source);
25
26     return source;
27 }
```

Listing 8.6: A custom atomic operation.

The loop incrementally updates `expected` with `source` (Line 11) and `source` itself with the value stored at `address` (Line 21) until we meet our assumption and thus have swapped the values (Line 24) or alternatively violated the constraint (Line 17). Concluding, every atomic function can be expressed in terms of CAS loops.

OUTLOOK

We have seen that warp intrinsics and atomic operations are powerful tools which enable us to write short yet efficient code. In general, they can be used to accelerate many traditional algorithms on the GPU (far too many to be discussed within the scope of this book). Nevertheless, we want to mention some noteworthy examples: The discussed wavefront relaxation scheme of DTW (see Section 7.4) which stores three diagonals could be implemented in registers for time series up to a length of 32. Moreover, we could design sorting networks such as Bitonic Sort or selection networks using intra-warp shuffles. Other examples include short length Fast Fourier Transforms (FFT), segmented and global prefix scans [15], or warp aggregation [1]. Furthermore, atomics could be used to dynamically assign block identifiers to enumerate the blocks according to their spawn order [11]. This can be used to model dependencies between thread blocks. A similar approach was used during the computation of prefix scans for inter-block synchronization without the need to terminate the kernel [15].

Finally, let us state a list of useful libraries that make excessive use of the discussed techniques. Thrust [10] is a high-level library bundled with CUDA which provides efficient parallel primitives such as prefix scans, reductions, histograms, and sorting algorithms for the GPU. It mimics the interface of STL vectors and thus provides a low barrier to entry especially for inexperienced users. CUDA Unbound (CUB) [2] is a yet more efficient low-level library which features highly configurable device-wide, block-wide and warp-wide parallel primitives individually optimized for several GPU generations. Moreover, CUB is a header-only library which can be easily integrated into existing projects. Fast Fourier Transforms of arbitrary length can be computed in single- and double-precision using the cuFFT library [6] which is also bundled with the CUDA toolkit. Other noteworthy examples include cuBLAS [3], cuRAND [7], cuDNN [5], and cuSOLVER [8]. Concluding, it is highly advisable to search for a potentially existing library before you try reinventing the wheel. These libraries are usually well documented, optimized for current GPU generations, and extensively tested which favors their use in production code.

8.2 UTILIZING MULTIPLE GPUS AND STREAMS (NEWTON ITERATION)

Up to this point, we have utilized a single GPU for the execution of kernels. In this section, you will learn how to concurrently harness multiple accelerator cards attached to the same compute node. Furthermore, we will discuss how you can hide the communication over the slow PCIe bus behind computation by interleaving memory transfers and kernel execution. Finally, we will combine both techniques to fully exploit the vast computational resources of single node workstations with multiple GPUs.

NEWTON'S ITERATION

In order to demonstrate the benefits of streamed computation on multiple GPUs, we need a compute-heavy kernel which rarely accesses global memory. A good representative for this class of kernels is the iterative fix point computation of zero crossings of arbitrary differentiable functions. For the sake of simplicity, we choose to compute the square root of a certain value using Newton's iteration method. Let $f : \mathbb{R} \to \mathbb{R}$, $x \mapsto f(x)$ be a differentiable function where $f'(x) \neq 0$ for all x and x_0 be an initial (not necessarily good) guess for $f(x) = 0$ then the recursive application of

$$x_{n+1} = x_n - \frac{f(x_n)}{f'(x_n)} \tag{8.4}$$

yields a better approximation for a zero crossing. Note that there could be multiple solutions of $f(x) = 0$, e.g., when investigating polynomials of higher degree – this approach merely computes one possible solution. The derivation of Eq. (8.4), its geometric interpretation, and further analysis of convergence speed and numerical stability can be found in any basic textbook about Univariate Analysis. In our case, we want to determine the positive square root of some given value α. Thus, we aim to solve the equation

$$f(x) = x^2 - \alpha \stackrel{!}{=} 0 \tag{8.5}$$

on the domain $\mathbb{R}^+ = (0, \infty)$. The first derivative $f(x)' = 2 \cdot x$ is greater than zero on the whole domain and thus the quotient in Eq. (8.4) is always well-defined. If we plug in the function into the iterative scheme, we obtain a simple recursive formula

$$x_{n+1} = x_n - \frac{x_n^2 - \alpha}{2 \cdot x_n} = \frac{1}{2}\left(x_n + \frac{\alpha}{x_n}\right) \quad . \tag{8.6}$$

Note that for positive values of x_0 and α all consecutive values x_n are positive and thus we compute a unique solution on $\mathbb{R}^+$. Hence, we can set $x_0 = \alpha$ in the initialization step. The iteration is usually performed until two consecutive results x_n and x_{n+1} agree within a certain error threshold or coincide in terms of floating point values. In our case, we pursue an orthogonal approach and perform the update procedure for a fixed number of iterations. Listing 8.7 shows an exemplary kernel which determines the square roots of all entries stored in a device array `value_t * Data`.

```
1  template <
2      typename index_t,
3      typename value_t,
4      index_t num_iters=256> __global__
5  void square_root_kernel(
6      value_t * Data,
7      index_t   length) {
8
9      const index_t thid = blockDim.x*blockIdx.x+threadIdx.x;
10
11     for (index_t i = thid; i < length; i += blockDim.x*gridDim.x){
12
13         value_t value = Data[i];
14         value_t root  = value;
15
16         # pragma unroll 32
17         for (index_t iter = 0; iter < num_iters && value; iter++)
18             root = 0.5*(root+value/root);
19
20        Data[i] = root;
21     }
22 }
```

Listing 8.7: Square root kernel based on Newton's iteration.

Note that this approach is inefficient due to the existence of the fast `sqrt()` call and the rather high number of iterations. However, we can easily alter the `num_iters` parameter to adjust the execution time in our experiments. Furthermore, there could be non-trivial tasks like solving equations like $f(x) = x \cdot \exp(\cdot x) - \alpha = 0$ that cannot be explicitly expressed in terms of x. In this case, CUDA does not provide a specialized instruction to efficiently compute the Lambert W function $x = W(\alpha)$.

The kernel can be called from within the main function. We further add timers to allow for the comparison of the execution time of the kernel with the time taken for memory transfers.

```
#include "../include/hpc_helpers.hpp"

int main () {

    typedef float    value_t;
    typedef uint64_t index_t;

    const index_t length = 1UL << 30;

    value_t * data = nullptr, * Data = nullptr;

    cudaMallocHost(&data, sizeof(value_t)*length);       CUERR
    cudaMalloc    (&Data, sizeof(value_t)*length);       CUERR

    for (index_t index = 0; index < length; index++)
        data[index] = index;

    TIMERSTART(overall)
    TIMERSTART(host_to_device)
    cudaMemcpy(Data, data, sizeof(value_t)*length,
               cudaMemcpyHostToDevice);                  CUERR
    TIMERSTOP(host_to_device)

    TIMERSTART(square_root_kernel)
    square_root_kernel<<<1024, 1024>>>(Data, length);   CUERR
    TIMERSTOP(square_root_kernel)

    TIMERSTART(device_to_host)
    cudaMemcpy(data, Data, sizeof(value_t)*length,
               cudaMemcpyDeviceToHost);                  CUERR
    TIMERSTOP(device_to_host)
    TIMERSTOP(overall)

    for (index_t index = 0; index < 10; index++)
        std::cout << index << " " << data[index] << std::endl;

    cudaFreeHost(data);                                  CUERR
    cudaFree(Data);                                      CUERR
}
```

Listing 8.8: Main function of Newton's iteration.

When executed on a single Pascal-based Titan X, we obtain the following output:

```
TIMING: 357.873 ms (host_to_device)
TIMING: 1346.27 ms (square_root_kernel)
TIMING: 325.595 ms (device_to_host)
TIMING: 2029.82 ms (overall)
```

It takes roughly 350 ms to copy the 4 GB of single-precision floating point values stored in `data` over the PCIe bus from the host to the device ($\approx$ 11.4 GB/s) and vice versa. The kernel itself runs approximately 1300 ms resulting in an overall time of two seconds for both the memory transfers and computation. Concluding, we spend roughly double the time taken for memory transfers on computation.

HARNESSING MULTIPLE GPUS

This subsection demonstrates the distribution of the aforementioned task among several GPUs attached to the same workstation. Note that throughout the rest of this section we will exclusively modify the main function and keep the kernel as is. We can reuse the same kernel by simply splitting the device array `Data` into chunks of size `length/num_gpus`. The amount of available CUDA-capable GPUs can be determined with the following two host-sided commands

```
int num_gpus;
cudaGetDeviceCount(&num_gpus);
```

You can set the environment variable `CUDA_VISIBLE_DEVICES` in a terminal if you want to explicitly mask certain GPUs. As an example, `CUDA_VISIBLE_DEVICES=0,2` would mask the devices 1 and 3 in a workstation with four GPUs. This might become handy if you use third party libraries such as Tensorflow and want to exclude the GPU which renders the Desktop in order to avoid occasional freezes of your GUI. Within a CUDA application we can choose the currently used device using `cudaSetDevice()` which accepts integral arguments ranging from `0` to `num_gpus-1`. Notably, this command is not a hardware switch which globally selects the used GPU like in a state machine: `cudaSetDevice()` acts within the currently defined scope and further is thread-safe. Consequently, allocation of memory and launching of kernels is bound to the selected device within the lifetime of that scope. At this point, we could optionally gather information about the attached devices such as provided VRAM, number of SMs, or clock rate in order to design proper distribution patterns of our data. This can be achieved as follows

```
cudaDeviceProp property;
cudaGetDeviceProperties(&property, gpu);
```

where `gpu` is an integer enumerating the devices. The `cudaDeviceProp` struct stores useful member variables such as the amount of VRAM (`size_t totalGlobalMem`), the number of SMs (`int multiProcessorCount`), or the frequency of the GPU (`int clockRate`) [4]. In our experiments, we will use two Pascal-based Titan X GPUs with 12 GB of VRAM. Hence, we can safely assume that both devices can store the data array of size 4 GB and further exhibit the same peak performance. Moreover, we assume that the length of `Data` is a multiple of the number of attached GPUs. In the following, we will split the host array `data` into chunks of size `length/num_gpus` and allocate `num_gpus` many device arrays of the same length on distinct GPUs, copy the partial arrays to the corresponding devices, launch the kernel operating on each partial device array, and finally copy the results back to the host.

```
1  #include "../include/hpc_helpers.hpp"
2
3  int main () {
4
5      typedef float    value_t;
6      typedef uint64_t index_t;
7
```

```
8       const index_t length = 1UL << 30;
9
10      // get number of GPUs
11      int num_gpus;
12      cudaGetDeviceCount(&num_gpus);
13      const index_t batch_size = length/num_gpus;
14
15      value_t * data = nullptr, * Data[num_gpus];
16
17      cudaMallocHost(&data, sizeof(value_t)*length);          CUERR
18
19      // for each GPU allocate partial data array
20      for (index_t gpu = 0; gpu < num_gpus; gpu++) {
21          cudaSetDevice(gpu);
22          cudaMalloc(&Data[gpu], sizeof(value_t)*batch_size); CUERR
23      }
24
25      for (index_t index = 0; index < length; index++)
26          data[index] = index;
27
28      TIMERSTART(overall)
29      // for each gpu copy partial array to GPUs
30      for (index_t gpu = 0; gpu < num_gpus; gpu++) {
31          const index_t offset = gpu*batch_size;
32          cudaSetDevice(gpu);                                 CUERR
33          cudaMemcpy(Data[gpu], data+offset,
34                     sizeof(value_t)*batch_size,
35                     cudaMemcpyHostToDevice);                 CUERR
36      }
37
38      // for each gpu execute the kernel on partial array
39      for (index_t gpu = 0; gpu < num_gpus; gpu++) {
40          cudaSetDevice(gpu);                                 CUERR
41          square_root_kernel<<<1024, 1024>>>
42                               (Data[gpu], batch_size);       CUERR
43      }
44
45      // for each gpu copy results back
46      for (index_t gpu = 0; gpu < num_gpus; gpu++) {
47          const index_t offset = gpu*batch_size;
48          cudaSetDevice(gpu);                                 CUERR
49          cudaMemcpy(data+offset, Data[gpu],
50                     sizeof(value_t)*batch_size,
51                     cudaMemcpyDeviceToHost);                 CUERR
52      }
53      TIMERSTOP(overall)
54
55      // some output of the result
56
57      //free memory for host and each of the devices
58      cudaFreeHost(data);                                     CUERR
59      for (index_t gpu = 0; gpu < num_gpus; gpu++) {
```

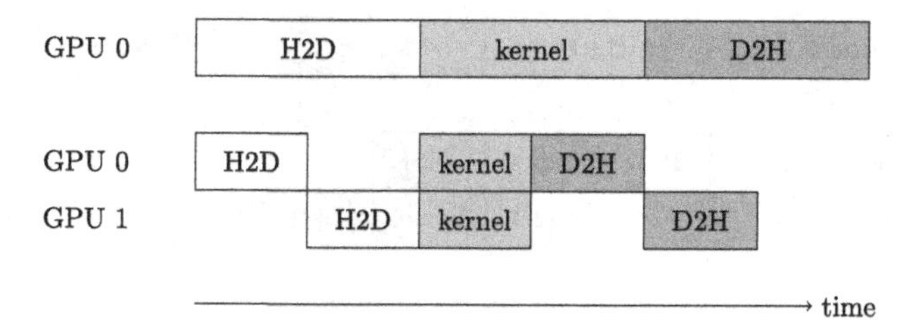

FIGURE 8.2

Schematic schedule of a task executed on a single GPU and the concurrent execution on two GPUs. In the latter case, we can save one-sixth of the overall execution time by cutting the execution time of the kernel in half. Note that the memory transfers take the same time since they share the same bandwidth of the PCIe bus.

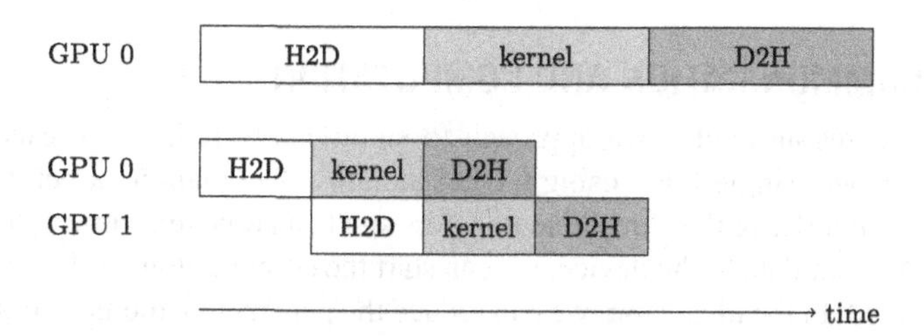

FIGURE 8.3

Schematic schedule of a task executed on a single GPU and the concurrent execution on two GPUs. In the latter case, we can save one-third by merging the for-loops of the memory transfers and kernel launches.

```
60             cudaSetDevice(gpu);
61             cudaFree(Data[gpu]);                                  CUERR
62         }
63     }
```

Listing 8.9: Main function of Newton's iteration using multiple GPUs.

When using two Pascal-based Titan X devices the GPU-related part of the program (Line 28 to Line 53) including memory transfers and kernel launches terminates after 1400 ms. This is slightly more than half of the 2000 ms overall execution time of the single GPU variant. Nevertheless, we can explain the 400 ms difference as follows: both GPUs block the sequential instruction flow until their respective memory transfer is accomplished and thus we can only cut the single GPU kernel execution time of 1300 ms in half, resulting in approximately 650 ms spent on computation and 750 ms on memory transfers (see Fig. 8.2). Consequently, we expect this approach to scale even worse when utilizing three or more devices. Note that we can slightly reduce the overall execution time of the multi-GPU approach to 1200 ms by merging the for-loops of the memory transfer to the device (Line 30) and the kernel launches (Line 39). Fig. 8.3 depicts the corresponding schedule.

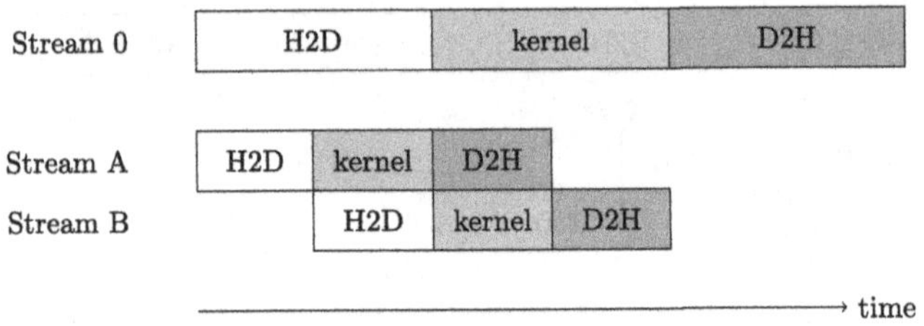

FIGURE 8.4

Schematic schedule of a task executed in the default zero-stream (Stream 0) and the interleaved execution of memory transfers (H2D: device to host, D2H: host to device) and kernel launches using two streams (Stream A and B). In the latter case, we can save one-third of the overall execution time by hiding communication behind computation.

INTERLEAVING COMMUNICATION AND COMPUTATION

This subsection discusses an alternative approach to significantly reduce the execution time of the `square_root_kernel` on a single GPU using CUDA streams. The main idea behind streaming relies on the fact that we can interleave the slow PCIe transfers and fast kernel executions. Hence, after having transferred a portion of our data to the device, we can start the computation while another portion of the data is still being copied. In the ideal case, we can reduce the perceivable memory transfers to the initial copy of the first chunk to the device and the last chunk to the host. The remaining memory transfers are executed during computation and thus do not contribute to the overall execution time. Unfortunately, this approach is only beneficial if the kernel execution time is in the same order-of-magnitude as the memory transfers. Note that for educational purposes we have chosen a compute-heavy kernel from the very beginning. Fig. 8.4 depicts an example using two CUDA streams.

Before we discuss the usage of user-defined streams let us briefly revisit CUDA's default behavior. Memory transfers and kernel launches are executed in the default *zero-stream* if we do not explicitly specify a user-defined stream. The zero-stream acts like expected: Several stacked kernels are executed in order without overlapping and synchronize with explicit global barriers (`cudaDeviceSynchronize()`) or implicitly defined ones (e.g. `cudaMalloc[Host]()`, `cudaMemset()`, and `cudaMemcpy()`). The only exception to this sequential behavior is the execution of host code which is not blocked by the asynchronous kernel invocations. Concluding, the zero-stream organizes the workflow of our program in a strict serial manner. Thus, a traditional kernel call is equivalent to a kernel invocation in the zero-stream (specified by the fourth argument in triple angle brackets):

```
kernel <<<num_blocks, num_threads, sh_mem>>>    (args);
kernel <<<num_blocks, num_threads, sh_mem, 0>>> (args);
```

In contrast, kernels launched in distinct user-defined streams act independently and thus do not synchronize with each other or kernels in the zero-stream. As a result, we can asynchronously spawn several kernels each using their own stream which are executed in parallel on the GPU. User-defined streams can be declared as follows: assume we want to create `num_streams` many streams then we can simply define an array of type `cudaStream_t *` and subsequently initialize it via:

```
cudaStream_t streams[num_streams];
```

```
for (int streamID = 0; streamID < num_streams; streamID++)
    cudaStreamCreate(&streams[streamID]);
```

Streams should be destroyed at the end of our program using `cudaStreamDestroy`:

```
for (int streamID = 0; streamID < num_streams; streamID++)
    cudaStreamDestroy(streams[streamID]);
```

The asynchronous invocation of several kernels in distinct stream is straightforward:

```
for (int streamID = 0; streamID < num_streams; streamID++)
     kernel <<<num_blocks, num_threads,
               sh_mem, streams[streamID]>>> (args);
```

Several kernel launches stacked within the same stream behave as usual: they are executed sequentially without overlap. Moreover, we can synchronize a certain stream with host code using `cudaStreamSynchronize(streams[streamID])` or alternatively enforce a global barrier affecting all user-defined streams and the zero-stream with a call to `cudaDeviceSynchronize()`. Furthermore, CUDA features a sophisticated event system which could be used to model complex dependencies between streams [4]. Finally, we have to discuss memory transfers which are now executed while other kernels perform computation. Unfortunately, we cannot use a traditional call to `cudaMemcpy()` since it imposes a global barrier across the whole device. Hence, we have to use `cudaMemcpyAsync()`, which accepts exactly the same arguments as `cudaMemcpy()` with the exception of an additional argument specifying the used stream:

```
cudaMemcpyAsync(target_ptr, source_ptr, size_in_bytes,
                transfer_mode, streams[streamID]);
```

As expected, asynchronous copies within the same stream stack in order and synchronize with kernel calls. There is only one important limitation: host memory should be allocated as *pinned memory* with `cudaMallocHost` in order to avoid swapping to disk. At the time of writing, it is not advisable to use host memory allocated with `new` or `malloc` in combination with streams. Note that CUDA further provides the asynchronous `MemsetAsync()` command. Unfortunately, there exists no asynchronous routine for memory allocation on the device that is callable from the host. Hence, we have to allocate device memory at the very beginning of our program.

Let us start coding. The workflow of our program in Listing 8.10 is straightforward. We partition the host array `value_t * data` into chunks of size `length/num_streams` and subsequently transfer each batch asynchronously to the device (Line 34). Afterwards, we launch `num_streams` kernels operating on their corresponding chunk (Line 40). Finally, we copy the result back chunk by chunk using streams again (Line 44). Looking for simplicity, the code assumes that the length of the array `Data` is a multiple of the number of streams.

```
1  #include "../include/hpc_helpers.hpp"
2
3  int main (int argc, char * argv[]) {
4
5      typedef float    value_t;
6      typedef uint64_t index_t;
7
```

```
8       const index_t length = 1UL << 30;
9
10      // get number of streams as command line argument
11      const index_t num_streams = atoi(argv[1]);
12      const index_t batch_size = length/num_streams;
13
14      // create streams
15      cudaStream_t streams[num_streams];
16      for (index_t streamID = 0; streamID < num_streams; streamID++)
17          cudaStreamCreate(streams+streamID);                  CUERR
18
19      value_t * data = nullptr, * Data = nullptr;
20
21      cudaMallocHost(&data, sizeof(value_t)*length);           CUERR
22      cudaMalloc    (&Data, sizeof(value_t)*length);           CUERR
23
24      for (index_t index = 0; index < length; index++)
25          data[index] = index;
26
27      TIMERSTART(overall)
28      for (index_t streamID = 0; streamID < num_streams; streamID++){
29
30          // compute global offset to local chunk
31          const index_t offset = streamID*batch_size;
32
33          // copy the data to the device using streams
34          cudaMemcpyAsync(Data+offset, data+offset,
35                          sizeof(value_t)*batch_size,
36                          cudaMemcpyHostToDevice,
37                          streams[streamID]);                  CUERR
38
39          // launch the kernel on each chunk
40          square_root_kernel<<<1024, 1024, 0, streams[streamID]>>>
41                            (Data+offset, batch_size);         CUERR
42
43          // copy the data back in chunks
44          cudaMemcpyAsync(data+offset, Data+offset,
45                          sizeof(value_t)*batch_size,
46                          cudaMemcpyDeviceToHost,
47                          streams[streamID]);                  CUERR
48      }
49
50      // synchronize all streams at once
51      cudaDeviceSynchronize();
52      TIMERSTOP(overall)
53
54      // some output of the result
55
56      // destroy the streams
57      for (index_t streamID = 0; streamID < num_streams; streamID++)
58          cudaStreamDestroy(streams[streamID]);                CUERR
59
```

```
60        // get rid of the memory
61        cudaFreeHost(data);                                         CUERR
62        cudaFree(Data);                                             CUERR
63    }
```

Listing 8.10: Main function of Newton's iteration using CUDA streams.

When executed on a Pascal-based Titan X, we measure the following runtimes depending on the number of used streams:

Streams	1	2	4	8	16	32	64
Time in ms	2030	1860	1770	1560	1450	1390	1350

Obviously, we can almost completely reduce the overall execution to the plain execution time of the kernel (1300 ms) by hiding the slow PCIe transfers behind computation. Notably, the streamed single GPU version is even faster than the naive non-streamed multi-GPU variant from the previous subsection (1400 ms).

STREAMED COMPUTATION ON MULTIPLE GPUS

Let us revisit what we have seen:

1. We can reduce the execution time of a kernel by utilizing several GPUs.
2. Memory transfers can be efficiently hidden by using streams.

It is obvious that we should combine both techniques in order to efficiently exploit the compute capabilities of all attached GPUs. We adapt the code shown in Listing 8.10 to use `num_streams` streams per GPU which each operate on a chunk of size `length/(num_streams*num_gpus)`. We again assume for simplicity that the length is a multiple of the chunk size.

```
1   #include "../include/hpc_helpers.hpp"
2
3   int main (int argc, char * argv[]) {
4
5       typedef float     value_t;
6       typedef uint64_t index_t;
7
8       const index_t length = 1UL << 30;
9       const index_t num_streams = atoi(argv[1]);
10
11      int num_gpus;
12      cudaGetDeviceCount(&num_gpus);
13      const index_t batch_size = length/(num_gpus*num_streams);
14
15      value_t * data = nullptr, * Data[num_gpus];
16      cudaStream_t streams[num_gpus][num_streams];
17
18      cudaMallocHost(&data, sizeof(value_t)*length);              CUERR
19
20      // malloc memory and create streams
```

```
21      for (index_t gpu=0; gpu<num_gpus; gpu++) {
22        cudaSetDevice(gpu);
23        cudaMalloc(&Data[gpu],
24                   sizeof(value_t)*batch_size*num_streams);    CUERR
25
26        for (index_t streamID=0; streamID<num_streams; streamID++)
27          cudaStreamCreate(&streams[gpu][streamID]);           CUERR
28      }
29
30      for (index_t index = 0; index < length; index++)
31          data[index] = index;
32
33      // asynchronous transfers and launches
34      TIMERSTART(overall)
35      for (index_t gpu=0; gpu<num_gpus; gpu++) {
36        const index_t offset = gpu*num_streams*batch_size;
37        cudaSetDevice(gpu);                                    CUERR
38
39        for (index_t streamID=0; streamID<num_streams; streamID++) {
40          const index_t loc_off = streamID*batch_size;
41          const index_t glb_off = loc_off+offset;
42
43          cudaMemcpyAsync(Data[gpu]+loc_off, data+glb_off,
44                          sizeof(value_t)*batch_size,
45                          cudaMemcpyHostToDevice,
46                          streams[gpu][streamID]);            CUERR
47
48          square_root_kernel
49                <<<1024, 1024, 0, streams[gpu][streamID]>>>
50                    (Data[gpu]+loc_off, batch_size);          CUERR
51
52          cudaMemcpyAsync(data+glb_off, Data[gpu]+loc_off,
53                          sizeof(value_t)*batch_size,
54                          cudaMemcpyDeviceToHost,
55                          streams[gpu][streamID]);            CUERR
56          }
57      }
58
59      // synchronize all devices
60      for (index_t gpu=0; gpu<num_gpus; gpu++) {
61        cudaSetDevice(gpu);                                   CUERR
62        cudaDeviceSynchronize();                              CUERR
63      }
64      TIMERSTOP(overall)
65
66      // some output of the result
67
68      // get rid of memory and streams
69      cudaFreeHost(data);                                     CUERR
70      for (index_t gpu=0; gpu<num_gpus; gpu++) {
71        cudaSetDevice(gpu);
72        cudaFree(Data[gpu]);                                  CUERR
```

```
73
74        for (index_t streamID=0; streamID<num_streams; streamID++)
75            cudaStreamDestroy(streams[gpu][streamID]);            CUERR
76    }
77 }
```

Listing 8.11: Main function of Newton's iteration using multiple GPUs and streams.

When executed on two Pascal-based Titan X GPUs, we measure the following runtimes depending on the number of used streams per GPU:

Streams per GPU	**1**	**2**	**4**	**8**	**16**	**32**	**64**
Time in ms	1020	930	880	770	710	690	670

We can reduce the overall runtime of the non-streamed single GPU version of 2000 ms to merely 670 ms which corresponds to approximately half of the plain execution time of the single GPU kernel (1300 ms). Note that even the single stream variant (1020 ms) performs better than the naive non-streamed multi-GPU version (1400 ms). As a result, the use of streams is highly advisable when utilizing several GPUs.

Finally, let us mention a useful tool for monitoring and debugging of CUDA applications. The NVIDIA visual profiler (nvvp) is a comprehensive application that allows for the investigation of stream scheduling on multiple GPUs and other important performance metrics in a user-friendly GUI. It can be used to analyze and improve the workflow of your application.

8.3 OUTLOOK

At this point, we have finished our journey. As you might have observed, CUDA is a mighty framework that allows for the massively parallel computation of huge data sets in short time. Notably, we can often outperform state-of-the-art multi-core CPUs as seen in Section 7.4 where we gain an additional speedup of roughly six over a dual socket Xeon CPU harnessing up to 32 hardware threads. However, speed comes with complexity: we have to properly access the attached memory, exploit local caches and employ sophisticated techniques such as warp intrinsics, atomics, and streams in order to yield reasonable performance. Although we have covered a significant fraction of CUDA's features, there are still some useful techniques left which cannot be discussed in detail within the scope of this book.

UNIFIED MEMORY

With the release of CUDA 8, NVIDIA introduced a *Unified Virtual Memory (UVM)* layer which treats the address spaces of host memory and device memory as one. UVM allows to design completely new distribution patterns especially when dealing with multiple GPUs. In the foreseeable future, we will see novel approaches such as lock-free hash maps distributed over multiple GPUs using system-wide atomics. However, this reduced complexity of source code might result in programs that are hard to analyze since the programmer has to pinpoint memory bottlenecks which are completely hidden behind syntactic sugar. As an example, the following code is a valid CUDA 8 program executed on the GPU despite the lack of explicit memory transfers:

```
#include <cstdint>
#include <iostream>

__global__ void iota_kernel(float * input, uint64_t size) {

    uint64_t thid = blockIdx.x*blockDim.x+threadIdx.x;
    for (uint64_t i = thid; i < size; i += gridDim.x*blockDim.x)
        input[i] = i;
}

int main () {

    uint64_t size = 1UL << 20;
    float * input = nullptr;
    cudaMallocHost(&input, sizeof(float)*size);
    iota_kernel<<<1024, 1024>>>(input, size);

    cudaDeviceSynchronize();

    for (uint64_t i = 0; i < 20; i++)
        std::cout << input[i] << std::endl;
}
```

Listing 8.12: An example for unified addressing in CUDA 8.

DYNAMIC PARALLELISM

All discussed examples in this book were executed on a static grid of thread blocks that had to be defined during kernel launch. However, an application might operate on an adaptive grid which is refined on demand. Examples are the integration of unsteady flows in the field of fluid dynamics, basic recursive applications such as Quicksort, Strassen's algorithms for fast matrix matrix multiplication, computation of fractals, or branch-and-bound algorithms. This adaptive refinement of the grid can be realized by calling kernels recursively from within a kernel. However, this comes with a reasonable overhead which renders *Dynamic Parallelism (DP)* impracticable in many cases. Nevertheless, DP should be considered especially when dealing with recursive refinements of spatial domains.

COOPERATIVE GROUPS

Up to CUDA version 8, warp intrinsics accomplishing intra-warp shuffles of register entries are limited to 32 consecutive threads. Moreover, we have learned that a warp is the de facto compute unit in the CUDA programming model that either applies 32 instructions in lock-step or serializes instruction divergences using masks. However, CUDA version 9 (release candidate of August 2017) introduces a new paradigm for the organization of threads: so-called cooperative groups. Loosely speaking, cooperative groups are user-defined teams of threads of flexible size providing convenient synchronization, communication and partition instructions. This paradigm shift in the organization of threads might have significant impact on the design of massively parallel algorithms in the future. As an example, the `__syncthreads()` call guarantees both a thread fence and a memory fence. Starting with CUDA 9, threads within a warp are not guaranteed to act in lock-step anymore (so-called independent thread scheduling) and thus we have to rethink intra-block communication using either shared memory or warp intrinsics. As a consequence, traditional software that exploits implicit warp synchronization, so-called warp-synchronous code, has to be rewritten or augmented with `__syncwarp()` statements. Among the new CUDA 9 features are persistent grids and multi-grids than can be synchronized across thread blocks and distinct GPUs. This allows for new computation schemes for iterative algorithms that rely on global barriers such as parallel reductions, parallel prefix sums, or sequence alignment algorithms.

TENSOR CORES

The Volta generation introduces novel compute units allowing for the ultra-efficient multiplication of small matrices: so-called tensor cores. Each of the 80 SMs consist of eight tensor core units for the computation of the affine mapping $D = A \cdot B + C$ where A and B are 4×4 FP16 matrices and C is an FP16 or FP32 matrix of the same shape. This certain operation is a fundamental building block in deep learning algorithms. The 640 tensor cores of a Tesla V100 card provide up to 120 tensor TFlop/s. Hence, a workstation equipped with eight V100 cards such as the DGX Volta box performs at 960 tensor TFlop/s for this specific task. Concluding, if you have to accelerate an application involving linear algebra and half-precision is sufficient (stochastic optimization, probabilistic machines, ensembles of weak linear classifiers), you should definitely have a look at tensor cores.

DISTRIBUTED COMPUTATION ON GPU CLUSTERS

In the CUDA chapters we have mainly discussed single GPU applications with the exception of the last section. Although it is straightforward to distribute a kernel over multiple GPUs attached to the same workstation, the communication between several compute nodes in a distributed memory architecture is far more complex. This can be achieved by interleaving CUDA with Message Passing Interface languages such as MPI (see Chapter 9) or PGAS-based languages such as UPC++ (see Chapter 10). In this context, a noteworthy technique is CUDA-aware MPI [13] which provides high-level wrappers for point-to-point primitives and global collectives which allow for the direct communication between compute nodes without stating explicit memory transfers over the PCIe bus.

8.4 ADDITIONAL EXERCISES

1. Cross-correlation, a minor modification of convolution, computes a weighted average of a signal f in a local neighborhood using the weight coefficients g. Assuming that f and g are sampled on a discrete grid and thus can be represented as arrays, we can compute the cross-correlation h of f and g as follows:

$$h[i] := (f \star g)[i] = \sum_{j=0}^{m-1} f[i+j] \cdot g[j] \quad \text{for all} \quad i \in \{0, \ldots, n-m\} \quad ,$$

where n is the length of the signal f and m is the number of weights stored in g. The final result h has $n - m + 1$ entries if $m \leq n$. In the following, assume $n = 2^{30}$ and $m = 2^5$.

(i) What is the time and memory complexity of the described computation pattern.

(ii) Implement an efficient kernel that stores g in constant memory and further evaluates the sum in shared memory in order to avoid redundant memory accesses to global memory. Do not forget to reserve storage for the halo values that are caused by the overlap of g with f:

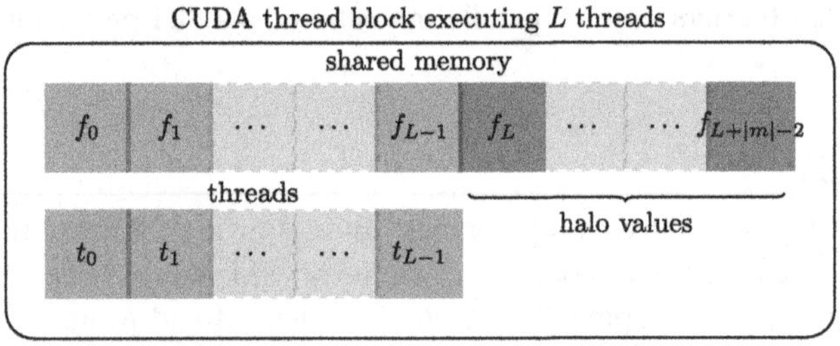

CUDA thread block executing L threads

shared memory

f_0 f_1 ⋯ ⋯ f_{L-1} f_L ⋯ ⋯ $f_{L+|m|-2}$

threads

halo values

t_0 t_1 ⋯ ⋯ t_{L-1}

(iii) Cross-correlation can also be computed by the pointwise multiplication of the corresponding Discrete Fourier Transforms (DFTs):

$$h = \mathcal{F}^{-1}\big(\mathcal{F}(f)^* \cdot \mathcal{F}(g)\big) \text{ where } \mathcal{F}(f)[k] = \frac{1}{n}\sum_{i=0}^{n-1} f[i] \cdot \exp\Big(2\pi\iota\frac{i \cdot k}{n}\Big) .$$

Here, $*$ corresponds to complex conjugation (time reversal in index space), $\iota = \sqrt{-1}$ denotes the imaginary solution of the equation $\iota^2 = -1$ and $\cdot$ is the index-wise complex multiplication of two numbers. DFTs and their inverses can be computed efficiently in $\mathcal{O}(n \log_2 n)$ time using so-called Fast Fourier Transforms (FFTs). How fast can you compute h using FFTs? Compare your result with the theoretical complexities in (i).

(iv) Implement the idea discussed in (iii) using NVIDIA's fast cuFFT library. First, embed g in a zero vector of length n. Second, compute the FFTs of both f and g. Third, evaluate the pointwise product using a custom kernel. Finally, compute the inverse transformation of your result.

(v) This task is for the mathematical reader: Prove the correlation theorem in (iii). First, show that a constant translation in index space results in a constant multiplicative phase $\exp(\iota\phi_0)$ in Fourier space. Second, apply a linear index substitution to the phase factors in the product $\mathcal{F}(f)^* \cdot \mathcal{F}(g)$. Finally, spot the inverse transformation by starring at the equation.

2. Revisit the wavefront kernel for the computation of the Dynamic Time Warping similarity measure in Listing 7.28. In the following we want to investigate further parallelization potential.

(i) Instead of using texture memory for the subject database, we could manually cache the subject sequences in shared memory in order to speed-up the random accesses during relaxation. Implement this approach.

(ii) During an update of a cell each thread t_l has to read two entries already processed by the previous thread t_{l-1} (left and diagonal) and one entry already processed by the same thread (above) as shown in Fig. 7.18. Implement a wavefront kernel for DTW that performs the intra warp communication using warp intrinsics instead of shared memory. For simplicity assume that the length of the time series is exactly $n = 31$.

3. The Hillis–Steele scan is a parallel computation scheme for the efficient calculation of prefix sums. It concurrently computes the inclusive prefix scan of a vector $x \in \mathbb{R}^n$ in $\log_2 n$ time using n compute units. The corresponding OpenMP code looks as follows:

```
1  #include <cmath>      // log
2  #include <assert.h>   // assert
3  #include <omp.h>      // openMP
4
5  template <
6      typename index_t,
7      typename value_t>
8  void hillis_steele(
9      value_t * x,
10     index_t   n) {
11
12     // make sure n is power of 2
13     assert((n & (n-1)) == 0);
14
15     // auxiliary memory
16     value_t *  y = new value_t[n];
17
18     # pragma omp parallel
19     for (index_t k = 0; k < std::log(n)/std::log(2); k++) {
20
21         // perform Hillis-Steele update
22         # pragma omp for
23         for (index_t i = 0; i < n; i++)
24             if (i >= (1<<k))
25                 y[i]  = x[i] + x[i-(1<<k)];
26             else
27                 y[i]  = x[i];
28
29         // copy auxiliary memory back to x
30         #pragma omp for
31         for (index_t i = 0; i < n; i++)
32             x[i] = y[i];
33     }
34
35     // free auxiliary memory
36     delete[] y;
37
38 }
```

(i) Draw a sketch that visualizes the computation pattern over the $\log_2 n$ iterations of the outer loop.

(ii) Implement a warp scan using warp shuffles.

(iii) Extend your implementation to work on segments with more than 32 threads within a block as follows. Compute the scans independently in each warp. Store the last entry of each prefix sum in shared memory and subsequently determine the prefix sum over these entries in registers. Adjust the overall prefix sum using the result from the previous step.

4. Let `uint32_t * Data` be an array of length $n = 2^{30}$ storing numbers from the set $S = \{0, \ldots, 9\}$ and `uint32_t * Counts` be an array of length $m = 10$ storing the number of occurrences for each element in S.

- **(i)** Implement a histogram kernel where each thread reads an entry from `Data` and atomically increments the corresponding slot in `Counts`.
- **(ii)** Improve that kernel by computing local histograms per CUDA thread block in shared memory and subsequently merge the partial histograms using atomic operations.
- **(iii)** Provide a register-only variant where each thread independently increments $m = 10$ registers. Subsequently the counts stored in registers have to be accumulated using warp intrinsics. Finally, the block-local histograms are written atomically to `Counts`.

Measure the execution times. Which approach performs best?

5. Revisit the shared memory-based kernel for the efficient computation of the covariance matrix in Listing 7.13

$$C_{jj'} = \frac{1}{m} \sum_{i=0}^{m-1} \overline{v}_j^{(i)} \cdot \overline{v}_{j'}^{(i)} \quad \text{for all } j, j' \in \{0, \ldots, n-1\} \quad .$$

Obviously, the sum can be calculated incrementally over the index i since addition is associative. Thus, one could determine C on half of the indices and later add the contributions of the other half.

- **(i)** Implement a multi-GPU variant that computes partial covariance matrices on distinct devices and subsequently merges the partial results.
- **(ii)** The same approach is applicable to CUDA streams. Make it so.
- **(iii)** Provide a streamed multi-GPU variant in order to max out the compute capability of your GPUs.
- **(iv)** Write a heterogeneous implementation that additionally harnesses the cores of your CPU to further improve performance.

REFERENCES

[1] NVIDIA Coporation, Optimized filtering with warp-aggregated atomics, https://devblogs.nvidia.com/parallelforall/cuda-pro-tip-optimized-filtering-warp-aggregated-atomics/, 2014 (visited on 01/20/2017).

[2] Duane Merril, NVIDIA Corporation, CUDA unbound, https://github.com/NVlabs/cub, 2017 (visited on 01/20/2017).

[3] NVIDIA Corporation, cuBLAS library, https://developer.nvidia.com/cublas, 2017 (visited on 01/20/2017).

[4] NVIDIA Corporation, CUDA programming guide version 8.0, https://docs.nvidia.com/cuda/cuda-c-programming-guide/, 2016 (visited on 09/25/2016).

[5] NVIDIA Corporation, cuDNN library, https://developer.nvidia.com/cudnn, 2017 (visited on 01/20/2017).

[6] NVIDIA Corporation, cuFFT library, https://developer.nvidia.com/cufft, 2017 (visited on 01/20/2017).

[7] NVIDIA Corporation, cuRAND library, https://developer.nvidia.com/curand, 2017 (visited on 01/20/2017).

[8] NVIDIA Corporation, cuSOLVER library, https://developer.nvidia.com/cusolver, 2017 (visited on 01/20/2017).

[9] NVIDIA Corporation, Faster parallel reductions on Kepler (blog), https://devblogs.nvidia.com/parallelforall/faster-parallel-reductions-kepler/, 2016 (visited on 09/25/2016).

[10] NVIDIA Corporation, Thrust parallel algorithms and data structures library, https://developer.nvidia.com/thrust (visited on 10/12/2015).

[11] Juan Gomez-Luna, et al., In-place data sliding algorithms for many-core architectures, in: Parallel Processing (ICPP), 2015 44th International Conference on, 2015, pp. 210–219.

[12] Sergey Ioffe, Christian Szegedy, Batch normalization: accelerating deep network training by reducing internal covariate shift, in: Proceedings of the 32nd International Conference on Machine Learning, ICML 2015, Lille, France, 6–11 July 2015, 2015, pp. 448–456, http://jmlr.org/proceedings/papers/v37/ioffe15.html.
[13] Jülich Supercomputing Centre (JSC), CUDA-aware MPI, https://www.fz-juelich.de/SharedDocs/Downloads/IAS/JSC/EN/slides/cuda/07-cuda-aware-MPI.pdf, 2016 (visited on 01/20/2017).
[14] W. Kahan, Pracniques: further remarks on reducing truncation errors, Communications of the ACM (ISSN 0001-0782) 8 (1) (Jan. 1965) 40–48, http://dx.doi.org/10.1145/363707.363723, http://doi.acm.org/10.1145/363707.363723.
[15] Yongchao Liu, Srinivas Aluru, LightScan: faster scan primitive on CUDA compatible manycore processors, in: CoRR, 2016, http://arxiv.org/abs/1604.04815.

CHAPTER

MESSAGE PASSING INTERFACE

9

Abstract

Up to now, we have studied how to develop parallel code for shared-memory architectures on multicore and manycore (GPU) systems. However, as explained in the first chapter, many HPC systems such as clusters or supercomputers consist of several compute nodes interconnected through a network. Each node contains its own memory as well as several cores and/or accelerators whose compute capabilities can be exploited with the techniques presented in the previous chapters. However, we need additional programming models to work with several nodes in the same program. The most common programming model for distributed-memory systems is message passing. The Message Passing Interface (MPI) is established as a *de facto* standard as it is based on the consensus of the MPI Forum, which has over 40 participating organizations, including vendors, researchers, software library developers, and users. MPI provides a portable, efficient, and flexible standard for message passing.
The goal of this chapter is to teach you how to develop parallel programs based on C++ according to the MPI standard. The code examples presented in this chapter address the most useful topics for new MPI programmers: point-to-point communication (blocking and nonblocking), collectives, derived datatypes and virtual topologies for complex communication.

Keywords

MPI, Message passing interface, Distributed memory, Cluster computing, SPMD, Two-sided communication, Deadlock, Nonblocking communication, Collective, Barrier, Broadcast, Scatter, Gather

CONTENTS

Parallel Programming. DOI: 10.1016/B978-0-12-849890-3.00009-5

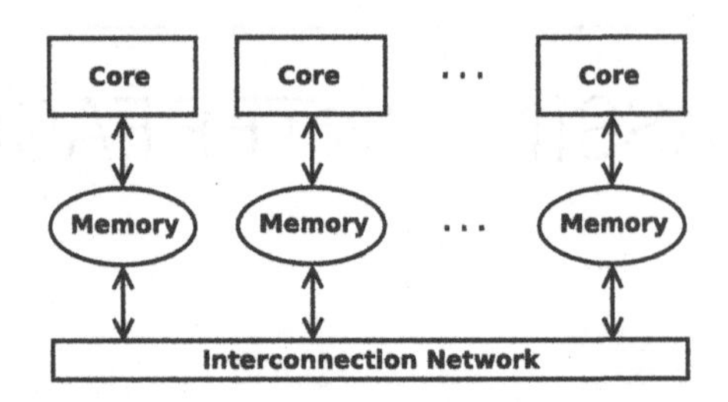

FIGURE 9.1

Abstraction of a traditional distributed-memory system with one CPU and one memory module per node.

9.1 INTRODUCTION TO MPI

Before the 1990s, writing parallel applications for different parallel architectures was a difficult and tedious task. Although many libraries could facilitate building parallel applications, there was no standardized and accepted way of doing it. Some parallel programmers realized that most of these libraries used the same message-passing model with only minor differences. Therefore they decided to work together in order to specify a general interface that allows programmers to write applications portable to different parallel architectures. This group of people, called the MPI Forum, finished the first interface specification (MPI-1) in 1994 [7]. The latest MPI version at the time of writing (v3.1) [8] was released in June 2015. Although there are other message-passing approaches as Parallel Virtual Machine (PVM) [15], MPI is so popular nowadays in the HPC community that it has been established as the *de facto* standard.

Note that MPI is only a definition of an interface that has been implemented by several developers for different architectures. Nowadays there exists a plethora of implementations whose routines or functions can be directly called from C, C++ and Fortran code. Some examples of MPI implementations are the open-source MPICH [19] and OpenMPI [20], as well as the commercial releases developed by vendors such as Intel [23], IBM [4] or HP [17]. Any parallel code that respects the MPI interface should work for all the aforementioned implementations, but the performance could vary [2, 21]. For instance, vendor implementations are usually optimized for their machines.

MPI follows the SPMD style, i.e., it splits the workload into different tasks that are executed on multiple processors. Originally, MPI was designed for distributed memory architectures, which were popular at that time. Fig. 9.1 illustrates the characteristics of these traditional systems, with several CPUs connected to a network and one memory module per CPU. A parallel MPI program consists of several processes with associated local memory. In the traditional point of view each process is associated with one core. Communication among processes is carried out through the interconnection network by using *send* and *receive* routines.

As architectural trends changed, the majority of current clusters contain shared-memory nodes that are interconnected through a network forming a hybrid distributed-memory/shared-memory system, as illustrated in Fig. 9.2. Modern clusters could even include manycore accelerators attached to the nodes. Nowadays MPI implementations are able to spawn several processes on the same machine. However, in order to improve performance, many parallel applications use the aforementioned hybrid approach:

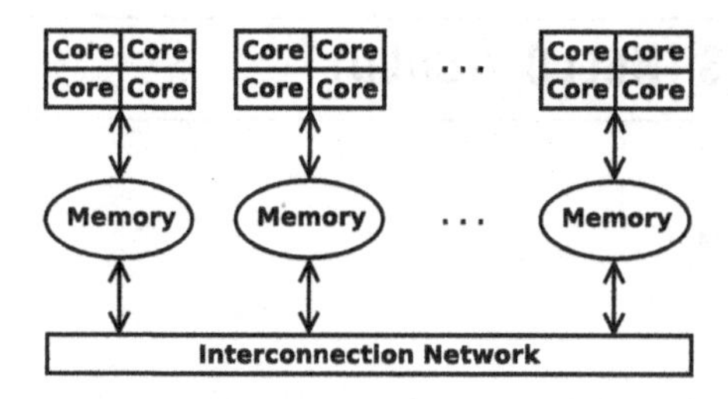

FIGURE 9.2

Abstraction of a modern distributed-memory system with several CPUs and one memory module per node.

one MPI process per node that calls multithreaded [3,10] or CUDA [1,13] functions to fully exploit the compute capabilities of the existing CPUs and accelerators cards within each node.

The commands to compile and execute MPI programs may vary depending on the used implementation. For instance, OpenMPI uses the command `mpic++` to compile code written in the C++ language. The command `mpirun` is used for the execution of our program together with the flag `-np` to specify the number of processes. The number of processes remains constant throughout the whole execution. Processes are mapped to different nodes as specified by the user in a configuration file. If any process fails, the whole application stops prematurely in an erroneous state.[1] Next, you will learn the commands necessary to compile and execute an MPI *Hello World* program utilizing four processes in OpenMPI (assuming that we can access the compiled binaries through a network file system and thus do not need to manually copy them to all nodes).

```
mpic++ -o hello hello.cpp
mpirun -np 4 ./hello
```

In this chapter we present the main features of MPI, the most popular interface for parallel programming on compute clusters. Analogous to previous chapters, we start with a simple *Hello World* program in Section 9.2 in order to demonstrate the basic setup and explain the concept of a communicator. We continue by describing the routines to perform blocking and nonblocking point-to-point communication (*send* and *receive*) in Sections 9.3 and 9.4, respectively, using two variants of *ping-pong* communication patterns. Collective routines, used to communicate data among all the processes of one communicator, are introduced in an example counting the amount of primes during Section 9.5. Further examples of collective routines are used all around the chapter. Section 9.6 presents two versions of parallel Jacobi iteration using both non-overlapping and overlapped communication patterns in order to demonstrate performance improvements obtained by the latter technique. The last two sections deal with parallel algorithms for matrix multiplication. In Section 9.7 derived datatypes are used to facilitate the distribution and final gathering of the matrices. Moreover, complex communicators are introduced in Section 9.8 in order to implement the communication pattern of SUMMA.

[1]Note, advanced techniques such as the checkpointing extension `openmpi-checkpoint` or the `dmtcp` tool can be deployed in order to allow the recovery from a fail-stop.

9.2 BASIC CONCEPTS (HELLO WORLD)

```
1  #include "mpi.h"
2
3  int main (int argc, char *argv[]){
4    // Initialize MPI
5    MPI::Init(argc,argv);
6
7    // Get the number of processes
8    int numP=MPI::COMM_WORLD.Get_size();
9
10   // Get the ID of the process
11   int myId=MPI::COMM_WORLD.Get_rank();
12
13   // Every process prints Hello
14   std::cout << "Process " << myId << " of "
15     << numP << ": Hello, world!" << std::endl;
16
17   // Terminate MPI
18   MPI::Finalize();
19   return 0;
20 }
```

Listing 9.1: MPI Hello World.

We start our journey into MPI programming with Listing 9.1, a version of the popular *Hello World*, which helps us to explain some basic features of the interface. First, the header `mpi.h` must be included to compile MPI code. In the `main()` function all processes are completely independent until the MPI initialization (`MPI::Init()`). From this point the processes can collaborate, send/receive messages or synchronize until reaching `MPI::Finalize()`. The finalization leads to freeing all the resources reserved by MPI. Two key concepts are further introduced in this example:

- **Communicator.** MPI uses communicator objects to define which collection of processes may communicate with each other. `MPI::Comm` is the abstract base communicator class, encapsulating the functionality common to all MPI communicators. For now, we use the predefined object `MPI::COMM_WORLD` that consists of all the MPI processes launched during our execution. The total amount of processes included in a communicator can be obtained with the method `Get_size()`.
- **Rank.** Within a communicator, every process has its own unique integer identifier assigned by the system when the process initializes. Ranks are contiguous, begin at zero, and can be obtained with the method `Get_rank()`.

This is a simple code without any communication among the involved processes. Every process just prints a message. The order of the print statements varies for different executions since we cannot control which process reaches the `std::cout` command first. A possible outcome is:

```
Process 3 of 4: Hello, world!
Process 0 of 4: Hello, world!
Process 1 of 4: Hello, world!
Process 2 of 4: Hello, world!
```

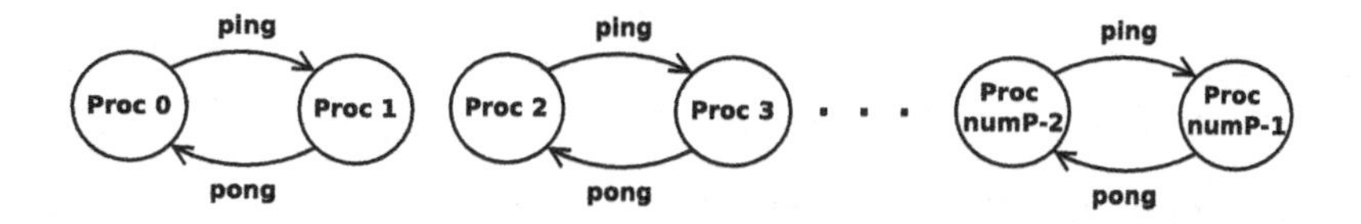

FIGURE 9.3

Pairs of processes involved in the *ping-pong* communication scheme.

9.3 POINT-TO-POINT COMMUNICATION (PING-PONG)

The concepts explained in the previous section enable us to develop MPI programs where the tasks of different processes are completely independent since they do not need to exchange information. Unfortunately, most applications cannot divide their workloads so easily and depend on communication between processes. The traditional MPI communication style is two-sided, i.e., the source and destination processes must be synchronized through *send* and *receive* methods.

A *ping-pong* between pairs of processes, sometimes referred to as *heartbeat* or *keep-alive message*, is used to present the concept of point-to-point communication. Assume we partition an even number of processes into pairs as illustrated in Fig. 9.3: $(0, 1)$, $(2, 3)$, $(4, 5)$, etc. Computation starts from the left process of the pair sending a message (*ping*) to the right one. Once the right process has received the *ping*, it immediately returns a *pong*. The number of *ping-pong* iterations shall be specified in the command line arguments of the program. For simplicity, the messages contain only one integer representing the current iteration number.

Listing 9.2 shows the first part of the program, which consists of concepts we are already familiar with:

1. initialization of MPI
2. obtaining information about the amount and rank of the processes
3. parsing the number of iterations from the command line arguments
4. checking if the number of processes is a multiple of two

If any of the parameters does not satisfy the aforementioned constraints, we call the routine `Abort()`, which makes a "best attempt" to terminate the tasks of all the processes in the specified `COMM_WORLD` communicator. It is usually implemented by sending `SIGTERM` to all processes. Here is the code associated with the initial part:

```
1  #include <stdlib.h>
2  #include "mpi.h"
3
4  int main (int argc, char *argv[]){
5    // Initialize MPI
6    MPI::Init(argc,argv);
7    // Get the number of processes
8    int numP=MPI::COMM_WORLD.Get_size();
9    // Get the ID of the process
10   int myId=MPI::COMM_WORLD.Get_rank();
```

```
11
12    if(argc < 2){
13      // Only the first process prints the output message
14      if(!myId)
15        std::cout << "ERROR: The syntax of the program is
16          ./ping-pong num_ping_pong" << std::endl;
17      MPI::COMM_WORLD.Abort(1);
18    }
19
20    if((numP%2) != 0){
21      // Only the first process prints the output message
22      if(!myId)
23        std::cout << "ERROR: The number of processes must be a "
24                  << "multiple of 2" << std::endl;
25      MPI::COMM_WORLD.Abort(1);
26    }
27
28    int num_ping_pong = atoi(argv[1]);
29    int ping_pong_count = 0;
```

Listing 9.2: Initial part of the *ping-pong* program.

At this point we can start sending *ping* and *pong* messages. The traditional communication scheme in MPI is two-sided, where both source and destination processes must indicate in the code that they are part of the communication. The *send* procedure in the source process involves storing the data into a buffer and notifying the communication device (which is often a network) that communication can start. This device is responsible for routing the message to the proper location. However, the destination process still has to acknowledge with a *receive* routine that it wants to receive the data. Once this happened, the data can be transmitted and both processes may continue their work. This approach is known as *blocking* communication since both processes do not proceed until communication has finished. We start explaining this mode because it is the default behavior of the functions `Send` and `Receive`. The nonblocking communication mode will be presented in Section 9.4. The C++ calls for *send* and *receive* are given as follows:

- `void Send(const void* buf, int count, const Datatype& datatype, int dest, int tag)`
- `void Recv(void* buf, int count, const Datatype& datatype, int source, int tag)`

The first argument specifies the data buffer, while the second one indicates the amount of elements involved in the communication. The third argument denotes an `MPI::Datatype` that describes the type of elements residing in the buffer. MPI provides several elementary datatypes e.g., `MPI::INT`, `MPI::FLOAT`, `MPI::DOUBLE`. Further, developers can define more complex types, as will be explained in Section 9.7. `Send` transmits the exact `count` of elements, and `Recv` will receive at most `count` elements. The remaining arguments specify the rank of the sending/receiving process and the tag of the message, respectively. Assume that the same process has to send many different types of messages to the same destination. Instead of having to go through extra measures to differentiate all these messages, MPI allows senders and receivers to additionally specify message IDs attached to the message (known as tags). When the destination process only requests a message with a certain tag number, messages with different tags will be buffered by the network until a corresponding receive has been posted. In general, note that for completion of the communication the `Send` and the `Recv` must be matching. This means, the ranks of the involved processes have to be specified correctly, the size constraints must be fulfilled, and the tags have to agree.

Back to our example, each of the *ping* and *pong* messages only contains one element representing the integer-encoded value of the current iteration step. Thus we already have almost all the information about the parameters to be specified in the calls. Listing 9.3 shows how to determine the partner identifiers for each pair:

```
31    int partner_id;
32    bool odd = myId%2;
33
34    if(odd){
35      partner_id = myId-1;
36    } else {
37      partner_id = myId+1;
38    }
```

Listing 9.3: Computation of pair ids in the *ping-pong* program.

If the rank of a process is even (`myId % 2 == 0`), the partner id can be obtained by incrementing its value by one. If the process is odd, the partner is the previous one. We complete the program with Listing 9.4. Processes with even rank start sending the *ping* message to their respective partners and wait for the reception of the pong. The behavior of odd processes is the opposite: they wait for the *ping* and, as soon as they have received it, they send *pong* to their partners.

```
39    while(ping_pong_count < num_ping_pong){
40      // First receive the ping and then send the pong
41      ping_pong_count++;
42
43      if(odd){
44        MPI::COMM_WORLD.Recv(&ping_pong_count, 1, MPI::INT,
45                             partner_id, 0);
46        MPI::COMM_WORLD.Send(&ping_pong_count, 1, MPI::INT,
47                             partner_id, 0);
48      } else {
49        MPI::COMM_WORLD.Send(&ping_pong_count, 1, MPI::INT,
50                             partner_id, 0);
51        MPI::COMM_WORLD.Recv(&ping_pong_count, 1, MPI::INT,
52                             partner_id, 0);
53      }
54    }
55
56    // Terminate MPI
57    MPI::Finalize();
58  }
```

Listing 9.4: Messages in the *ping-pong* program.

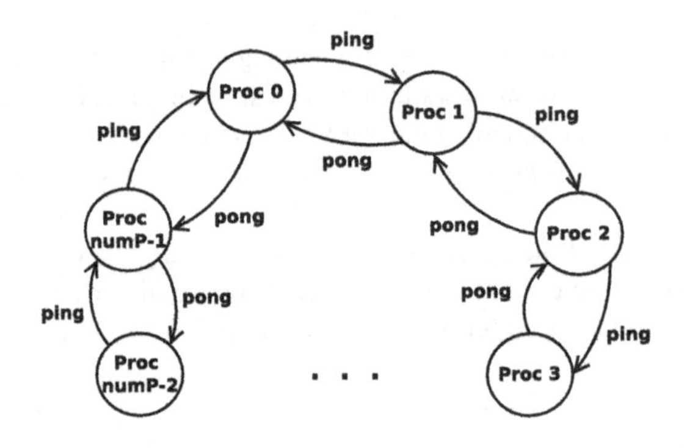

FIGURE 9.4

Abstraction of the *ping-pong* messages on an ordered ring of processes.

9.4 NONBLOCKING COMMUNICATION (PING-PONG IN A RING OF PROCESSES)

We have presented the function `Send()` as an example of a routine for initiating point-to-point communication. This is the basic blocking send operation and it may be implemented differently on different systems. Although we mentioned in the previous section that it is a blocking routine, this is an oversimplified statement. The formal definition states that it returns only after the application buffer in the sending task is free for reuse. The MPI standard permits the use of a system buffer but does not necessarily require it. Some implementations may use a synchronous send (`Ssend()`) to implement the basic blocking send, where the process is actually blocked until the destination process has started to receive the message. Therefore, while `Recv()` is always blocking, `Send()` is usually blocking only for large messages.

Blocking communication could lead to deadlocks among processes. A deadlock is defined as a specific condition when two or more processes are each waiting for the other one to release a resource, or more than two processes are waiting for resources in a circular chain. We use the code shown in Listing 9.5, with a variation of the *ping-pong* program, to illustrate this situation. Instead of grouping the processes in pairs for the communication, we structure them with an ordered ring as depicted in Fig. 9.4. In this case we need the number of the previous and next process for communication. For almost any Process `i`, the next process has rank $i + 1$. The only exception is Process `numP-1`, whose next process is `0`. The approach to determine the previous process is similar: `i-1` in all cases but process `0`, whose predecessor is `numP-1`. Here we show a preliminary version of the new *ping-pong* program:

```
1  #include <stdlib.h>
2  #include "mpi.h"
3
4  int main (int argc, char *argv[]){
5    // Initialize MPI
6    MPI::Init(argc,argv);
7    // Get the number of processes
8    int numP=MPI::COMM_WORLD.Get_size();
```

```
9      // Get the ID of the process
10     int myId=MPI::COMM_WORLD.Get_rank();
11
12     if(argc < 2){
13       // Only the first process prints the output message
14       if(!myId)
15         std::cout << "ERROR: The syntax of the program is
16           ./ping-pong num_ping_pong" << std::endl;
17       MPI::COMM_WORLD.Abort(1);
18     }
19
20     int num_ping_pong = atoi(argv[1]);
21     int ping_pong_count = 0;
22     int next_id = myId+1, prev_id=myId-1;
23
24     if(next_id >= numP)
25       next_id = 0;
26
27     if(prev_id < 0)
28       prev_id = numP-1;
29
30     while(ping_pong_count < num_ping_pong){
31       ping_pong_count++;
32
33       // Send the ping
34       MPI::COMM_WORLD.Send(&ping_pong_count,1,MPI::INT,next_id,0);
35
36       // Wait and receive  the ping
37       MPI::COMM_WORLD.Recv(&ping_pong_count,1,MPI::INT,prev_id,0);
38
39       // Send the pong
40       MPI::COMM_WORLD.Send(&ping_pong_count,1,MPI::INT,prev_id,0);
41
42       // Wait and receive the pong
43       MPI::COMM_WORLD.Recv(&ping_pong_count,1,MPI::INT,next_id,0);
44     }
45
46     // Terminate MPI
47     MPI::Finalize();
```

Listing 9.5: Simulation of *ping-pong* on a ring of processes with blocking communication.

Unfortunately, this code generates a circular chain of `Send()` calls. Surprisingly, the above listed code will actually not deadlock in many scenarios. As stated before, although `Send()` is a blocking call, the MPI specification states that this routine blocks until the send buffer can be reclaimed, i.e., when the network can buffer the message. If the messages eventually cannot be buffered by the network, they will block until a matching receive has been posted. In our case, messages are small enough such that almost all networks will be able to buffer them. However, what would happen if the messages were larger and could not be buffered by the network? In general, as the characteristics of the networks might be very different, it is not a good practice to rely on the buffer capacity. One solution to prevent unexpected deadlocks is the use of nonblocking communication calls:

- `MPI::Request Isend(const void* buf, int count, const Datatype& datatype, int dest, int tag)`
- `MPI::Request Irecv(void* buf, int count, const Datatype& datatype, int source, int tag)`

The only difference in the syntax compared to the basic `Send()` and `Recv()` calls is that the nonblocking routines return an object of the class `MPI::Request`. This object contains information about the state of the messages. Both `Isend()` and `Irecv()` return immediately without blocking the computation, and we must interact with the `MPI::Request` object to synchronize the messages. MPI provides two methods to synchronize nonblocking communication. First, the method `Wait()` (in the class `MPI::Request`) blocks the computation until the message has been sent or received. Additionally, `Test()` returns a Boolean that indicates if the operation has finished, but never blocks the process.

We modify the loop of Listing 9.6 so that it now uses nonblocking communication routines. As `Isend()` does not block the computation, there is no risk that all processes stay blocked in the `Send()` routine. The following code is safe of deadlocks:

```
1   #include <stdlib.h>
2   #include "mpi.h"
3
4   int main (int argc, char *argv[]){
5     // Initialize MPI
6     MPI::Init(argc,argv);
7     // Get the number of processes
8     int numP=MPI::COMM_WORLD.Get_size();
9     // Get the ID of the process
10    int myId=MPI::COMM_WORLD.Get_rank();
11
12    if(argc < 2){
13      // Only the first process prints the output message
14      if(!myId)
15        std::cout << "ERROR: The syntax of the program is
16          ./ping-pong num_ping_pong" << std::endl;
17      MPI::COMM_WORLD.Abort(1);
18    }
19
20    int num_ping_pong = atoi(argv[1]);
21    int ping_pong_count = 0;
22    int next_id = myId+1, prev_id=myId-1;
23
24    if(next_id >= numP)
25      next_id = 0;
26
27    if(prev_id < 0)
28      prev_id = numP-1;
29
30    MPI::Request rq_send, rq_recv;
31
32    while(ping_pong_count < num_ping_pong){
33      // First receive the ping and then send the pong
34      ping_pong_count++;
```

```
35        rq_send = MPI::COMM_WORLD.Isend(&ping_pong_count, 1, MPI::INT,
36                                        next_id, 0);
37        rq_recv = MPI::COMM_WORLD.Irecv(&ping_pong_count, 1, MPI::INT,
38                                        prev_id, 0);
39        rq_recv.Wait();
40
41        rq_send = MPI::COMM_WORLD.Isend(&ping_pong_count, 1, MPI::INT,
42                                        prev_id, 0);
43        rq_recv = MPI::COMM_WORLD.Irecv(&ping_pong_count, 1, MPI::INT,
44                                        next_id, 0);
45        rq_recv.Wait();
46      }
47
48      // Terminate MPI
49      MPI::Finalize();
```

Listing 9.6: Simulation of *ping-pong* on a ring of processes with nonblocking communication.

Note that the deadlock situation would have been solved by replacing only the `Send()` methods by the corresponding `Isend()`. We have also modified the *receive* functions to provide an example using `Irecv()`. Besides avoiding unexpected deadlocks, nonblocking communication is usually employed to overlap computation and communication, as will be explained in Section 9.6.

9.5 COLLECTIVES (COUNTING PRIMES)

Most communication patterns can be designed with the already explained point-to-point communication. However, MPI provides a set of routines for communication patterns that involve all the processes of a certain communicator, so-called collectives. The two main advantages of using collectives are:

- Reduction of the programming effort. We can reuse code already implemented by the MPI developers instead of programming handcrafted versions of complex communication patterns.
- Performance optimization, as the implementations are usually efficient, especially if optimized for specific architectures [14,22].

A new program is used as an example for illustrating collectives in MPI. We use parallel programming to determine the total number of primes in a range between 0 and certain upper bound n. The value of n is provided as command line argument to the program. The sequential reference code of a naive prime search is given by:

```
int totalPrimes = 0;
bool prime;
for(int i=2; i<=n; i++){
  prime = true;
  for(int j=2; j<i; j++){
    if((i%j) == 0){
      prime = false;
```

```
            break;
        }
    }
    totalPrimes += prime;
}
```

The outer loop brute-forces all numbers from 2 to n (0 and 1 are not prime). For each number j, we determine if i is a multiple of j. If this multiple exists, we finish the probing of i since it is obviously not prime. If no divisors are found, i.e. i is prime, we increment the variable `totalPrimes` by one.

The beginning of the parallel program is very similar to the previous examples (see Listing 9.7): MPI initialization and parsing of the relevant parameters. However, we introduce a slight modification. Only Process 0 reads the number n from the standard input, so we will need to send it to the other processes before starting the computation. Note, this example is artificial since all processes actually have access to the parameters of an MPI program. However, it will be useful to explain how to efficiently send the same data from one process to the others. Thus the initialization reads as follows:

```
1  #include <stdlib.h>
2  #include "mpi.h"
3
4  int main (int argc, char *argv[]){
5    // Initialize MPI
6    MPI::Init(argc,argv);
7    // Get the number of processes
8    int numP=MPI::COMM_WORLD.Get_size();
9    // Get the ID of the process
10   int myId=MPI::COMM_WORLD.Get_rank();
11
12   if(argc < 2){
13     // Only the first process prints the output message
14     if(!myId)
15       std::cout << "ERROR: The syntax of the program is ./primes n"
16         << std::endl;
17     MPI::COMM_WORLD.Abort(1);
18   }
19
20   int n;
21   if(!myId)
22     n = atoi(argv[1]);
23
24   // Barrier to synchronize the processes before measuring time
25   MPI::COMM_WORLD.Barrier();
26
27   // Measure the current time
28   double start = MPI::Wtime();
```

Listing 9.7: Initial part of the program to count the number of primes.

Two new MPI methods that need further explanation have been included in the previous code. `MPI::Wtime()` is a method that returns the current time as a `double`. It is employed to measure the execution time of the program. However, in parallel programs it is desirable that all processes start measuring the execution time at the same moment. This is the reason why we have included the method

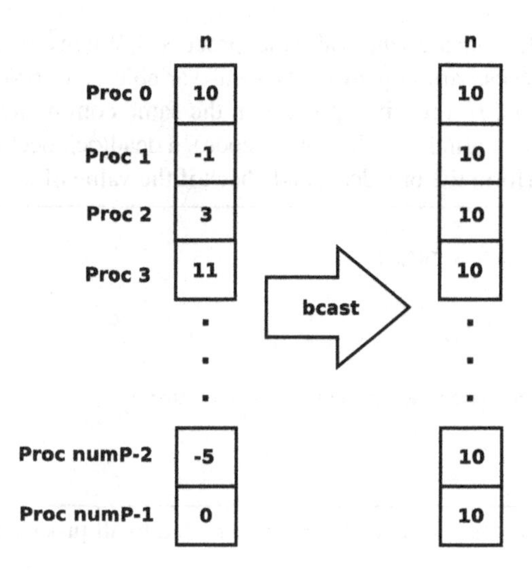

FIGURE 9.5

Broadcast of one element from Process 0.

`Barrier()` in the code. This is the simplest collective available in MPI that synchronizes the processes of a specified communicator: no process can continue until all have reached the barrier.

Once the timer has been correctly initialized, we broadcast the value of the upper bound n, i.e., we send it from Process 0 to all other processes in the specified communicator, as shown in Fig. 9.5. `Bcast()` is another collective since all processes of a communicator are involved. Of course, the same can be achieved with simple point-to-point communication:

```
if(!myId)
  for(i=1; i<numP; i++)
    MPI::COMM_WORLD.Send(&n, 1, MPI::INT, i, 0);
else
  MPI::COMM_WORLD.Recv(&n, 1, MPI::INT, 0, 0);
```

However, we can reduce the programming effort by using the methods provided by MPI implementations for common collectives like the aforementioned broadcast. Moreover, these methods usually obtain better performance than our ad hoc implementations, as they are usually designed with trees or other types of communication patterns that are optimized for the characteristics and topologies of the underlying interconnection networks. The syntax of the broadcast method is:

- `void Bcast(void* buffer, int count, const MPI::Datatype& datatype, int root)`

This routine must be called by all the processes of the communicator. Most arguments have already been explained in the section dealing with point-to-point communication. The only new parameter is `root`: it specifies the source process holding the data. Although the root process and receiver processes perform different jobs, they all call the same method. When the root process (in our example, it is

Process 0) calls `Bcast()`, the data are sent to all other processes. When a receiver process calls `Bcast()`, the data from the root process are copied to its local variable. The broadcast is finished when all processes received their data. If a receiver process in the same communicator does not participate in the broadcast (by simply not stating it in the source code), a deadlock occurs. We include the fragment shown in Listing 9.8 to perform the broadcast and check if the value of n is correct.

```
31  // Send the value of n to all processes
32  MPI::COMM_WORLD.Bcast(&n, 1, MPI::INT, 0);
33
34  if(n < 1){
35    // Only the first process prints the output message
36    if(!myId)
37      std::cout << "ERROR: The parameter 'n' must be higher than 0"
38        << std::endl;
39    MPI::COMM_WORLD.Abort(1);
40  }
```

Listing 9.8: Broadcast upper bound to all processes.

Now that n is available to all processes we can start the counting of primes. We divide the workload such that each process probes different integers, i.e., distributing the elements i of the outer loop among them. Concretely, we apply a cyclic distribution where the numbers are assigned to different processes one by one. As can be seen in Listing 9.9, the inclusion of the variables `myId` and `numP` in the syntax of the loop allows each process to work with more than one value of i.

```
41  // Perform the computation of the number of primes
42  // between 0 and n in parallel
43  int myCount = 0;
44  int total;
45  bool prime;
46
47  // Each process analyzes only part of the numbers below n
48  // The distribution is cyclic for better workload balance
49  for(int i=2+myId; i<=n; i=i+numP){
50    prime = true;
51    for(int j=2; j<i; j++){
52      if((i%j) == 0){
53        prime = false;
54        break;
55      }
56    }
57    myCount += prime;
58  }
```

Listing 9.9: Calculation of the number of primes.

At this point of the program every process has accumulated a different value in `myCount`, which indicates the number of found primes in its specific range of indices. The total amount of primes is the sum of the all `myCount` values stored in the processes. Similarly to the broadcast, we could perform this summation by sending the *myCount* values from all processes to Process 0 and finally adding up the values in a sequential manner:

```
if(!myId){
  total = myCount;
  for(int i=1; i<numP; i++){
    MPI::COMM_WORLD.Recv(&myCount, 1, MPI::INT, i, 0);
    total += myCount;
  }
}
else
  MPI::COMM_WORLD.Send(&myCount, 1, MPI::INT, 0, 0);
```

This is a correct but inefficient implementation of the reduction using point-to-point communication. An important drawback is that the code of the destination process requires ordered *receives*. Assume that Process 2 sends the message some seconds earlier than Process 1. An efficient out-of-order implementation would use these seconds to update the variable `total` with the value from Process 2. However, in the previous implementation `total` must be updated first with the value from Process 1, then from Process 2, etc. We now modify our previous reduction to support out-of-ordered updates:

```
MPI::Status status;
if(!myId){
  total = myCount;
  for(int i=1; i<numP; i++){
    MPI::COMM_WORLD.Recv(&myCount, 1, MPI::INT,
                         MPI::ANY_SOURCE, 0, status);
    total += myCount;
  }
}
else {
  MPI::COMM_WORLD.Send(&myCount, 1, MPI::INT, 0, 0);
}
```

Simply specifying `MPI::ANY_SOURCE` in the `root` value, the destination process updates `total` as soon as it receives one message. It does not matter which process sends first. `MPI::ANY_TAG` provides a similar functionality for the `tag` parameter. Furthermore, we can specify in `Recv()` an object of the class `MPI::Status`, which includes the methods `Get_source()`, `Get_tag()`, and `Get_error()` to obtain information about the state of the message.

In general, it is more efficient to employ the collective `Reduce()`, explicitly designed for this type of operations (Fig. 9.6). The syntax of the MPI reduction collective is:

- `void Reduce(const void* sendbuf, void* recvbuf, int count, const MPI::Datatype& datatype, const MPI::Op& op, int root)`

The only argument that needs explanation is `op`, which represents the operation that will be performed during the reduction. In our program we need to employ addition (`MPI::SUM`) but there are other operations defined by MPI such as product, logical and/or, min/max, etc. Additionally, we can work with custom operations defined by the user. Note, ensure that your custom operation is an associative map, otherwise your results may be incorrect. The remaining part of our program (Listing 9.10) includes:

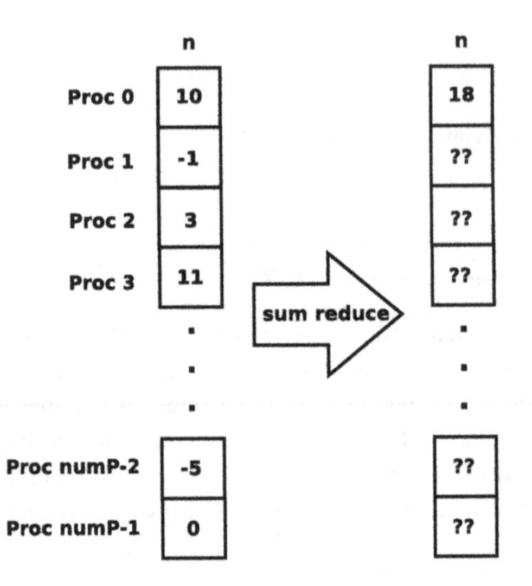

FIGURE 9.6

Sum reduction to Process 0.

1. the collective for reduction
2. the measure of the final time
3. printing out the result
4. MPI finalization

```
59  // Reduce the partial counts into 'total' in the Process 0
60  MPI::COMM_WORLD.Reduce(&myCount, &total, 1,
61                         MPI::INT, MPI::SUM, 0);
62
63  // Measure the current time
64  double end = MPI::Wtime();
65
66  if(!myId){
67    std::cout << total << " primes between 1 and "
68      << n << std::endl;
69    std::cout << "Time with " << numP << " processes: "
70      << end-start << " seconds" << std::endl;
71  }
72
73  // Terminate MPI
74  MPI::Finalize();
```

Listing 9.10: Reduction of the number of primes obtained by each process.

The program presented in this section made use of three common blocking collectives (`Barrier()`, `Bcast()` and `Reduce()`) as representative examples. Although the programs of the next sections will use more examples, we want to mention a brief list of useful MPI collectives.

- `Allreduce()`. Combination of reduction and a broadcast so that the output is available for all processes.
- `Scatter()`. Split a block of data available in a root process and send different fragments to each process.
- `Gather()`. Send data from different processes and aggregate it in a root process.
- `Allgather()`. Similar to `Gather()` but the output is aggregated in buffers of all the processes.
- `Alltoall()`. All processes scatter data to all processes.

Furthermore, there exist variants of these routines for nonblocking communications, as well as variable block size per process. More information about these collectives can be found in the MPI manual.

9.6 OVERLAPPING COMPUTATION AND COMMUNICATION (JACOBI ITERATION)

A common task in scientific applications is the iterative computation of steady-state solutions of the Poisson equation $\Delta\phi(p) = f(p)$ over the rectangular domain Ω with Dirichlet boundary condition $\phi(p) = g(p)$ for all points p on the boundary $\partial\Omega$. The sum of second derivatives

$$\Delta = \sum_{k=0}^{d-1} \frac{\partial^2}{\partial x_k^2} = \frac{\partial^2}{\partial x_0^2} + \cdots + \frac{\partial^2}{\partial x_{d-1}^2}$$

is called Laplace operator in Euclidean space and locally measures the amount of curvature of a real-valued function ϕ defined on a d-dimensional domain. Loosely speaking, if one substitutes the linear map Δ with a matrix A and the functions ϕ and f with vectors x and b, the problem simplifies to finding a solution of the equation $A \cdot x = b$ for given A and b. We will see that in the special case of $A = \Delta$ a solution can be computed by repeatedly averaging local patches on a discretized representation of ϕ.

The presented algorithm falls in a broader class of so-called *stencil code* or *tensor convolution algorithms*[2] which iteratively apply a small-sized mask (stencil) to matrices or higher dimensional arrays (tensors), respectively. For the sake of simplicity, we fix the dimension to $d = 2$, i.e., ϕ can be interpreted as continuous image and thus its discretized version as a matrix `data` with finite shape `rows` × `cols`. Further, we set the heterogeneous term f to zero. However, the derivation for non-vanishing f is straightforward and left to the reader as an exercise.

A discrete approximation of the Laplacian Δ using a finite differences approach with positive step size h reads as follows:

$$(\Delta\phi)(x, y) \approx \frac{\phi(x+h, y) + \phi(x-h, y) + \phi(x, y+h) + \phi(x, y-h) - 4 \cdot \phi(x, y)}{h^2}$$

[2]Nonlinear tensor convolution such as max-pooling is an important filter in deep neural networks.

Since $\Delta\phi = 0$ per construction, we can solve the equation for $\phi(x, y)$. Further we set $h = 1$ to unit step size, which is a reasonable choice if we want to enumerate grid vertices with index tuples (i, j).[3] Thus the discretized update rule on the matrix representation `data` of the continuous image ϕ is given by the ordinary average over the 4-neighborhood of a pixel $(i, j) \in \Omega \setminus \partial\Omega$:

$$\texttt{data}[i, j] \leftarrow \frac{\texttt{data}[i+1, j] + \texttt{data}[i-1, j] + \texttt{data}[i, j+1] + \texttt{data}[i, j-1]}{4}.$$

Due to the fact that each tuple (i, j) can be updated independently and because of the massive amount of grid vertices that are needed for a decent approximation of the continuous domain, tensor convolution algorithms are often parallelized [6,18]. The update step is repeated in an inherent sequential loop until convergence of the matrix. Fig. 9.7 illustrates the so-called Jacobi algorithm for an arbitrary matrix.

Next you can see the sequential code of a single Jacobi iteration step using two arrays (`data` and `buffer`) of shape `rows`$\times$`cols`. The first array stores the matrix data while the second is an auxiliary array for updating the values.

```
for(int i=1; i<rows-1; i++)
  for(int j=1; j<cols-1; j++)
    // calculate discrete Laplacian by averaging 4-neighborhood
    buff[i*cols+j] = 0.25f*(data[(i+1)*cols+j]+data[i*cols+j-1]+
                      data[i*cols+j+1]+data[(i-1)*cols+j]);

// this copy could be avoided by swapping pointers
memcpy(data, buff, rows*cols*sizeof(float));
```

The sum of squared residues between the values stored in `data` and `buffer` will be employed in our example as convergence measure:

```
float error = 0.0;
for(int i=1; i<rows-1; i++)
  for(int j=1; j<cols-1; j++)
    // determine difference between 'data' and 'buff'
    error += (data[i*cols+j]-buff[i*cols+j])*
             (data[i*cols+j]-buff[i*cols+j]);
```

Our parallel version of Jacobi iteration uses a 1D block distribution, as illustrated in Fig. 9.8. The shape of the matrix, as well as the convergence threshold, are given as program parameters. For the sake of simplicity, we assume that the number of rows must be a multiple of the number of processes. Listing 9.11 shows the initialization of the parameters and the matrix. Only Process 0 calls the function `readInput` that initializes the matrix (it can be implemented to read it from a file, or to generate the matrix ad hoc).

```
1  #include <stdlib.h>
2  #include <stdio.h>
3  #include <iostream>
```

[3] Here we suppress unnecessary noise in the equations by neglecting the affine transformation between the image domain and the matrix shape.

Initial data

-1	-1	-1	-1	-1	-1	-1	-1	-1	-1	-1	-1
3	3	3	3	3	3	3	3	3	3	3	3
3	3	3	3	3	3	3	3	3	3	3	3
2	2	2	2	2	2	2	2	2	2	2	2
2	2	2	2	2	2	2	2	2	2	2	2
2	2	2	2	2	2	2	2	2	2	2	2
1	1	1	1	1	1	1	1	1	1	1	1
1	1	1	1	1	1	1	1	1	1	1	1
1	1	1	1	1	1	1	1	1	1	1	1
0	0	0	0	0	0	0	0	0	0	0	0
0	0	0	0	0	0	0	0	0	0	0	0
-1	-1	-1	-1	-1	-1	-1	-1	-1	-1	-1	-1

data after 1 iteration

-1	-1	-1	-1	-1	-1	-1	-1	-1	-1	-1	-1
3	2	2	2	2	2	2	2	2	2	2	3
3	2.8	2.8	2.8	2.8	2.8	2.8	2.8	2.8	2.8	2.8	3
2	2.3	2.3	2.3	2.3	2.3	2.3	2.3	2.3	2.3	2.3	2
2	2	2	2	2	2	2	2	2	2	2	2
2	1.8	1.8	1.8	1.8	1.8	1.8	1.8	1.8	1.8	1.8	2
1	1.3	1.3	1.3	1.3	1.3	1.3	1.3	1.3	1.3	1.3	1
1	1	1	1	1	1	1	1	1	1	1	1
1	0.8	0.8	0.8	0.8	0.8	0.8	0.8	0.8	0.8	0.8	1
0	0.3	0.3	0.3	0.3	0.3	0.3	0.3	0.3	0.3	0.3	0
0	-0.3	-0.3	-0.3	-0.3	-0.3	-0.3	-0.3	-0.3	-0.3	-0.3	0
-1	-1	-1	-1	-1	-1	-1	-1	-1	-1	-1	-1

data after 25 iterations

-1	-1	-1	-1	-1	-1	-1	-1	-1	-1	-1	-1
3	1.0	0.2	-0.1	-0.3	-0.3	-0.3	-0.3	-0.1	0.2	1.0	3
3	1.6	0.9	0.5	0.4	0.3	0.3	0.4	0.5	0.9	1.6	3
2	1.6	1.2	0.9	0.8	0.7	0.7	0.8	0.9	1.2	1.6	2
2	1.6	1.3	1.1	1.0	0.9	0.9	1.0	1.1	1.3	1.6	2
2	1.5	1.3	1.1	1.0	1.0	1.0	1.0	1.1	1.3	1.5	2
1	1.1	1.0	1.0	0.9	0.9	0.9	0.9	1.0	1.0	1.1	1
1	0.9	0.8	0.7	0.7	0.7	0.7	0.7	0.7	0.8	0.9	1
1	0.6	0.5	0.4	0.3	0.3	0.3	0.3	0.4	0.5	0.6	1
0	0.1	0.0	0.0	-0.1	-0.1	-0.1	-0.1	0.0	0.0	0.1	0
0	-0.3	-0.5	-0.5	-0.5	-0.5	-0.5	-0.5	-0.5	-0.5	-0.3	0
-1	-1	-1	-1	-1	-1	-1	-1	-1	-1	-1	-1

data after 75 iterations

-1	-1	-1	-1	-1	-1	-1	-1	-1	-1	-1	-1
3	0.9	0.1	-0.3	-0.4	-0.5	-0.5	-0.4	-0.3	0.1	0.9	3
3	1.5	0.7	0.3	0.1	0.0	0.0	0.1	0.3	0.7	1.5	3
2	1.5	1.0	0.6	0.4	0.3	0.3	0.4	0.6	1.0	1.5	2
2	1.5	1.1	0.8	0.6	0.5	0.5	0.6	0.8	1.1	1.5	2
2	1.4	1.0	0.8	0.6	0.5	0.5	0.6	0.8	1.1	1.4	2
1	1.0	0.8	0.6	0.5	0.5	0.5	0.5	0.6	0.8	1.0	1
1	0.8	0.6	0.4	0.4	0.3	0.3	0.4	0.4	0.8	0.8	1
1	0.5	0.3	0.2	0.1	0.1	0.1	0.1	0.2	0.3	0.5	1
0	0.0	-0.1	-0.2	-0.2	-0.3	-0.3	-0.2	-0.2	-0.1	0.0	0
0	-0.4	-0.5	-0.6	-0.6	-0.6	-0.6	-0.6	-0.6	-0.5	-0.4	0
-1	-1	-1	-1	-1	-1	-1	-1	-1	-1	-1	-1

FIGURE 9.7

Repetitive application of the update step to the upper left input matrix.

```
#include <string.h>
#include "mpi.h"

int main (int argc, char *argv[]){
  // Initialize MPI
  MPI::Init(argc,argv);
  // Get the number of processes
  int numP=MPI::COMM_WORLD.Get_size();
  // Get the ID of the process
  int myId=MPI::COMM_WORLD.Get_rank();

```

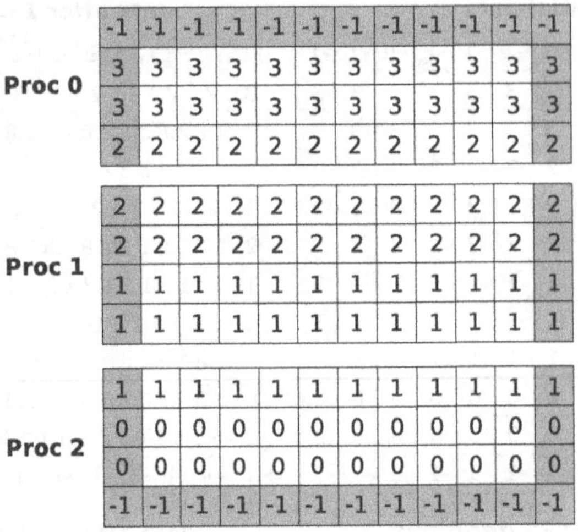

Proc 0	-1	-1	-1	-1	-1	-1	-1	-1	-1	-1	-1	-1
	3	3	3	3	3	3	3	3	3	3	3	3
	3	3	3	3	3	3	3	3	3	3	3	3
	2	2	2	2	2	2	2	2	2	2	2	2
Proc 1	2	2	2	2	2	2	2	2	2	2	2	2
	2	2	2	2	2	2	2	2	2	2	2	2
	1	1	1	1	1	1	1	1	1	1	1	1
	1	1	1	1	1	1	1	1	1	1	1	1
Proc 2	1	1	1	1	1	1	1	1	1	1	1	1
	0	0	0	0	0	0	0	0	0	0	0	0
	0	0	0	0	0	0	0	0	0	0	0	0
	-1	-1	-1	-1	-1	-1	-1	-1	-1	-1	-1	-1

FIGURE 9.8

An example of a 1D distribution using three processes. The rows of the matrix are distributed using a block approach.

```
15   if(argc < 4){
16         // Only the first process prints the output message
17         if(!myId)
18       std::cout << "ERROR: The syntax of the program is
19                 ./jacobi rows cols errThreshold" << std::endl;
20
21          MPI::COMM_WORLD.Abort(1);
22   }
23
24   int rows = atoi(argv[1]); int cols = atoi(argv[2]);
25   float errThres = atof(argv[3]);
26
27   if((rows < 1) || (cols < 1)){
28     // Only the first process prints the output message
29     if(!myId)
30       std::cout << "ERROR: The number of rows and columns must
31                    be higher than 0" << std::endl;
32
33     MPI::COMM_WORLD.Abort(1);
34   }
35
36   if(rows%numP){
37     // Only the first process prints the output message
38     if(!myId)
39       std::cout << "ERROR: The number of rows must be multiple
40                    of the number of processes" << std::endl;
41
```

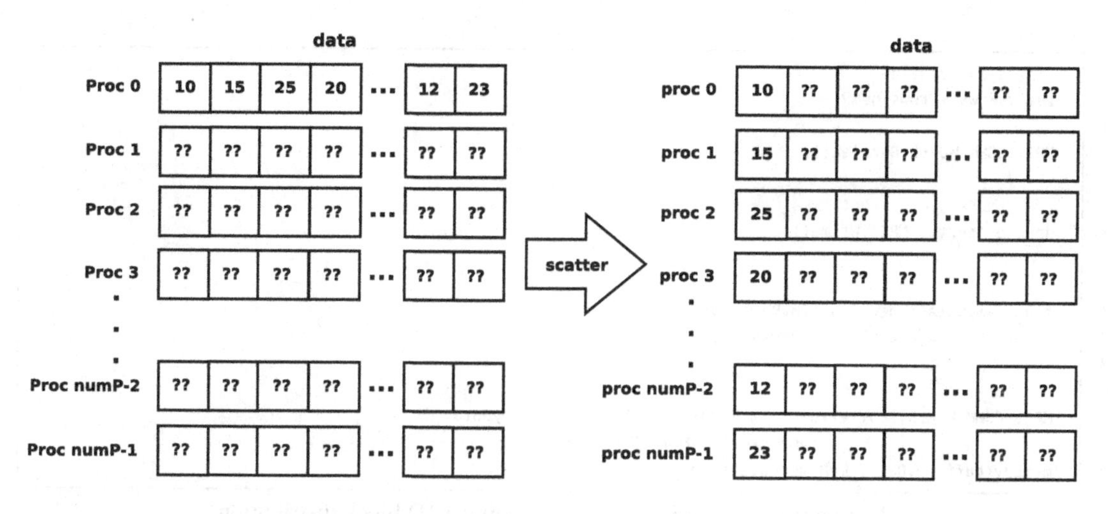

FIGURE 9.9

Scattering of an array from Process 0.

```
42        MPI::COMM_WORLD.Abort(1);
43      }
44
45      float *data;
46
47      // Only one process reads the data of the initial matrix
48      if(!myId){
49        data = new float[rows*cols];
50        readInput(rows, cols, data);
51      }
```

Listing 9.11: Initialization of variables and matrix for parallel Jacobi iteration.

The next step is distributing the matrix among the processes. We use the collective `Scatter()` which, as presented in Section 9.5, splits a block of data available in the root process and sends different fragments to each process. Its behavior is illustrated in Fig. 9.9 and its syntax is:

- `Scatter(const void* sendbuf, int sendcount, const MPI::Datatype& sendtype, void* recvbuf, int recvcount, const MPI::Datatype& recvtype, int root)`

Apart from the parameter to specify the `root` process, we must specify three arguments for both the source and destination processes: the source/destination buffer for the data, the datatype, and the number of elements. Note that `sendcount` is the number of elements received by a process, not the length of the scattered array. Listing 9.12 shows how to distribute the matrix as shown in Fig. 9.8 using the `Scatter()` collective. This code fragment also includes the initialization of the timer and the auxiliary array `buff`.

```
52  // The computation is divided by rows
53  int myRows = rows/numP;
54
55  MPI::COMM_WORLD.Barrier();
56
57  // Measure the current time
58  double start = MPI::Wtime();
59
60  // Arrays for the chunk of data to work
61  float *myData = new float[myRows*cols];
62  float *buff = new float[myRows*cols];
63
64  // Scatter the input matrix
65  MPI::COMM_WORLD.Scatter(data, myRows*cols, MPI::FLOAT, myData,
66                          myRows*cols, MPI::FLOAT, 0);
67  memcpy(buff, myData, myRows*cols*sizeof(float));
```

Listing 9.12: Scattering of a matrix using a 1D block distribution.

At this point the processes have almost all the necessary values to initially update their rows in the first iteration. For instance, in Fig. 9.8, Process 1 has all the necessary information to update rows 5 and 6 (i.e., its central rows) but it cannot update rows 4 and 7 yet. However, in order to update their boundary rows, the processes need the values of rows stored in other processes. In our example, Process 1 needs row 3 (stored in Process 0) to update row 4, and it also needs row 8 (stored in Process 2) to update row 7. As a result, we need to share the boundary rows among processes at the beginning of each iteration.

Under these assumptions, the Jacobi iteration steps are implemented in the body of the `while`-loop listed in Listing 9.13. After sending and receiving the boundary rows, each process performs the updating of cells and afterwards calculates its local error. The iteration finishes with an extended version of the *reduction* collective called `Allreduce()` that stores the sum of the local errors in all processes. Note, this combined reduction and broadcast of the global error is crucial to avoid a deadlock that may be caused by a mixed evaluation of the checked condition in the `while`-loop.

```
68  float error = errThres+1.0;
69  float myError;
70
71  // Buffers to receive the boundary rows
72  float *prevRow = new float[cols];
73  float *nextRow = new float[cols];
74
75  while(error > errThres){
76    if(myId > 0)
77      // Send the first row to the previous process
78      MPI::COMM_WORLD.Send(myData, cols, MPI::FLOAT, myId-1, 0);
79
80    if(myId < numP-1){
81      // Receive the next row from the next process
82      MPI::COMM_WORLD.Recv(nextRow, cols, MPI::FLOAT, myId+1, 0);
83
```

```
84          // Send the last row to the next process
85          MPI::COMM_WORLD.Send(&myData[(myRows-1)*cols], cols,
86                               MPI::FLOAT, myId+1, 0);
87        }
88
89        if(myId > 0)
90          // Receive the previous row from the previous process
91          MPI::COMM_WORLD.Recv(prevRow, cols, MPI::FLOAT, myId-1, 0);
92
93        // Update the first row
94        if((myId > 0) && (myRows>1))
95          for(int j=1; j<cols-1; j++)
96            buff[j] = 0.25*(myData[cols+j]+myData[j-1]+
97              myData[j+1]+prevRow[j]);
98
99        // Update the main block
100       for(int i=1; i<myRows-1; i++)
101         for(int j=1; j<cols-1; j++)
102           // calculate discrete Laplacian by average 4-neighborhood
103           buff[i*cols+j]=
104               0.25f*(myData[(i+1)*cols+j]+myData[i*cols+j-1]+
105               myData[i*cols+j+1]+myData[(i-1)*cols+j]);
106
107       // Update the last row
108       if((myId < numP-1) && (myRows > 1))
109         for(int j=1; j<cols-1; j++)
110           buff[(myRows-1)*cols+j] =
111               0.25*(nextRow[j]+myData[(myRows-1)*cols+j-1]+
112               myData[(myRows-1)*cols+j+1]+myData[(myRows-2)*cols+j]);
113
114       // Calculate the error of the block
115       myError = 0.0;
116       for(int i=0; i<myRows; i++)
117         for(int j=1; j<cols-1; j++)
118           // determine difference between 'data' and 'buff'
119           myError += (myData[i*cols+j]-buff[i*cols+j])*
120                      (myData[i*cols+j]-buff[i*cols+j]);
121
122       memcpy(myData, buff, myRows*cols*sizeof(float));
123
124       // Sum the error of all the processes
125       // Output is stored in the variable 'error' of all processes
126       MPI::COMM_WORLD.Allreduce(&myError, &error, 1, MPI::FLOAT,
127                                 MPI::SUM);
128     }
```

Listing 9.13: Parallel Jacobi iteration.

Once the matrix has converged, all processes must send their final portion of the matrix to the root, so it can print the result. This operation is performed with the `Gather()` collective whose behavior is the opposite of `Scatter()` (see Fig. 9.10).

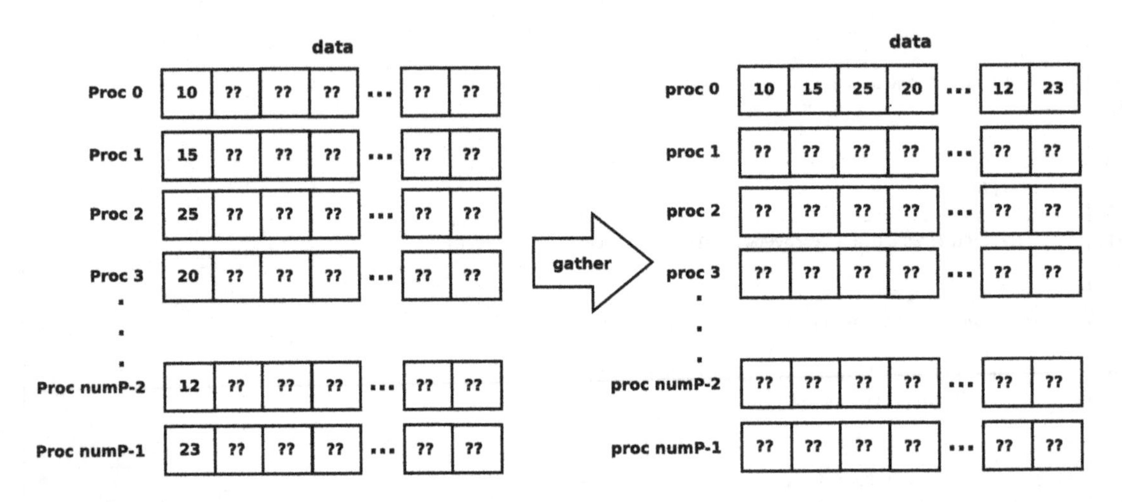

FIGURE 9.10

Gathering of an array to Process 0.

- `Gather(const void* sendbuf, int sendcount, const MPI::Datatype& sendtype, void* recvbuf, int recvcount, const MPI::Datatype& recvtype, int root)`

Gathering and printing the results (calling an external function `printOutput()`), measuring the execution time and MPI finalization are included in Listing 9.14.

```
129  // Only Process 0 writes
130  // Gather the final matrix to the memory of Process 0
131  MPI::COMM_WORLD.Gather(myData, myRows*cols, MPI::FLOAT, data,
132                         myRows*cols, MPI::FLOAT, 0);
133
134  // Measure the current time
135  double end = MPI::Wtime();
136
137  if(!myId){
138    std::cout << "Time with " << numP << " processes: "
139              << end-start << " seconds" << std::endl;
140    printOutput(rows, cols, data);
141    delete [] data;
142  }
143
144  delete [] myData;
145  delete [] buff;
146  delete [] prevRow;
147  delete [] nextRow;
148
```

```
149    // Terminate MPI
150    MPI::Finalize();
151  }
```

Listing 9.14: Gathering of a matrix using a 1D block distribution.

Unfortunately, the described parallelization scheme has a performance issue: all processes must wait to update the rows until the communication has finished. However, most of the computation in each process (i.e. updating all internal rows) is independent of the communication since we only access rows that are stored in the same process.

Could we overlap communication and computation by updating the internal rows while the messages are still transmitted? The answer is yes. Overlapping is a frequently used technique to improve the performance of MPI programs [5,12].

Listing 9.15 shows how to modify the presented algorithm by overlapping computation and communication using nonblocking communication presented in Section 9.4. Similarly to the previous version, each iteration starts with the communication. However, now we use `Isend()` and `Irecv()` instead of their blocking counterparts. Therefore processes can update their internal rows while the messages are being communicated. Before updating the boundary rows, we wait for the completion of the communication with the method `Wait()` defined in the class `MPI::Request`.

```
68    float error = errThres+1.0;
69    float myError;
70
71    // Buffers to receive the boundary rows
72    float *prevRow = new float[cols];
73    float *nextRow = new float[cols];
74    MPI::Request request[4];
75
76    while(error > errThres){
77      if(myId > 0){
78        // Send the first row to the previous process
79        request[0] = MPI::COMM_WORLD.Isend(myData, cols, MPI::FLOAT,
80                                           myId-1, 0);
81        // Receive the previous row from the previous process
82        request[1] = MPI::COMM_WORLD.Irecv(prevRow, cols, MPI::FLOAT,
83                                           myId-1, 0);
84      }
85
86      if(myId < numP-1){
87        // Send the last row to the next process
88        request[2] = MPI::COMM_WORLD.Isend(&myData[(myRows-1)*cols],
89                          cols, MPI::FLOAT, myId+1, 0);
90        // Receive the next row from the next process
91        request[3] = MPI::COMM_WORLD.Irecv(nextRow, cols,
92                          MPI::FLOAT, myId+1, 0);
93      }
94
95      // Update the main block
96      for(int i=1; i<myRows-1; i++)
```

```
97        for(int j=1; j<cols-1; j++)
98            // Discrete Laplacian by averaging 4-neighborhood
99            buff[i*cols+j]=
100               0.25f*(myData[(i+1)*cols+j]+myData[i*cols+j-1]+
101               myData[i*cols+j+1]+myData[(i-1)*cols+j]);
102
103     // Update the first row
104     if(myId > 0){
105       request[1].Wait(status);
106       if(myRows > 1)
107         for(int j=1; j<cols-1; j++)
108           buff[j] = 0.25*(myData[cols+j]+myData[j-1]+
109                     myData[j+1]+prevRow[j]);
110     }
111
112     // Update the last row
113     if(myId < numP-1){
114       request[3].Wait(status);
115       if(myRows > 1)
116         for(int j=1; j<cols-1; j++)
117           buff[(myRows-1)*cols+j] =
118               0.25*(nextRow[j]+myData[(myRows-1)*cols+j-1]+
119               myData[(myRows-1)*cols+j+1]+myData[(myRows-2)
120               *cols+j]);
121     }
122
123     memcpy(myData, buff, myRows*cols*sizeof(float));
124
125     // Sum the error of all the processes
126     // Output is stored in the variable 'error' of all processes
127     MPI::COMM_WORLD.Allreduce(&myError, &error, 1, MPI::FLOAT,
128                             MPI::SUM);
129   }
```

Listing 9.15: Parallel Jacobi iteration with nonblocking communication.

In order to test whether nonblocking communication is useful, we have executed our blocking and nonblocking versions on a system with four 16-core AMD Opteron 6272 processors (i.e., 64 cores at 2.10 GHz). Fig. 9.11 shows the speedups of both codes computing a matrix of dimensions 4096×4096 floats (strong scaling) and error threshold of 0.1. For a small number of processes, the speedups are nearly the same because the blocks are large enough such that the computational time on each iteration step is much longer than the communication time. Thus, hiding communication time does not impact performance. However, the implementation with nonblocking communication on 64 processes is significantly better. In this case communication and computation time in each iteration step are more similar (as blocks are smaller). Thus, hiding the communication time has a great influence on performance. Note that, due to the high communication requirements, the blocking approach does not scale for 64 processes.

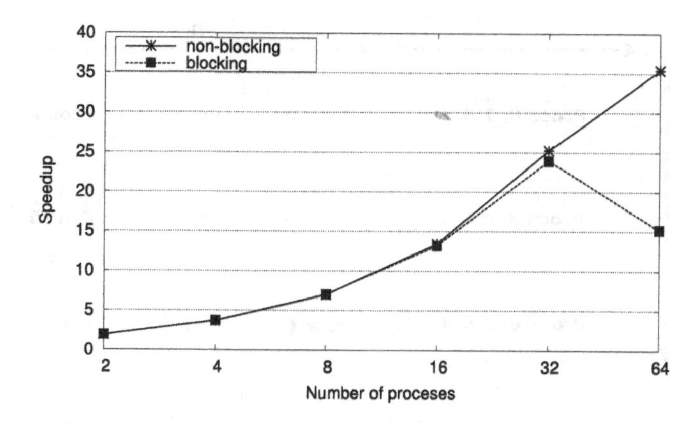

FIGURE 9.11

Performance of the Jacobi parallelization with blocking and nonblocking communications on a system with 64 cores. Matrix dimensions are 4096 × 4096 and error threshold is 0.1.

9.7 DERIVED DATATYPES (MATRIX MULTIPLICATION WITH SUBMATRIX SCATTERING)

As we have seen in the previous sections, all communication routines (both point-to-point and collectives) have arguments of the class `MPI::Datatype` to specify the type of the data. Up to now we have always used predefined MPI types such as `MPI::INT` or `MPI::FLOAT`, which allows us to send data of the same type that are contiguous in memory.

However, what happens if we want to send data that are noncontiguous in memory? Should we create one message for each contiguous block (with the performance drawback of sending several small messages)? What happens in the case that we want to send data of different type? Using the predefined datatypes as building blocks, MPI allows programmers to define their own derived types using a series of MPI methods.

There are three kinds of derived datatypes in MPI. The methods are part of any class that extends `MPI::Datatype`:

- `MPI::Datatype Create_contiguous(int count)`. It creates a new datatype defined as `count` contiguous elements of another existing type.
- `MPI::Datatype Create_vector(int count, int blocklength, int stride)`. It can be used to reference equally-spaced, fixed-sized blocks of data. Each block is simply a concatenation of `blocklength` elements of the old datatype and the spacing `stride` between blocks is a multiple of the extent of the old datatype to skip. The number of blocks in the datatype is specified by `count`.
- `MPI::Datatype Create_struct(int count, const int array_of_blocklengths[], const MPI::Aint array_of_displacements[],`

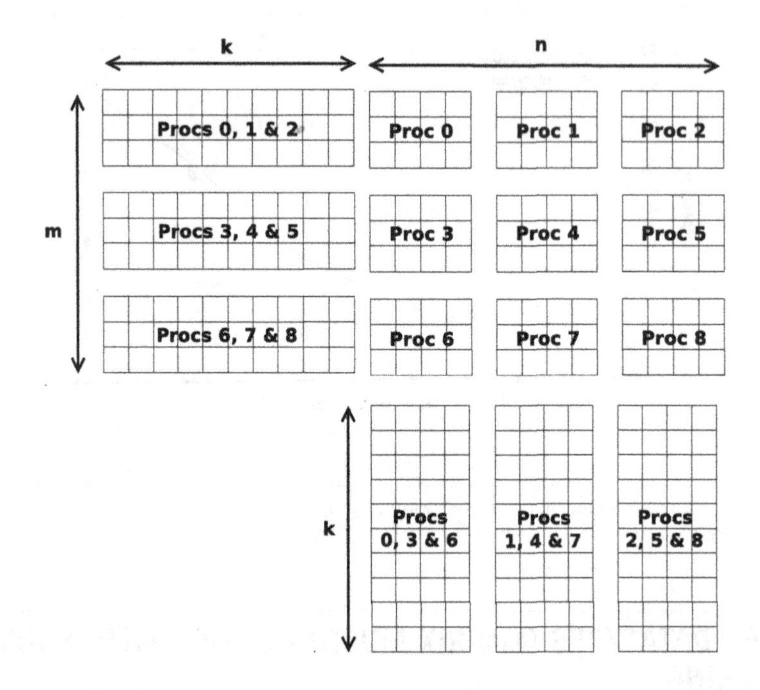

FIGURE 9.12

Example distribution of the three involved matrices: A (left) $\cdot B$ (bottom) $= C$ (right). We use nine processes and the dimensions are: $m = 9$, $k = 10$, $n = 12$.

`const MPI::Datatype array_of_types[])`. This is the more general datatype. It allows for replication of an old type in noncontiguous blocks. However, block sizes and strides can vary.

We will use the three kinds of derived datatypes in a parallel implementation of the scaled matrix product $\alpha \cdot A \cdot B = C$ where A, B and C are matrices of dimensions $m \times k$, $k \times n$ and $m \times n$, respectively. Further, α is a float to scale the matrix product.

Each process is responsible for a 2D tile of C. We distribute the rows of A and the columns of B so that each process has the necessary rows and columns to calculate its block of C. Fig. 9.12 illustrates an example of this distribution for nine processes. For the sake of simplicity, both m and n must be multiples of the number of processes.

In this case the shape of the matrices, as well as the value of the scalar α, are read from a configuration file by Process 0 and afterwards broadcast to all processes. As can be seen in Listing 9.16, we start designing a *struct* `params` in order to store the parameters. After the MPI initialization, we create an MPI struct datatype (`paramsType`) to send these parameters with only one message. Note that we also save in the variable `dimGrid` the dimension of the process grid and we check that we can create a square grid (Line 24).

The creation of the datatype consists of two steps. First, we declare it with the proper `Create_struct()` routine. The first argument specifies that our struct has two blocks. Therefore the other arguments must be arrays of two entries, one per block. Filling the length of the blocks (3, 1) and their datatypes (`MPI::INT`, `MPI::FLOAT`) is straightforward. However, determining the displacements or offsets is less obvious. Although we know that the first block does not have displacement (value 0), we need the routine `Get_extent()`, that provides the size of an MPI datatype, to indicate the displacement of the second block. Once the datatype is defined, we must confirm the creation of the type with the method `Commit()`. Listing 9.16 finishes by broadcasting the structure, the checking of the dimensions and the initialization of the matrices by Process 0 (using a function that reads the matrix values from input files).

```
1   #include <stdlib.h>
2   #include <stdio.h>
3   #include <iostream>
4   #include <string.h>
5   #include <math.h>
6   #include "mpi.h"
7
8   struct params{
9     int m, k, n;
10    float alpha;
11  };
12
13  int main (int argc, char *argv[]){
14    // Initialize MPI
15    MPI::Init(argc,argv);
16
17    // Get the number of processes
18    int numP=MPI::COMM_WORLD.Get_size();
19    int gridDim = sqrt(numP);
20
21    // Get the ID of the process
22    int myId=MPI::COMM_WORLD.Get_rank();
23
24    if(gridDim*gridDim != numP){
25      // Only the first process prints the output message
26      if(!myId)
27        std::cout << "ERROR: the number of processes must be square"
28                  << std::endl;
29
30      MPI::COMM_WORLD.Abort(1);
31    }
32
33    // Arguments for the datatype
34    params p;
35    int blockLengths[2] = {3, 1};
36    MPI::Aint lb, extent;
37    MPI::INT.Get_extent(lb, extent);
38    MPI::Aint disp[2] = {0, 3*extent};
39    MPI::Datatype types[2] = {MPI::INT, MPI::FLOAT};
40
41    // Create the datatype for the parameters
42    MPI::Datatype paramsType =
43        MPI::INT.Create_struct(2, blockLengths, disp, types);
```

```
44    paramsType.Commit();
45
46    // Process 0 reads the parameters from a configuration file
47    if(!myId)
48      readParams(&p);
49
50    MPI::COMM_WORLD.Barrier();
51    double start = MPI::Wtime();
52
53    // Broadcast of all the parameters using one message
54    MPI::COMM_WORLD.Bcast(&p, 1, paramsType, 0);
55
56    if((p.m < 1) || (p.n < 1) || (p.k<1)){
57      // Only the first process prints the output message
58      if(!myId)
59        std::cout << "ERROR: 'm', 'k' and 'n' must be higher than 0"
60                  << std::endl;
61
62      MPI::COMM_WORLD.Abort(1);
63    }
64
65    if((p.m%gridDim) || (p.n%gridDim)){
66      // Only the first process prints the output message
67      if(!myId)
68        std::cout << "ERROR: 'm', 'n' must be multiple of the grid
69                    dimensions" << std::endl;
70
71      MPI::COMM_WORLD.Abort(1);
72
73    }
74
75    float *A, *B, *C, *myA, *myB, *myC;
76    // Only one process reads the data from the files
77    if(!myId){
78      A = new float[p.m*p.k];
79      B = new float[p.k*p.n];
80      readInput(p.m, p.k, p.n, A, B);
81    }
```

Listing 9.16: Initialization and broadcasting of variables for parallel matrix multiplication.

As only Process 0 has initialized the matrices, we must distribute the necessary fragments among the processes. Note that Fig. 9.12 illustrates an example of how we distribute the matrices for our product. In our example, as the grid dimensions are 3×3, the number of rows and columns per block are $\frac{9}{3} = 3$ and $\frac{12}{3} = 4$, respectively. The first two lines of Listing 9.17 show how to determine the number of rows per block of A and C (`blockRows`) and the number of columns per block of B and C (`blockCols`).

Listing 9.17 continues with the distribution of the rows of A. As the elements of a matrix in C++ are consecutively stored by rows, we can represent a block of rows with the derived datatype "contiguous." In the declaration of the datatype (Lines 86–87) we must only specify the number of elements, i.e., the number of rows per block (`blockRows`) multiplied by the length of each row (k). After the `Commit()`,

Process 0 sends the corresponding block of rows to all processes, which store it in their auxiliary array myA.

This simple example of a derived datatype demonstrates the usefulness of the command Create_contiguous(). However, the row distribution of A in this example could have been accomplished by simply sending $blockRows \cdot k$ many MPI::FLOATs. Nevertheless, Create_contiguous() is useful for defining concatenated types from either existing fundamental datatypes or programmer-defined derived types.

```
82  int blockRows = p.m/gridDim;
83  int blockCols = p.n/gridDim;
84  MPI::Request req;
85
86  // Create the datatype for a block of rows of A
87  MPI::Datatype rowsType =
88       MPI::FLOAT.Create_contiguous(blockRows*p.k);
89  rowsType.Commit();
90
91  // Send the rows of A that needs each process
92  if(!myId)
93    for(int i=0; i<gridDim; i++)
94      for(int j=0; j<gridDim; j++)
95        req = MPI::COMM_WORLD.Isend(&A[i*blockRows*p.k], 1,
96                                    rowsType, i*gridDim+j, 0);
97
98  myA = new float[blockRows*p.k];
99  MPI::COMM_WORLD.Recv(myA, 1, rowsType, 0, 0);
```

Listing 9.17: Distribution of the rows of A among processes.

Listing 9.18 shows the distribution of the columns of B among processes. The first lines are dedicated to creating the new datatype. In this case the elements of the block of columns are not consecutive in memory thus we use Create_vector(). Fig. 9.13 illustrates the meaning of the parameters of Create_vector() when creating a datatype for blocks of columns. count is the number of blocks, i.e., the number of matrix rows. The length of the block has been previously calculated in blockCols. stride indicates the number of elements between the first position of each block, i.e., the length of a row (n in B).

After calling Commit() Process 0 sends the corresponding columns with only one message per process. Blocks are not consecutive in the original matrix. Thus without the new datatype we would have needed k messages of $blockCols$ many floats per process, increasing the complexity and the runtime of our code. Regarding the Recv(), we want to store the elements associated with each process consecutively in myB. We cannot accomplish it using colsType, as it would leave empty strides between blocks. Therefore we indicate in Recv() that we want to save $k \cdot blockCols$ many float values.

Up to now, all calls to *send* and *receive* participating in the same point-to-point communication used identical datatypes. However, as we have shown in this example, this is not compulsory (and sometimes even not advisable). The number of transmitted bytes must agree within the constraints of *send* and *receive*, but the elements can be represented with different types.

```
100 // Create the datatype for a block of columns of B
101 MPI::Datatype colsType =
```

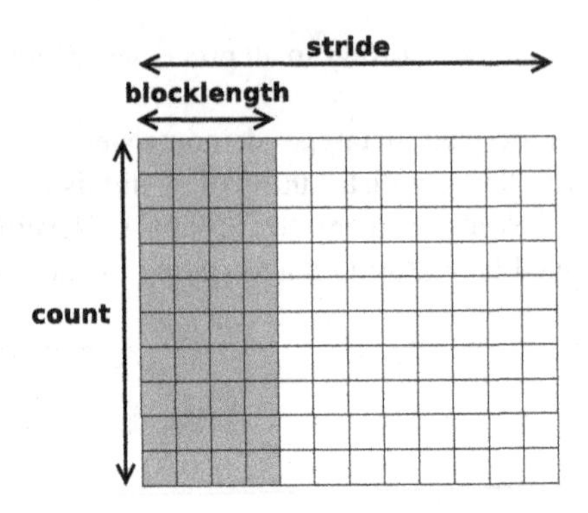

FIGURE 9.13

Representation of the arguments of `Create_vector()` to design a datatype for blocks of columns. In this example there are four columns per block.

```
102            MPI::FLOAT.Create_vector(p.k, blockCols, p.n);
103   colsType.Commit();
104
105   // Send the columns of B that needs each process
106   if(!myId)
107     for(int i=0; i<gridDim; i++)
108       for(int j=0; j<gridDim; j++)
109         req = MPI::COMM_WORLD.Isend(&B[blockCols*j], 1, colsType,
110                                    i*gridDim+j, 0);
111
112   myB = new float[p.k*blockCols];
113   MPI::COMM_WORLD.Recv(myB, p.k*blockCols, MPI::FLOAT, 0, 0);
```

Listing 9.18: Distribution of the columns of B among processes.

Once all processes have received their necessary rows of A and columns of B, they compute their partial matrix product stored in the array `myC`. Listing 9.19 shows the corresponding portion of the program:

```
114   // Array for the chunk of data to work
115   myC = new float[blockRows*blockCols];
116
117   // The multiplication of the submatrices
118   for(int i=0; i<blockRows; i++)
119     for(int j=0; j<blockCols; j++){
120       myC[i*blockCols+j] = 0.0;
121         for(int l=0; l<p.k; l++)
122         myC[i*blockCols+j] += p.alpha*myA[i*p.k+l]*
123                              myB[l*blockCols+j];
124   }
```

Listing 9.19: Partial multiplication on each process.

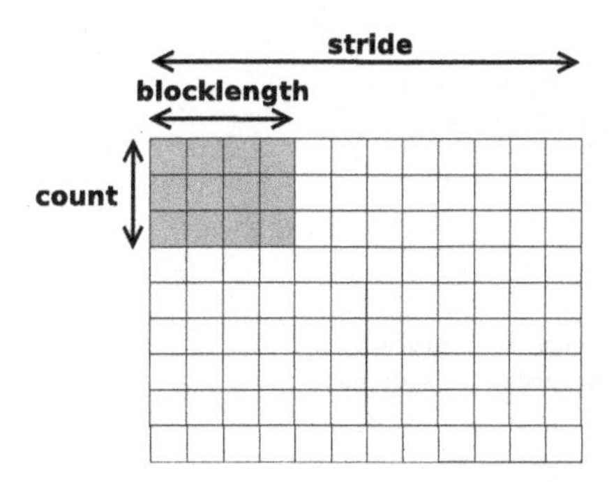

FIGURE 9.14

Representation of the arguments of `Create_vector()` to design a datatype for 2-dimensional blocks. In this example there are three rows and four columns per block.

After the partial computation of the matrix product in each node, we must gather the final result in matrix C stored in Process 0. However, we encounter a problem: consecutive rows of the auxiliary matrix `myC` are not stored in consecutive positions in C. Going back to the example of Fig. 9.12, in the bottom matrix we can see that the first and the second row of the 2D block of C in Process 0 must be separated in C by eight elements (the first row of the blocks in Process 1 and 2). In Listing 9.20 the partial results are sent to Process 0 using the predefined type `MPI::FLOAT` since values are stored consecutively in `myC`. However, a new datatype `block2DType` is designed using `Create_vector()` in order to store the values of the 2D blocks with only one call to `Recv()`. The only difference when using `colsType` is the proper choice of `count` ($nRows$ instead of k). Fig. 9.14 illustrates the meaning of the vector type arguments. The program finishes with the deallocation of arrays and datatypes (using the `Free()` routine).

```
125  // Only Process 0 writes
126  // Gather the final matrix to the memory of Process 0
127  // Create the datatype for a block of columns
128  MPI::Datatype block2DType =
129                 MPI::FLOAT.Create_vector(blockRows,
130                                          blockCols, p.n);
131  block2DType.Commit();
132
133  if(!myId){
134    C = new float[p.m*p.n];
135
136    for(int i=0; i<blockRows; i++)
137      memcpy(&C[i*p.n], &myC[i*blockCols],
138             blockCols*sizeof(float));
139
140      for(int i=0; i<gridDim; i++)
141        for(int j=0; j<gridDim; j++)
142          if(i || j)
143            MPI::COMM_WORLD.Recv(&C[i*blockRows*p.n+j*blockCols],
144                          1, block2DType, i*gridDim+j, 0);
```

```
145   } else
146       MPI::COMM_WORLD.Send(myC, blockRows*blockCols,
147                            MPI::FLOAT, 0, 0);
148
149   // Measure the current time and print by Process 0
150   double end = MPI::Wtime();
151
152   if(!myId){
153     std::cout << "Time with " << numP << " processes: "
154               << end-start << " seconds" << std::endl;
155     printOutput(p.m, p.n, C);
156     delete [] A;
157     delete [] B;
158     delete [] C;
159   }
160
161   // Delete the types and arrays
162   rowsType.Free();
163   colsType.Free();
164   block2DType.Free();
165
166   delete [] myA;
167   delete [] myB;
168   delete [] myC;
169
170   // Terminate MPI
171   MPI::Finalize();
172   return 0;
173 }
```

Listing 9.20: Distribution of the columns of B among processes.

9.8 COMPLEX COMMUNICATORS (MATRIX MULTIPLICATION USING SUMMA)

A drawback of the algorithm presented in the previous section is related to the fact that some data of A and B are stored redundantly. For instance, the first block of rows is replicated in processes 0, 1, and 2 (see Fig. 9.12). Memory is not infinite in our clusters. Thus we have to reduce memory overhead when multiplying large-scale matrices. Moreover, the initial distribution of data is more expensive if we replicate data. In this section we will use the *Scalable Universal Matrix Multiplication Algorithm* (SUMMA) [9] as an example of a parallel algorithm where the three matrices are distributed among processes without redundancy.

Fig. 9.15 shows how the three matrices are partitioned and which process is associated with each block. The advantage of this distribution compared to the previous section is that no single element of any matrix is replicated in more than one process. However, not all the data necessary to compute the partial products are stored in the local memory of the processes. For instance, Process 0 computes the C_{00} block. Thus it needs to calculate the product $A_{00} \cdot B_{00}$ using the factors stored in its local

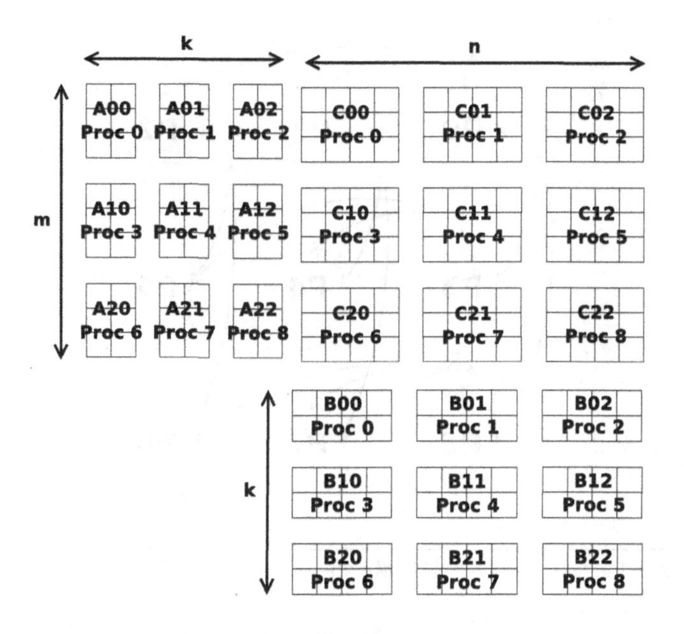

FIGURE 9.15

Example of the distribution of the three matrices involved in the SUMMA algorithm: A (left) $\cdot B$ (bottom) $= C$ (right). We use nine processes and the dimensions are: $m = 9$, $k = 6$, $n = 12$.

memory but also has to perform $A_{01} \cdot B_{10}$ and $A_{02} \cdot B_{20}$. However, the submatrices A_{01}, A_{02}, B_{10} and B_{20} are not stored in the local memory of Process 0. Similarly, Process 1 computes the contribution $C_{01} = A_{00} \cdot B_{01} + A_{01} \cdot B_{11} + A_{02} \cdot B_{21}$, but four of the involved blocks (A_{00}, A_{02}, B_{11} and B_{21}) are stored in other processes.

SUMMA realizes the communication using calls to broadcast along the rows and columns of a 2D grid of processes in order to distribute the blocks. The computation is performed in a loop of $\sqrt{\texttt{numP}}$ iterations. During each iteration one process in each row broadcasts its block A_{ij} to the other processes in the same row. Similarly, the block B_{ij} is broadcast along the columns of the grid. Figs. 9.16, 9.17, and 9.18 illustrate this procedure using nine processes on a 3×3 grid. In each iteration step the involved processes multiply the received blocks of A and B and afterwards accumulate the contributions in their corresponding block of C. As an example, Process 0 multiplies the pairs of blocks (A_{00}, B_{00}), (A_{01}, B_{10}) and (A_{02}, B_{20}) in consecutive iterations. At the end of the loop the expected result $A_{00} \cdot B_{00} + A_{01} \cdot B_{10} + A_{02} \cdot B_{20}$ is stored in block C_{00}.

Listing 9.21 shows the first part of our MPI SUMMA implementation. All concepts have been explained in the previous sections. After the MPI initialization, all processes parse the matrix dimensions from the command line arguments of the program (Lines 28–30). The code continues with parameter checking: the processes can be structured in a square 2D grid (Lines 34–41), the matrix dimensions are multiples of the grid dimensions (Lines 43–51), and the matrix dimensions are nonzero (Lines 53–60). If all parameters satisfy the stated constraints, Process 0 reads the input matrices from

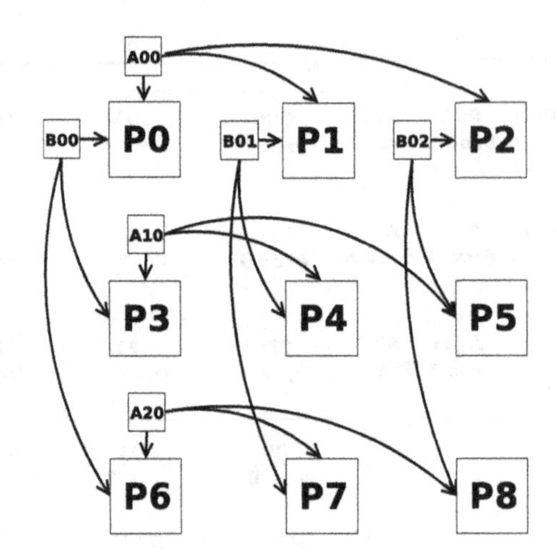

FIGURE 9.16

Broadcasting scheme in the first iteration of SUMMA on 3 × 3 grid.

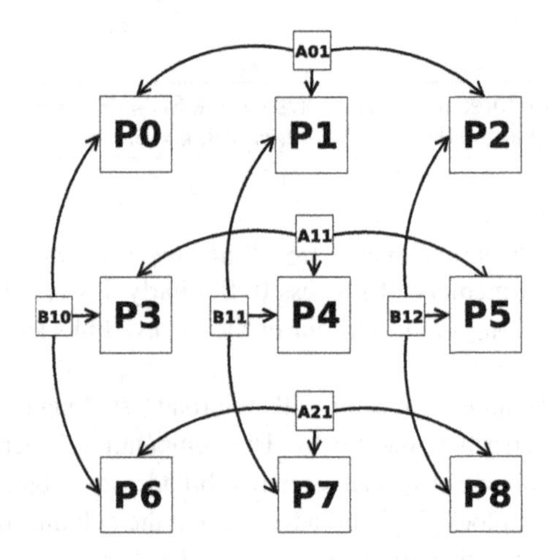

FIGURE 9.17

Broadcasting scheme in the second iteration of SUMMA on 3 × 3 grid.

a file (Lines 65–71) and distributes the 2D blocks with new datatypes, as explained in the previous section (Lines 73–112).

```
1  #include <stdlib.h>
2  #include <stdio.h>
3  #include <iostream>
4  #include <string.h>
5  #include <math.h>
```

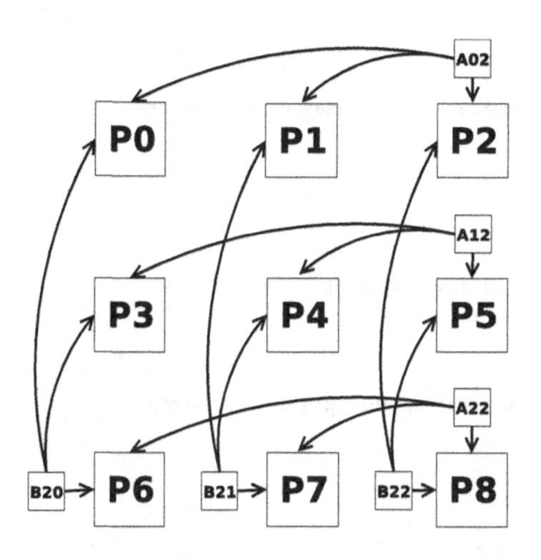

FIGURE 9.18

Broadcasting scheme in the third iteration of SUMMA on 3 × 3 grid.

```
6   #include "mpi.h"
7
8   int main (int argc, char *argv[]){
9     // Initialize MPI
10    MPI::Init(argc,argv);
11
12    // Get the number of processes
13    int numP=MPI::COMM_WORLD.Get_size();
14
15    // Get the ID of the process
16    int myId=MPI::COMM_WORLD.Get_rank();
17
18    if(argc < 4){
19      // Only the first process prints the output message
20      if(!myId)
21        std::cout <<
22        "ERROR: The syntax of the program is ./summa m k n"
23        << std::endl;
24
25      MPI::COMM_WORLD.Abort(1);
26    }
27
28    int m = atoi(argv[1]);
29    int k = atoi(argv[2]);
30    int n = atoi(argv[3]);
31
32    int gridDim = sqrt(numP);
33    // Check if a square grid could be created
34    if(gridDim*gridDim != numP){
```

```
35      // Only the first process prints the output message
36      if(!myId)
37        std::cout << "ERROR: The number of processes must be square"
38                  << std::endl;
39
40      MPI::COMM_WORLD.Abort(1);
41    }
42
43    if((m%gridDim) || (n%gridDim) || (k%gridDim)){
44      // Only the first process prints the output message
45      if(!myId)
46        std::cout
47        << "ERROR: 'm', 'k' and 'n' must be multiple of sqrt(numP)"
48        << std::endl;
49
50      MPI::COMM_WORLD.Abort(1);
51    }
52
53    if((m < 1) || (n < 1) || (k<1)){
54      // Only the first process prints the output message
55      if(!myId)
56        std::cout << "ERROR: 'm', 'k' and 'n' must be higher than 0"
57        << std::endl;
58
59      MPI::COMM_WORLD.Abort(1);
60    }
61
62    float *A, *B, *C;
63
64    // Only one process reads the data from the files
65    if(!myId){
66      A = new float[m*k];
67      readInput(m, k, A);
68      B = new float[k*n];
69      readInput(k, n, B);
70      C = new float[m*n];
71    }
72
73    // The computation is divided by 2D blocks
74    int blockRowsA = m/gridDim;
75    int blockRowsB = k/gridDim;
76    int blockColsB = n/gridDim;
77
78    // Create the datatypes of the blocks
79    MPI::Datatype blockAType = MPI::FLOAT.Create_vector(blockRowsA,
80                                     blockRowsB, k);
81    MPI::Datatype blockBType = MPI::FLOAT.Create_vector(blockRowsB,
82                                     blockColsB, n);
83    MPI::Datatype blockCType = MPI::FLOAT.Create_vector(blockRowsA,
84                                     blockColsB, n);
85    blockAType.Commit(); blockBType.Commit(); blockCType.Commit();
86
```

```
87  float* myA = new float[blockRowsA*blockRowsB];
88  float* myB = new float[blockRowsB*blockColsB];
89  float* myC = new float[blockRowsA*blockColsB]();
90  float* buffA = new float[blockRowsA*blockRowsB];
91  float* buffB = new float[blockRowsB*blockColsB];
92
93  // Measure the current time
94  MPI::COMM_WORLD.Barrier();
95  double start = MPI::Wtime();
96
97  MPI::Request req;
98
99  // Scatter A and B
100 if(!myId){
101   for(int i=0; i<gridDim; i++)
102     for(int j=0; j<gridDim; j++)
103       req = MPI::COMM_WORLD.Isend(A+i*blockRowsA*k+j*blockRowsB,
104                                  1, blockAType, i*gridDim+j, 0);
105       req = MPI::COMM_WORLD.Isend(B+i*blockRowsB*n+j*blockColsB,
106                                  1, blockBType, i*gridDim+j, 0);
107 }
108
109 MPI::COMM_WORLD.Recv(myA, blockRowsA*blockRowsB,
110                      MPI::FLOAT, 0, 0);
111 MPI::COMM_WORLD.Recv(myB, blockRowsB*blockColsB,
112                      MPI::FLOAT, 0, 0);
```

Listing 9.21: Initialization and distribution of the matrices in SUMMA.

After matrix distribution we can start the SUMMA loop. We need one broadcast per iteration, row and column of the grid. We cannot directly use the broadcast presented in Section 9.5 since we do not need to send each block to all processes, but to a subsets of all processes. Up to now, we only know the universal communicator `MPI::COMM_WORLD` that includes all processes. However, MPI provides the routine `Split()` to create new communicators as subset of an existing one.

- `Split(int color, int key)`. It returns a new communicator that is a subset of the previous one. `color` identifies the subset where the process is included. `key` indicates the rank of the process within the communicator.

Table 9.1 lists the necessary values to divide the processes in `COMM_WORLD` into communicators that reassemble the discussed communication scheme. Each communicator is created with a simple call to `Split()`, as demonstrated in Listing 9.22.

```
113 // Create the communicators
114 MPI::Intercomm rowComm = MPI::COMM_WORLD.Split(myId/gridDim,
115                                               myId%gridDim);
116 MPI::Intercomm colComm = MPI::COMM_WORLD.Split(myId%gridDim,
117                                               myId/gridDim);
```

Listing 9.22: Creation of the communicators.

Table 9.1 **`color` and `key` for each process to create the communicators for rows and columns over a 3 × 3 grid.**

Process ↓	Row Communicator		Column Communicator	
	Color	*Key*	*Color*	*Key*
0	0	0	0	0
1	0	1	1	0
2	0	2	2	0
3	1	0	0	1
4	1	1	1	1
5	1	2	2	1
6	2	0	0	2
7	2	1	1	2
8	2	2	2	2

Listing 9.23 shows the main SUMMA loop. Thanks to the new communicators, broadcasting in each iteration is performed with only two lines of code (Lines 129–130). After the communication of the proper block, each process updates the values of block C_{ij} with the corresponding product (Lines 133–137).

```
118  // The main loop
119  for(int i=0; i<gridDim; i++){
120    // The owners of the block to use must copy it to the buffer
121    if(myId%gridDim == i){
122      memcpy(buffA, myA, blockRowsA*blockRowsB*sizeof(float));
123    }
124    if(myId/gridDim == i){
125      memcpy(buffB, myB, blockRowsB*blockColsB*sizeof(float));
126    }
127
128    // Broadcast along the communicators
129    rowComm.Bcast(buffA, blockRowsA*blockRowsB, MPI::FLOAT, i);
130    colComm.Bcast(buffB, blockRowsB*blockColsB, MPI::FLOAT, i);
131
132    // The multiplication of the submatrices
133    for(int i=0; i<blockRowsA; i++)
134      for(int j=0; j<blockColsB; j++)
135        for(int l=0; l<blockRowsB; l++)
136          myC[i*blockColsB+j] += buffA[i*blockRowsB+l]*
137                                 buffB[l*blockColsB+j];
138  }
```

Listing 9.23: SUMMA loop.

Finally, the blocks of C must be sent to Process 0 so it can print the output. The remaining part of the program (Listing 9.24) is similar to the end of the matrix multiplication shown in the previous section.

```
139    // Only Process 0 writes
140    // Gather the final matrix to the memory of Process 0
141    if(!myId){
142      for(int i=0; i<blockRowsA; i++)
143        memcpy(&C[i*n], &myC[i*blockColsB],
144               blockColsB*sizeof(float));
145
146        for(int i=0; i<gridDim; i++)
147          for(int j=0; j<gridDim; j++)
148            if(i || j)
149              MPI::COMM_WORLD.Recv(&C[i*blockRowsA*n+j*blockColsB],
150                                   1, blockCType, i*gridDim+j, 0);
151    } else
152      MPI::COMM_WORLD.Send(myC, blockRowsA*blockColsB,
153                           MPI::FLOAT, 0, 0);
154
155    // Measure the current time
156    double end = MPI::Wtime();
157
158    if(!myId){
159      std::cout << "Time with " << numP << " processes: "
160                << end-start << " seconds" << std::endl;
161      printOutput(m, n, C);
162      delete [] A;
163      delete [] B;
164      delete [] C;
165    }
166
167    MPI::COMM_WORLD.Barrier();
168
169    delete [] myA;
170    delete [] myB;
171    delete [] myC;
172    delete [] buffA;
173    delete [] buffB;
174
175    // Terminate MPI
176    MPI::Finalize();
177    return 0;
178  }
```

Listing 9.24: End of MPI SUMMA.

We provide a performance comparison of the two presented MPI implementations for matrix multiplication (matrix scattering and SUMMA). The results indicate that SUMMA is a better option since we use less memory and additionally obtain a better runtime. This advantage in favor of SUMMA can be explained by its superior data distribution scheme. The data replication of matrix scattering in the first algorithm renders the communication more expensive. Moreover, as the blocks in SUMMA are smaller, we additionally improve computation performance thanks to better usage of CPU caches (see Fig. 9.19).

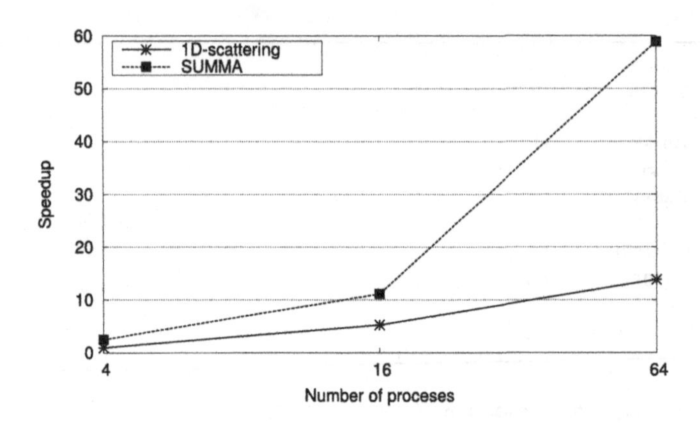

FIGURE 9.19

Performance comparison of matrix multiplication algorithms using matrix scattering and SUMMA on a system with 64 cores. The shape of the three matrices are fixed to 8192 × 8192.

Although `Split()` is the most common communicator creation function, there are other alternatives. For instance, `Dup()` creates a copy of the original communicator. Additionally, we can use the class MPI::Group to associate several processes and later create a group with them. A group that includes all the processes of a communicator is obtained with `Get_group()`. The opposite function, to create one communicator with all the processes of a group is `MPI::Intercomm::Create`. The procedure to construct a communicator that cannot be created with the basic `Split()` and `Dup()` routines consists in creating the desired MPI::Group and then call to `Create()`. The memory allocated by the groups is released with `Free()`. There exist several routines to manipulate groups:

- `Union()`. Produces a group by combining all the processes of two original groups.
- `Intersection()`. Takes two groups as input and creates a new group with those processes that are included in both original groups.
- `Difference()`. Creates a group with those processes included in the first input group but not in the second one.
- `Incl()`. It receives one group and a list of rank as inputs. The new group includes the processes of the original group with the ranks included in the list.
- `Excl()`. Similar to the previous one, but the new group includes all processes within the original group except the ones of the list.

9.9 OUTLOOK

Additional features, such as nonblocking collective routines and parallel I/O routines, included in the MPI standard v3.1 [8] have not been addressed in this chapter. One additional and important improvement included in MPI since that version are the Remote Memory Accesses (RMA) or one-sided communication. However, we advise the reader to skip this topic until finishing this book, as MPI's

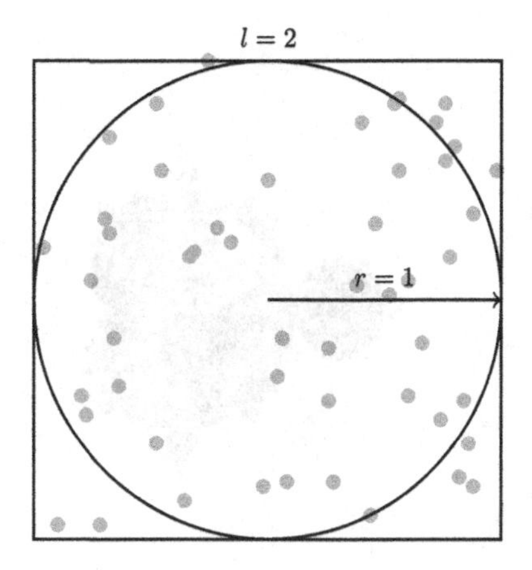

FIGURE 9.20

Approximation of the inscribed areas $A(D)$ and $A(S)$ by uniformly drawing $n = 50$ random samples in the interval $[-1, 1] \times [-1, 1]$.

one-sided communications calls [8] are an adaptation of the remote copies available in PGAS languages such as UPC++ which will be explained in detail during the next chapter.

9.10 ADDITIONAL EXERCISES

1. Let S be a solid square of length $l = 2$ and D be a solid disk of radius $r = 1$ both centered at the origin $(0, 0)$. The quotient of their inscribed areas is proportional to π since

$$Q = \frac{A(D)}{A(S)} = \frac{\pi \cdot r^2}{l^2} = \frac{\pi}{4} .$$

   As a result, a reasonable approximation of π can be determined by computing the ratio of points that fall into D when uniformly drawing n random samples in S. In the limit $n \to \infty$ we reassemble the inscribed areas $A(D)$ and $A(S)$ (see Fig. 9.20). Implement the described Monte Carlo algorithm using MPI where each process draws $\frac{n}{\#\text{processes}}$ many samples.
2. `Allreduce` is an MPI collective that performs a reduction and an additional broadcast of the result to all processes. Implement a function that achieves the same exclusively using `send` and `recv` routines in a ring of processes as shown in Fig. 9.4. Concretely, design the reduction for the operation `MPI::SUM` and the datatype `MPI::INT` working on an array of length `n` compliant to the call: `MPI::COMM_WORLD.Allreduce(myA, A, n, MPI::INT, MPI::SUM)`.
3. A complex number $c = x + i \cdot y \in \mathbb{C}$ where $x, y \in \mathbb{R}$ is part of the Mandelbrot set if and only if the recursively defined complex-valued sequence $z_{n+1} = z_n^2 + c$ with initial condition $z_0 = 0$ converges i.e. $\lim_{n\to\infty} z_n$ exists (is a finite value). Without proof, one can show that every complex number c with squared radius $|c|^2 = x^2 + y^2 > 4$ generates a diverging sequence. Thus we

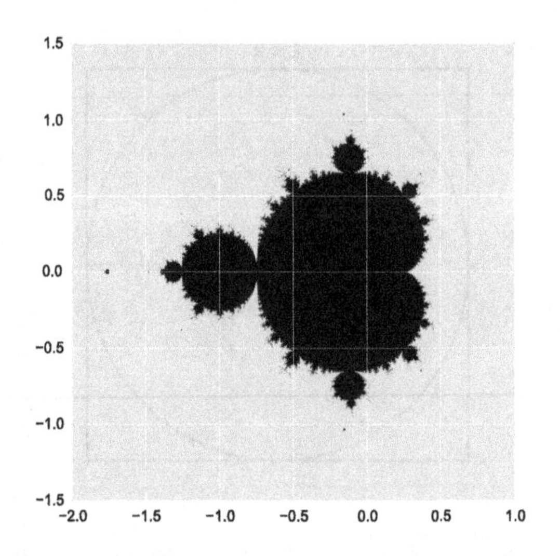

FIGURE 9.21

Mandelbrot set (black area) sampled with 1920 samples per axis.

approximate the Mandelbrot set by the set of all complex numbers c whose radius $|c|$ is smaller than two after a fixed number of iterations. It can be represented as a 2D image where x and y coordinates represent the real and imaginary part of the complex number (see Fig. 9.21). Implement an MPI program that computes an approximation of the Mandelbrot set over the domain $[-2, 1] \times [-1.5, 1.5]$ receiving as input the number of samples per axis.

a. Distribute the pixels with a static distribution (the same number of pixels per process).
b. Measure the scalability of your program. What are the disadvantages of a static distribution?
c. Improve the scalability with a master-slave approach. Hint: use `Recv()` with `MPI::ANY_SOURCE` in the master.

4. The standard collective routines are blocking. However, there exist also nonblocking counterparts for some of them, e.g., the nonblocking broadcast:

- `MPI::Request Ibcast(void *buffer, int count, MPI::Datatype & type, int root);`

Similarly to `Isend()` and `Irecv()`, they can be used in combination with `Wait()` and `Test()`. Improve the performance of the SUMMA algorithm presented in Section 9.8 by overlapping computation and communication. Before the processes perform the partial matrix product of iteration k, start the broadcasting of the blocks necessary for the next iteration $k + 1$.
Hint: Use `MPI_Ibcast()` from the C interface of MPI.

5. As mentioned in Section 9.1, many modern compute clusters consist of several nodes each providing several CPU cores. Hybrid implementations using MPI for inter-node communication and OpenMP for shared-memory computations usually obtain better performance than pure MPI approaches.

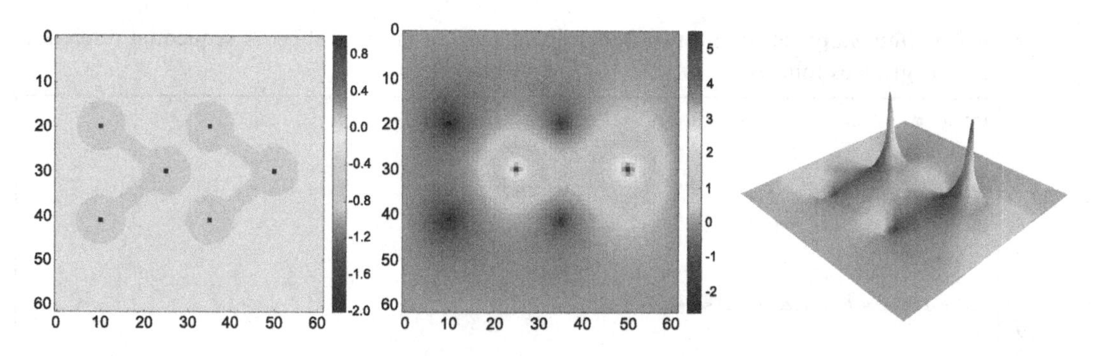

FIGURE 9.22

An example for the solution of the Poisson equation. The first panel illustrates the distribution f of two positively charged hydrogen atoms and a negatively charged oxygen atom for each of the two water molecules (gray silhouettes). The second panel depicts the corresponding electric potential ϕ. The last panel provides the graph of ϕ embedded in 3D space for a better visualization.

Develop a hybrid OpenMP/MPI implementation of SUMMA where each process utilizes n threads in order to parallelize the partial matrix products.

6. Due to the increasing popularity of GPUs, there has been a trend in HPC to attach CUDA-enabled accelerators to the compute nodes of modern clusters. These clusters are also called heterogeneous platforms.
Develop a hybrid CUDA/MPI implementation of SUMMA where the partial matrix products are performed in parallel using either the cuBLAS library or handcrafted CUDA kernels.
Hint: Many MPI implementations can be configured to be CUDA-aware. Alternatively, you can interleave CUDA-specific memcopies and the collectives provided by MPI.

7. Let $\Delta\phi(p) = f(p)$ be the Poisson equation with non-vanishing heterogeneous term $f(p) \neq 0$ analogous to the section about Jacobi iteration (see Fig. 9.22).

a. Derive the corresponding update rule for the image matrix `data`:

$$\texttt{data}[i,j] \leftarrow \frac{\texttt{data}[i+1,j] + \texttt{data}[i-1,j]}{4} + \frac{\texttt{data}[i,j+1] + \texttt{data}[i,j-1]}{4} - \frac{\texttt{rho[i, j]}}{4}$$

where `rho` is the discretized version of f with the same shape as `data`.

b. Design an MPI algorithm that distributes the matrix `data` on a two-dimensional grid of tiles. Make use of derived datatypes in order to efficiently communicate columns between neighboring blocks.

c. Compare the runtime of the algorithm that employs a one-dimensional distribution of tiles and your novel two-dimensional partition for a reasonable number of processes. Can you back up your experimental measurements with theoretical ideas?

8. The Smith–Waterman algorithm is used in bioinformatics to align a pair of proteins or nucleotide sequences [16] of lengths m and n (*pairwise sequence alignment*). The algorithm consists of two major steps:

- **Relaxation Step:** relaxing a two-dimensional scoring matrix `table`. A sequential reference code is given as follows:

```
int n, m; // specified by the user

// char arrays of length m and n
char * seq0 = read_seq0(m);
char * seq1 = read_seq1(n);

// the relaxation table
int * table = new size[(m+1)*(n+1)];

// initialization of borders with zeros
for (int i = 0; i < m+1; i++)
    table[i*(n+1)] = 0;
for (int j = 1; j < n+1; j++)
    table[j] = 0;

// start the relaxation
for (int i = 1; i < m+1, i++) {

    // get i-1 th symbol in sequence 0
    char symb0 = seq0[i-1];

    for (int j = 1; j < n+1; j++) {
        // get j-1 th symbol in sequence 1
        char symb1 = seq1[j-1];

        // get diagonal upper and left entry
        int diag = table[(i-1)*(n+1)+j-1];
        int abve = table[(i-1)*(n+1)+j];
        int left = table[i*(n+1)+j-1];

        // here we use a simplified scoring scheme
        int score = (symb0 == symb1) ? 1 : -1;

        // determine the best predecessor
        int contribution = score + max(diag,
                                  max(abve, left));

        // update the cell: values cannot be negative
        table[i*(n+1)+j] = max(contribution, 0);
    }
}

// proceed with finding the greatest score value in table
// ...

// free resource
// ...
```

Listing 9.25: Smith–Waterman reference code.

- **Traceback:** Determining an optimal (possibly not unique) alignment path

The relaxation step is computationally more demanding ($\mathcal{O}(m \cdot n)$) than the traceback step ($\mathcal{O}(\max(n, m))$). Thus we focus on the parallelization of the relaxation. Write an MPI program that relaxes the scoring table for two given sequences. The program will distribute the columns of the table such that each process will be responsible for calculating the values of a block of columns. For simplicity, assume that the length of the sequences is a multiple of the number of processes.
Some hints to optimize your MPI code:

- Note that Process p needs the values of the last column of Process $p-1$. If Process p waits until Process $p-1$ has finishing the computation of its whole column, the computation will be serialized. That should be avoided at any cost. Thus send the columns in fragments as soon as these fragments have finished the relaxation.
- The use of derived datatypes is advisable.
- Send the columns with nonblocking routines in order to overlap computation and communication.

9. Cannon's algorithm [11] is an alternative to SUMMA for matrix multiplication without replication of data among processes. It is based on shifting partial blocks. Implement this algorithm using MPI. Use communicators for rows and columns in your implementation.
   a. Look up and understand the sequential algorithm.
   b. Design an efficient implementation of Cannon's algorithm with MPI. Hint: use the same communicators and derived datatypes as in SUMMA.
10. k-Means is a popular data mining algorithm to group n real-valued data points x in d-dimension into k clusters. Assume that the points are stored in a data matrix D of shape $n \times d$. Lloyd's algorithm is usually applied to determine an approximate solution of the objective function:

$$\min_{\{C_l\}} \sum_{l=0}^{k-1} \sum_{x \in C_l} (x - m_l)^2 \quad \text{with} \quad m_l = \frac{1}{|C_l|} \sum_{x \in C_l} x \quad ,$$

where l enumerate the clusters C_l and m_l denotes their centers obtained by ordinary averaging of data points within each cluster. The algorithm can be split into two steps that are repeatedly interleaved until convergence (i.e. the clusters do not change anymore).

- **Assignment Step:** For each data point x we compute the corresponding cluster index $l = \operatorname{argmin}_{l'}(x - m_{l'})^2$, i.e., the nearest cluster center to x under Euclidean distance is determined by bruteforcing all cluster indices $0 \leq l' < k$. This can be achieved in $\mathcal{O}(n)$ time.
- **Update Step:** After having computed the corresponding cluster index l for each data point x, i.e. $x \in C_l$, we update the cluster centers m_l in linear time by ordinary averaging.

Concluding, if τ is the number of iterations Lloyd's algorithm has an overall time complexity of $\mathcal{O}(\tau \cdot n)$. Unfortunately, k-Means can only be used to separate spherical clusters due to the used

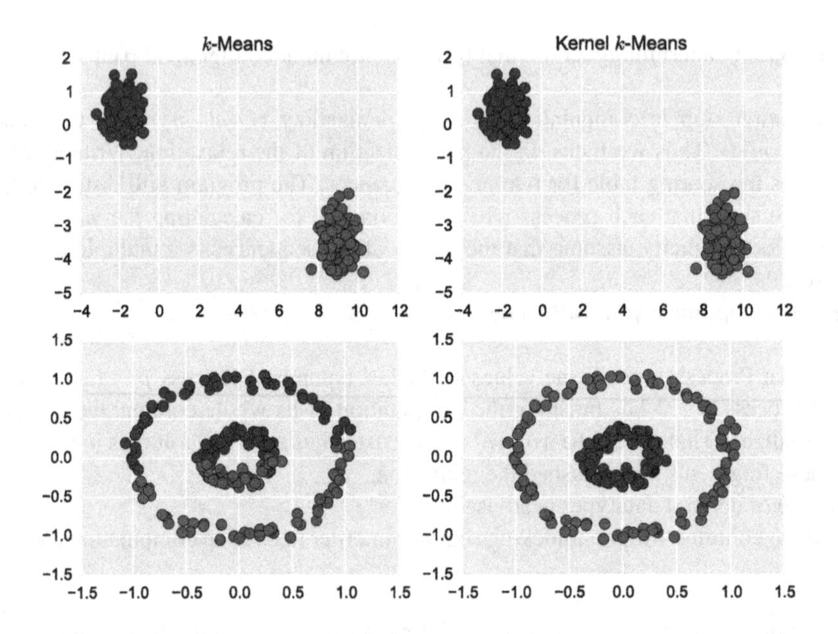

FIGURE 9.23

Clustering of two data sets each consisting of two clusters with 100 data points each. k-Means computes the correct result on spherical Gaussians (upper left) but fails to predict the clusters on the nonlinearly separable circles (lower left). However, Kernel k means (plots on the right) predicts the correct clusters on both data sets using a radial basis function kernel matrix $K(x, y) = \exp\big(-0.5 \cdot (x - y)^2\big)$. The circles can be untangled in higher dimensional feature space (simply move the inner circle out of the x–y plane).

Euclidean distance measure (see Fig. 9.23). This drawback can be fixed using a kernelized variant of the described algorithm. Assume we could map the data points $x \mapsto \phi(x)$ by some function ϕ to a high (possibly infinite) dimensional space, then clusters might be separable by planes. The key idea of Kernel k-Means is to apply k-Means in feature space instead of the data domain. Substituting the feature map and the equation for the centers in the objective function

$$\min_{\{C_l\}} \sum_{l=0}^{k-1} \sum_{x \in C_l} \Big(\phi(x) - \frac{1}{|C_l|} \sum_{y \in C_l} \phi(y)\Big)^2$$

$$= \min_{\{C_l\}} \sum_{l=0}^{k-1} \sum_{x \in C_l} \Big(\phi(x)^2 - \frac{2}{|C_l|} \sum_{y \in C_l} \phi(x) \cdot \phi(y) + \frac{1}{|C_l|^2} \sum_{y \in C_l} \sum_{z \in C_l} \phi(y) \cdot \phi(z)\Big)$$

the whole expression can be rewritten exclusively in terms of scalar products $K(x, y) := \phi(x) \cdot \phi(y)$ in feature space. Surprisingly, we do not even need to know the feature map ϕ if someone provides us a valid (symmetric and positive definite) kernel matrix K of shape $n \times n$.[4]

[4]Note, the choice $\phi(x) = x$, $K(x, y) = x \cdot y$ is the special case of traditional k-Means.

a. Analyze the time and memory requirements of Kernel k-Means.

b. Can you lower the memory footprint if someone provides you a function to compute the entries of K on-the-fly, e.g. $K(x, y) = \exp(-(x - y)^2)$? Implement this idea using multiple processes in MPI.

c. The kernel matrix K may be huge depending on the number of points n. Design a communication pattern for distributed-memory architectures in order to efficiently compute the cluster assignments. Write an efficient implementation in MPI with minimal data redundancy by distributing K across several compute nodes. Note, you only know K, ϕ is unknown.

REFERENCES

[1] Tejaswi Agarwal, Michela Becchi, Design of a hybrid MPI-CUDA benchmark suite for CPU-GPU clusters, in: Procs. 23rd Intl. Conf. on Parallel Architectures and Compilation Techniques (PACT'14), Edmonton, Canada, 2014.

[2] Ahmed Bukhamsin, Mohamad Sindi, Jallal Al-Jallal, Using the Intel MPI benchmarks (IMB) to evaluate MPI implementations on an Infiniband Nehalem Linux cluster, in: Procs. Spring Simulation Multiconference (SpringSim'10), Orlando, FL, USA, 2010.

[3] Martin J. Chorley, David W. Walker, Performance analysis of a hybrid MPI/OpenMP application on multi-core clusters, Journal of Computational Science 1 (3) (2010) 168–174.

[4] IBM Platform Computing, IBM Platform MPI home page, http://www-03.ibm.com/systems/platformcomputing/products/mpi/ (visited on 08/20/2015).

[5] Anthony Danalis, et al., Transformations to parallel codes for communication-computation overlap, in: Procs. ACM/IEEE Conf. on Supercomputing (SC'05), Seattle, WA, USA, 2005.

[6] Hikmet Dursun, et al., A multilevel parallelization framework for high-order stencil computations, in: Procs. 15th Euro-Par Conf. (Euro-Par'09), Delft, The Netherlands, 2009.

[7] Message Passing Interface Forum, MPI: A Message-Passing Interface Standard, report, University of Tennessee, Knoxville, Tennessee, May 1994, http://www.mpi-forum.org/docs/mpi-1.0/mpi10.ps (visited on 08/20/2015).

[8] Message Passing Interface Forum, MPI: A Message-Passing Interface Standard Version 3.1, report, University of Tennessee, Knoxville, Tennessee, June 2015, http://www.mpi-forum.org/docs/mpi-3.1/mpi31-report.pdf (visited on 08/20/2015).

[9] Robert A. van de Geijn, Jerrel Watts, SUMMA: scalable universal matrix multiplication algorithm, Concurrency and Computation: Practice and Experience 9 (4) (1997) 255–274.

[10] Haoqiang Jina, et al., High performance computing using MPI and OpenMP on multi-core parallel systems, Parallel Computing 37 (9) (2011) 562–575.

[11] H. Lee, J.P. Robertson, J.A. Fortes, Generalized Cannon's algorithm for parallel matrix multiplication, in: Procs. 11th Intl. Conf. on Supercomputing (ICS '97), Vienna, Austria, 1997.

[12] G. Liu, T. Abdelrahman, Computation-communication overlap on network-of-workstation multiprocessor, in: Procs. Intl. Conf. on Parallel and Distributed Processing Techniques and Applications (PDPTA '98), Las Vegas, NV, USA, 1998.

[13] Yongchao Liu, Bertil Schmidt, Douglas L. Maskell, DecGPU: distributed error correction on massively parallel graphics processing units using CUDA and MPI, BMC Bioinformatics 12 (85) (2011).

[14] Amith R. Mamidala, et al., MPI collectives on modern multicore clusters: performance optimizations and communication characteristics, in: Procs. 8th Intl. Symp. on Cluster Computing and the Grid (CCGRID'08), Lyon, France, 2008.

[15] PVM Project Members, Parallel Virtual Machine (PVM) home page, http://www.csm.ornl.gov/pvm/ (visited on 08/20/2015).

[16] Saul B. Needleman, Christian D. Wunsch, A general method applicable to the search for similarities in the amino acid sequence of two proteins, Journal of Molecular Biology 48 (3) (1970) 443–453.

[17] Hewlett Packard, MPI for HP ProLiant systems, https://h20392.www2.hp.com/portal/swdepot/displayProductInfo.do?productNumber=MPISW (visited on 08/20/2015).

[18] Jian Tao, Marek Blazewicz, Steven R. Brandt, Using GPU's to accelerate stencil-based computation kernels for the development of large scale scientific applications on heterogeneous systems, in: Procs. 17th ACM SIGPLAN Symp. on Principles and Practice of Parallel Programming (PPoPP'12), New Orleans, LA, USA, 2012.

[19] MPICH Team, MPICH home page, https://www.mpich.org/ (visited on 08/20/2015).

[20] OpenMPI Team, Open MPI: open source high performance computing home page, http://www.open-mpi.org/ (visited on 08/20/2015).
[21] Rajeev Thakura, William Gropp, Test suite for evaluating performance of multithreaded MPI communication, Parallel Computing 35 (12) (2009) 608–617.
[22] Bibo Tu, et al., Performance analysis and optimization of MPI collective operations on multi-core clusters, The Journal of Supercomputing 60 (1) (2012) 141–162.
[23] Intel Developer Zone, Intel MPI library home page, https://software.intel.com/en-us/intel-mpi-library/ (visited on 08/20/2015).

CHAPTER

UNIFIED PARALLEL C++ 10

Abstract

Although MPI is commonly used for parallel programming on distributed-memory systems, Partitioned Global Address Space (PGAS) approaches are gaining attention for programming modern multi-core CPU clusters. They feature a hybrid memory abstraction: distributed memory is viewed as a shared memory that is partitioned among nodes in order to simplify programming. In this chapter you will learn about Unified Parallel C++ (UPC++), a library-based extension of C++ that gathers the advantages of both PGAS and Object Oriented paradigms.
The examples included in this chapter will help you to understand the main features of PGAS languages and how they can simplify the task of programming parallel source code for clusters and supercomputers. In particular, we study UPC++ examples addressing the topics of memory affinity, privatization, remote memory accesses, asynchronous copies, and locks.

Keywords

Partitioned global address space, PGAS, Distributed memory, Cluster computing, Unified parallel C++, UPC++, Memory affinity, Remote memory, Shared array, Pointer privatization, One-sided communication

CONTENTS

10.1 INTRODUCTION TO PGAS AND UPC++

As explained in the previous chapter, in an attempt to improve performance on modern architectures that mix shared and distributed memory, several applications favor a hybrid MPI+OpenMP approach. The development of such hybrid parallel code can be challenging since programmers must be aware of the characteristics and performance issues of both approaches. An alternative approach to increase

Parallel Programming. DOI: 10.1016/B978-0-12-849890-3.00010-1

the programmability of modern clusters and supercomputers without compromising performance and portability is the usage of the PGAS paradigm. PGAS is a parallel programming model that combines the advantages of traditional message-passing and shared-memory models about which you have learned in the previous chapters. PGAS languages expose to the programmer a global shared address space, which is logically divided among processes, so that each process is associated with or presents affinity to a part of the shared memory. PGAS languages explicitly expose the non-uniform nature of memory access times: operations on local data (i.e., the portion of the address space to which a particular processor has affinity) are much faster than operations on remote data (i.e., any part of the shared address space to which a processor does not have affinity). Some of the attractive features of this approach are:

- The global shared address space facilitates the development of parallel programs, allowing all processes to directly read and write remote data and avoiding the error-prone data movements of the message-passing paradigm [9,5].
- Accesses to shared memory allow for the development of efficient one-sided communication that can outperform traditional two-sided communication, as a process can directly read and write on remote memory without the explicit cooperation of the process on the remote core [1,7].
- Compared to the shared memory paradigm, the performance of the codes can be increased by taking into account data affinity as typically the accesses to remote data will be much more expensive than the accesses to local data.
- PGAS languages provide a programming model that can be used across the whole system instead of relying on two distinct programming models that must be combined, as in the hybrid OpenMP+MPI solution. Thus PGAS languages aim to deliver the same performance as hybrid approaches using a uniform programming model.

Examples of PGAS languages include Unified Parallel C (UPC) [14,13], Co-Array Fortran (CAF, included in Fortran98 specification) [4], Titanium [10], Chapel [2], X10 [15] and Fortress [8]. UPC++ [16] is a compiler-free approach based on UPC that uses C++ templates and runtime libraries to provide PGAS features (see Fig. 10.1). As it is a C++ library, we just need a C++ compiler simplifying the installation procedure (no ad hoc compiler is necessary). Furthermore, this compiler-free idea benefits from better portability among platforms, more flexible interoperability with other parallel language extensions, and significant savings in development and maintenance cost (programmability).

The execution model of UPC++ is SPMD and each independent execution unit is called a thread, which can be implemented as an OS process or a Pthread. The number of UPC++ threads is fixed during program execution. In UPC++ (and, in general, all PGAS approaches) there is a global address space accessible to all threads (see Fig. 10.2). This space is logically partitioned among all threads. Each thread is associated with or presents *affinity* to one fragment of memory. When translating this memory model to an actual distributed-memory system, UPC++ runtime guarantees that memory resides in the same node as the thread with affinity, and thus we know that accesses to this part of the memory are relatively fast. Accesses using addresses without affinity might need additional communication and thus take longer time.

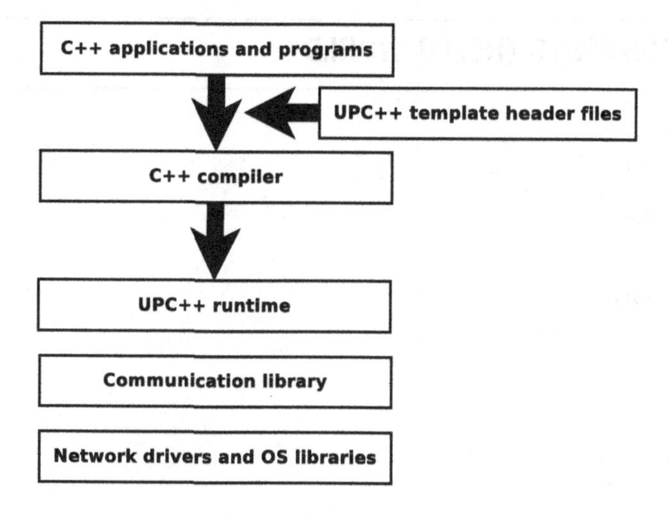

FIGURE 10.1

UPC++ software stack.

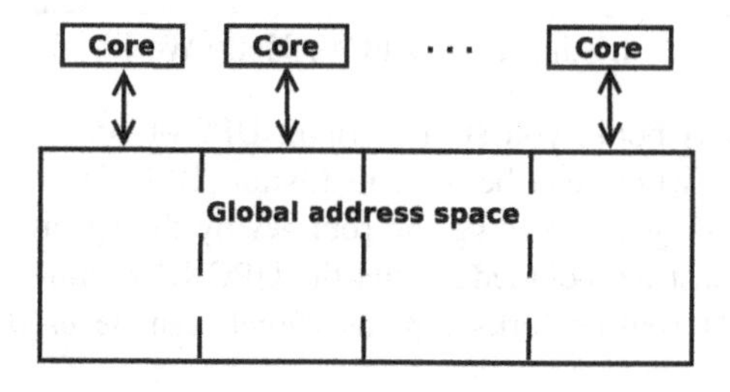

FIGURE 10.2

Abstraction of the memory model in PGAS languages.

At the time of writing this book UPC++ is a novel library and its mechanism for installation and execution depends on the hardware and software of the system. We recommend to follow the guidelines of the UPC++ community for updated information [12].

The basic ideas about how setting up UPC++ code are presented in Section 10.2. We continue by describing how to efficiently work with global partitioned memory in Section 10.3, where the concepts of shared arrays and pointer privatization are used to develop a parallel vector update routine. Additional features to efficiently manage the memory model such as global pointers and collectives are introduced in Section 10.4 as part of a code to count the number of occurrences of a letter in a text in parallel. Section 10.5 presents UPC++ locks and uses them for parallel image histogramming. Finally, remote function invocation is employed in Section 10.6 to accelerate the drawing of Mandelbrot sets.

10.2 BASIC CONCEPTS (HELLO WORLD)

```
1  #include <upcxx.h>
2
3  int main (int argc, char *argv[]){
4    // Initialize UPC++
5    upcxx::init(&argc, &argv);
6
7    // Get the number of threads
8    int numT=upcxx::ranks();
9
10   // Get the ID of the thread
11   int myId=upcxx::myrank();
12
13   // Every thread prints Hello
14   std::cout << "Thread " << myId << " of "
15     << numT << ": Hello, world!" << std::endl;
16
17   // Terminate UPC++
18   upcxx::finalize();
19   return 0;
20 }
```

Listing 10.1: UPC++ Hello World.

As it is a tradition in this book, you start learning UPC++ programming with the corresponding version of *Hello World*, which can be seen in Listing 10.1. This simple code is similar to the MPI variant (see Listing 9.1), just replacing the routines by the specific syntax of this new library. First, the header `upcxx.h` must be included so that the UPCXX routines are available. In the `main()` function the `upcxx::init()` routine forks threads, which can be used until they are joined with `upcxx::finalize()`.

A conceptual difference compared to the MPI variant of the code is that UPC++ does not work with communicators. Thus the identifier and the total number of threads obtained by `upcxx::myrank()` and upcxx::ranks() are always related to the total number of threads in the execution. A possible outcome of Listing 10.1 is:

```
Thread 3 of 4: Hello, world!
Thread 1 of 4: Hello, world!
Thread 0 of 4: Hello, world!
Thread 2 of 4: Hello, world!
```

10.3 MEMORY AFFINITY AND PRIVATIZATION (VECTOR UPDATE)

Now that we know how to create our first UPC++ program, let us continue showing you how to distribute data and tasks among threads. First, we introduce the concepts of local and remote. According to the abstraction shown in Fig. 10.2, the shared global address space is logically partitioned among threads. We define the *remote memory* of Thread i as the subspace that does not present affinity to

Thread i. Furthermore, the *local memory* of Thread i is defined as the subspace with affinity to it. Note that, differently from MPI, local is not equivalent to private, i.e., any thread can directly read data from the local memory of other threads without the need for explicitly calling communication routines.

The global address space is a logical abstraction created by UPC++, but it does not directly correspond to any particular parallel hardware. The same UPC++ program can work on different parallel architectures (either shared- or distributed-memory systems). The runtime just adapts the mapping of the different portions of the global address space to the underlying hardware memory. For instance, when working on a shared-memory system, the global memory will be symmetrically partitioned among threads. In this case, the performance of accesses to remote and local memory is similar, as all data are stored on the same memory module. In contrast, when a UPC++ program is executed on a distributed-memory system, where each node has its own memory module, the logical global space gathers the memory of the whole system, and each partition corresponds to the memory module of each node. In this case, local memory is mapped onto the module of the node where the thread is executed (thus, shared-memory accesses). Remote memory accesses might need to transfer data from other nodes, being significantly slower as they require network communications. However, such communications do not affect UPC++ programmability because they are directly performed by the runtime system and the complexity is hidden to the programmer.

```
1  #include <upcxx.h>
2
3  int main (int argc, char *argv[]){
4    // Initialize UPC++
5    upcxx::init(&argc, &argv);
6
7    // Private C++ array
8    float p[3];
9
10   // Shared UPC++ arrays
11   upcxx::shared_array<float, 3> a(18);
12   upcxx::shared_array<float> c(18);
13   float *privC = (float *) &c[upcxx::myrank()];
14
15   // Terminate UPC++
16   upcxx::finalize();
17   return 0;
18 }
```

Listing 10.2: Array definitions in UPC++.

Listing 10.2 illustrates different ways to allocate memory in the UPC++ global address space. First, we declare a traditional C++ array p in Line 8, that will create a private copy of this array in the different subspaces. As can be seen in Fig. 10.3 data is replicated in the different parts of the memory (one copy per thread). Each thread has its own pointer to the beginning of its array and threads can only access the memory allocated in the part with affinity. We can also use the C++ dynamic memory allocation methods with the same results (e.g., using the `new[]` function).

However, we know that working in parallel often requires to distribute data among threads, instead of replicating. The UPC++ shared array is a useful structure to represent distributed data. When creating a shared array we must specify the datatype and the block factor (i.e., the length of each block) as C++

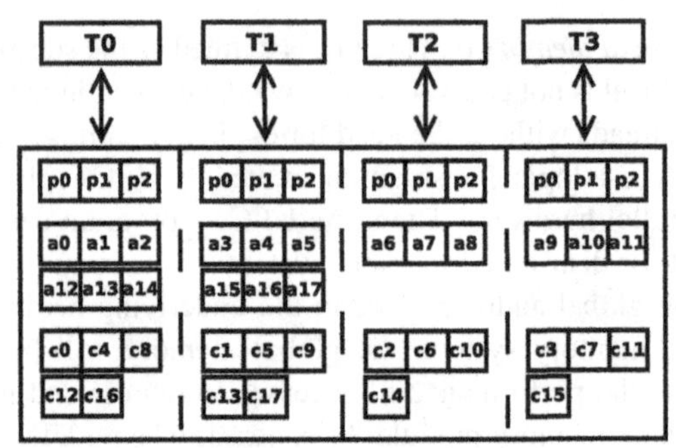

FIGURE 10.3

Different examples of arrays in UPC++.

template parameters. For instance, array *a* has 18 elements, divided into three contiguous elements per block. The first three elements are allocated in the part of the global memory with affinity to Thread 0; the fourth, fifth, and sixth elements in the part with affinity to Thread 1, and so on. The assignment is round-robin, i.e., after allocating one block in the part with affinity to the last thread, the next block is allocated in the portion of memory local to the first thread. The default block factor is one, i.e., pure cyclic distribution of data, as used in array *c* in Line 12. Note that when working with shared arrays no data is replicated and all threads can use the same pointer to access the same array.

Thanks to shared arrays we can easily work with distributed data as we only need to know the global pointer. On the contrary, in MPI we need to know in which private memory the data resides and translate from global to local indices. Furthermore, in UPC++ all threads can directly access any part of the array to implement remote memory accesses. Fig. 10.4 illustrates the UPC++ behavior to access shared arrays. The complexity of remote accesses is hidden to the users, which simplifies the programming task.

Before we study the example for this section, we want to remark that we can also use traditional C++ pointers to access local data allocated in a shared array. An example is the pointer $privC$ in Listing 10.2. Each thread has its own copy of $privC$ accessing different parts of the shared array c. For instance, Thread 0's $privC$ is initialized to $c[0]$, while for Thread 1 it points to $c[1]$. This private pointer iterates only through the data stored in the part of the global address space with affinity to the thread owner. For instance, the next element pointed by $privC$ in Thread 0 is $c[4]$ ($privC[1] == c[4]$) instead of $c[1]$ (it has affinity to Thread 1). The mechanism of using traditional C+++ pointers to access data of shared arrays is called *privatization*. We must be careful and not initialize private pointers to parts of global memory without affinity (e.g., $privC = c[0]$ in all threads). Privatization forces us to pay attention to data with affinities and to work with local indices (similarly to MPI), increasing programming complexity. However, as C++ pointers do not need to save information about block factor and memory affinities, pointer arithmetic is simpler and performance usually increases.

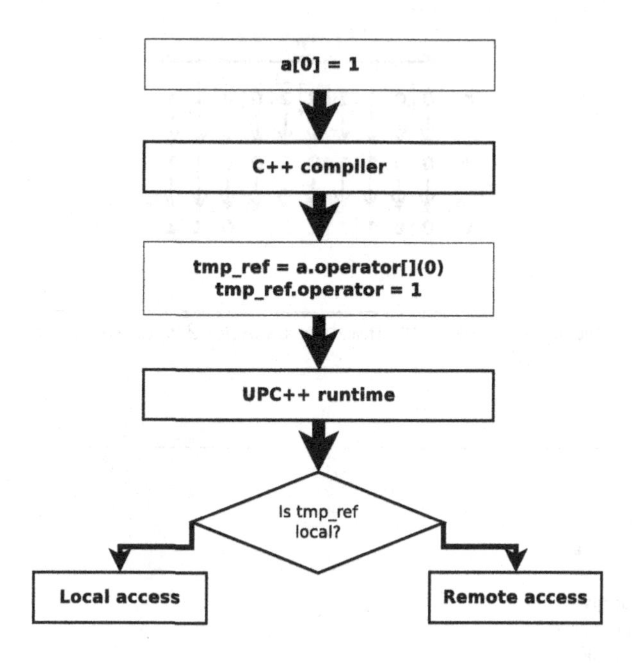

FIGURE 10.4

Behavior of UPC++ to access shared arrays.

We use vector update (sometimes called AXPY) in this section as the example to show how to exploit global address space to distribute data and workload among threads. This numerical routine receives as input two vectors x and y of length n, as well as a scalar value α, and it updates all the elements i of vector y ($0 \leq i < n$) as follows:

```
for (i=0; i<n; i++) y[i] = alpha * x[i] + y[i];
```

As the computation cost of each element i is the same, the workload can be distributed among threads so that each one processes the same number of elements. Even though different types of distribution could be used, it is important to maximize memory affinity. For example, consider a block-cyclic distribution where each block has two elements and the vector length is a multiple of two. Fig. 10.5 illustrates an example with three threads, arrays of length 10, and each thread computing the elements with affinity.

As previously mentioned, UPC++ shared arrays are useful structures to represent distributed data, as blocks are allocated into different memory parts. Listing 10.3 shows the initialization of the shared arrays used for the AXPY routine. This code starts with the initialization of the UPC++ environment (Line 6), the extraction of the number of threads and the id (Lines 7 and 8), and the parameter checking (Lines 10 to 35). The two shared arrays are created in Lines 38 and 39 by specifying that blocks have two floats (in the template) and the total size of the array (in parenthesis) is n, previously received as parameter. In the loop between Lines 47 and 52 each thread initializes the elements with affinity within the arrays.

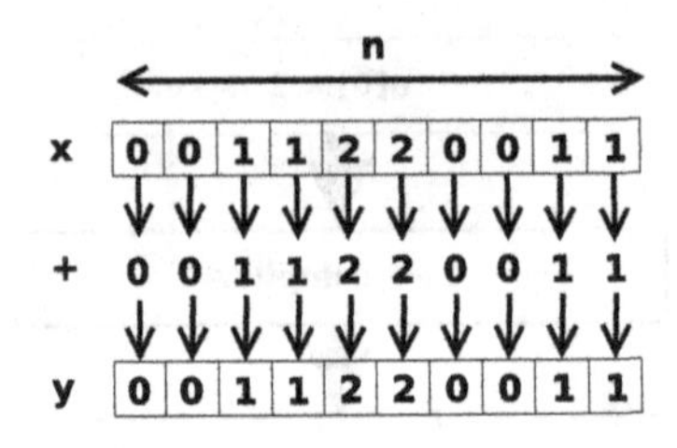

FIGURE 10.5

Abstraction of a block-cyclic workload distribution with block factor 2 when AXPY is executed with three threads and the vector length is 10.

```
1  #include <upcxx.h>
2  #include <timer.h>
3
4  int main (int argc, char *argv[]){
5    // Initialize UPC++
6    upcxx::init(&argc, &argv);
7    int numT = upcxx::ranks();
8    int myId = upcxx::myrank();
9
10   if(argc < 3){
11     // Only the first process prints the output message
12     if(!MYTHREAD)
13       std::cout << "ERROR: The syntax of the program is
14                    ./axpy n alpha" << std::endl;
15     exit(1);
16   }
17
18   int n = atoi(argv[1]);
19   float alpha = atof(argv[2]);
20
21   if(n < 1){
22     // Only the first process prints the output message
23     if(!myId)
24       std::cout << "ERROR: 'n' must be higher than 0"
25                 << std::endl;
26     exit(1);
27   }
28
29   if(n\%2){
30     // Only the first process prints the output message
31     if(!myId)
32       std::cout << "ERROR: The blocks (of size 2) must
33                    be complete" << std::endl;
34     exit(1);
35   }
36
37   // Declare the shared arrays
```

```
38  upcxx::shared_array<float, 2> x(n);
39  upcxx::shared_array<float, 2> y(n);
40
41  // To measure time
42  upcxx::timer t;
43  upcxx::barrier();
44  t.start();
45
46  // Initialize arrays
47  for(int i=2*myId; i<n; i+=2*numT){
48    x[i] = i;
49    y[i] = numT;
50    x[i+1] = i+1;
51    y[i+1] = numT;
52  }
```

Listing 10.3: Initialization of vectors for AXPY.

This code also introduces the UPC++ class to measure time: `timer`. It is included in the file `timer.h` and provides the methods `start()`, `stop()` and `reset()` to control the timer, as well as methods to output the measured time in different units (from seconds to nanoseconds). In order to correctly measure performance we need to synchronize all threads before starting the timer (Line 44). This is done in Line 43 by `barrier()`, a simple UPC++ collective. More UPC++ collectives will be presented in the next section.

Once the vectors are initialized, the computation of the AXPY routine itself shown in Listing 10.4 is straightforward. Threads do not need to calculate their local indices as they can directly traverse the whole shared vector with a simple for-loop (Line 54). Each thread starts discarding those elements with affinity to previous threads and pointing to the first element with affinity ($2 \cdot myId$). In each iteration it calculates the values for two elements and then proceeds to the next block that belongs to it by jumping over the blocks of the other threads (stride $2 \cdot numT$). The program finishes by stopping the timer (Line 60) after the corresponding barrier and with one thread writing the output into a file and printing the measured time.

```
53  // Compute AXPY with block factor 2
54  for(int i=2*myId; i<n; i+=2*numT){
55    y[i] += alpha*x[i];
56    y[i+1] += alpha*x[i+1];
57  }
58
59  upcxx::barrier();
60  t.stop();
61  if(!myId){
62    std::cout << "Time with " << numT << " processes: "
63              << t.secs() << " seconds" << std::endl;
64    printOutput(n, y);
65  }
```

Listing 10.4: Numerical computation of AXPY.

Note that all memory accesses during the iterations of the loop (Lines 55 and 56) are performed to local memory or memory with affinity. This is beneficial in terms of performance as we can guarantee

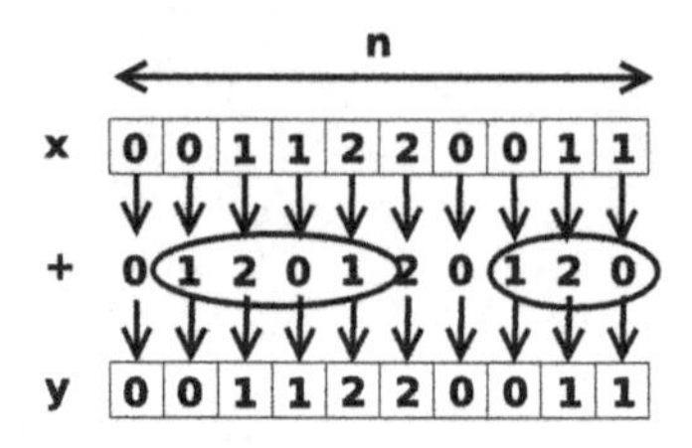

FIGURE 10.6

Abstraction of an inefficient workload distribution when AXPY is executed with three threads and the vector length is 10. Arrays present block factor of 2 while workload is distributed using a pure cyclic approach (block factor 1). Remote accesses are highlighted in ovals.

that threads are mapped to cores with direct access to these memory addresses. Nevertheless, the variation of the main loop presented in Listing 10.5 is completely valid although it requires several remote accesses. In this case the workload among threads (Line 54) follows a pure cyclic distribution (block factor 1), which is different from the data distribution of the shared array (block factor 2). Although the vector y written into the output file (function `printOutput()`) is the same in Listings 10.4 and 10.5, the latter will be slower because threads might require network communication to access these remote data. Fig. 10.6 illustrates the problem of using different distributions for data and workload, where three of the ten calculations require two remote accesses. As an example, we have executed these two versions in a cluster with InfiniBand network using only two nodes and one UPC++ thread per node and the second version is almost two times slower.

```
53    // Compute AXPY with block factor 1
54    for(int i=myId; i<n; i+=numT){
55      y[i] += alpha*x[i];
56    }
57
58    upcxx::barrier();
59    t.stop();
60    if(!myId){
61      std::cout << "Time with " << numT << " processes: "
62                << t.secs() << " seconds" << std::endl;
63      printOutput(n, y);
64    }
```

Listing 10.5: Numerical computation of AXPY using several remote accesses.

Listing 10.6 shows a similar computation but using private pointers to access the memory fragment with affinity. As the shared arrays are declared with block factor 2, the private pointers initially point at position $2 \cdot myId$, i.e., the first element with affinity (see Lines 54 and 55). Then each thread only has to consecutively access all the elements within its local memory with the loop of Line 58. Private pointers will never access addresses without affinity. Additionally, performance increases thanks to the lower complexity of C++ traditional pointers compared to shared arrays. In fact, in the previously mentioned scenario with two nodes and one thread per node the code with privatization (Listing 10.6) is five times faster than the original AXPY version (Listing 10.4).

```
53    // Private pointers
54    float *privX = (float *) &x[myId*2];
55    float *privY = (float *) &y[myId*2];
56
57    // Compute AXPY with private pointers
58    for(int i=0; i<myBlocks*2; i++){
59      privY[i] += alpha*privX[i];
60    }
61
62    upcxx::barrier();
63    t.stop();
64    if(!myId){
65      std::cout << "Time with " << numT << " processes: "
66                << t.secs() << " seconds" << std::endl;
67      printOutput(n, y);
68    }
```

Listing 10.6: Numerical computation of AXPY using privatization.

10.4 GLOBAL POINTERS AND COLLECTIVES (LETTER COUNT)

In the previous section we have learned about the `shared_array` structure as an easy way to distribute data within the global address space. However, there are two important limitations regarding the block factor: it must be the same for all blocks and it must be known at compile time. Assume that we want to distribute an array in a pure block way (i.e., each thread has only one block of consecutive elements) and the size of the array n is specified through a parameter. The block factor of such array should be $\frac{n}{numT}$. However, as this is not a constant known at compile time, we cannot use it as template parameter of the `shared_array` structure.

How can we distribute our data in UPC++ with a dynamic block factor? This can be achieved by global pointers (`global_ptr` structure). These pointers can allocate and access memory in any part of the global address space. Therefore we can use one global pointer per block of our distributed array. Each global pointer can dynamically allocate memory with an independent call to the method `allocate(int rank, size_t bytes)` (`rank` being the thread with affinity to the memory subspace where data is stored and `bytes` being the allocation size), so each block can have a different size and this size can be specified at runtime. Note that the UPC++ routine `allocate()` allows remote memory allocation (i.e., one thread can allocate memory with affinity to any other thread). This is not possible in other PGAS languages such as UPC. Once the memory is not going to be used again, we should call the function `deallocate()` to free it.

We illustrate how to use global pointers to create a distributed array using a UPC++ program that counts the number of occurrences of a letter in a text. We distribute the text in a pure block way, i.e., blocks of equal size and one block per thread. Each thread counts the occurrence of the letter in its corresponding text fragment The partial counts are added to compute the final result. The letter to find and the text length are specified as command line parameters. For simplicity, we assume that the length of the text must be a multiple of the number of threads. The input text is read from a file with an auxiliary function `readText()`.

```
1  #include <upcxx.h>
2
3  int main (int argc, char *argv[]){
4    // Initialize UPC++
5    upcxx::init(&argc, &argv);
6
7    int numT = upcxx::ranks();
8    int myId = upcxx::myrank();
9
10   if(argc < 3){
11     // Only the first thread prints the output message
12     if(!MYTHREAD)
13       std::cout << "ERROR: The syntax of is ./letter l n"
14                 << std::endl;
15     exit(1);
16   }
17
18   char l = *argv[1];
19   int n = atoi(argv[2]);
20
21   if(n < 0){
22     // Only the first thread prints the output message
23     if(!myId)
24       std::cout <<"ERROR: 'n' must be higher than 0" << std::endl;
25     exit(1);
26   }
27
28   if(n%numT){
29     // Only the first thread prints the output message
30     if(!myId)
31       std::cout << "ERROR: 'n' must be a multiple of the number of
32                 threads" << std::endl;
33     exit(1);
34   }
35
36   // Create the array of global pointers
37   upcxx::shared_array<upcxx::global_ptr<char>> p(numT);
38
39   // Each thread allocates the memory of its subspace
40   int blockFactor = n/numT;
41   p[myId] = upcxx::allocate(myId, blockFactor*sizeof(char));
42
43   // Thread 0 must wait until all threads have allocated memory
44   upcxx::barrier();
45
46   // Thread 0 reads the text and copies the fragments
47   if(!myId){
48     char *text = new char[n];
49     readText(n, text);
50
51     for(int i=0; i<numT; i++)
```

```
52        upcxx::copy<char>(&text[blockFactor*i], p[i], blockFactor);
53
54      delete text;
55    }
56
57    // Threads must wait until Thread 0 has copied all the fragments
58    upcxx::barrier();
```

Listing 10.7: Initialization of structures for the parallel letter counting.

Listing 10.7 summarizes the initialization of the text and its distribution in global address space. After checking the parameters, we declare the structures that will help us to represent the distributed text. First, a shared array `p` of global pointers in Line 37. As the input text will be distributed using one block per thread, we need $numT$ global pointers (one per block). Pointers are cyclically distributed among the parts of the global address space since no specific block factor is specified for the `p` structure. Each thread uses the pointer that is mapped to the memory with affinity (`p[myId]`) to dynamically allocate memory in its subspace. This is accomplished in Line 41 by the `allocate()` function previously described.

Once all threads have allocated the block in their local memory, Thread 0 will use the global pointers to copy the corresponding fragments of the input. `barrier()` in Line 44 is necessary to guarantee that all blocks have been already allocated before they are used as copy destinations. Then, the input is read by Thread 0 (we have not included the code of the `readText()` function for simplicity) and the portions of the text are copied to the corresponding blocks of memory. For each block we have decided not to directly assign the elements one-by-one with a `for` loop, which could lead to several small messages congesting the network when the memory modules are in different nodes of a cluster. Instead, we use the function `copy<T>(T *src, T *dst, size_t count)` to copy a complete block `count` elements of datatype `T` at once from `src` to `dst` (see Line 52). One copy per block is performed. Another call to `barrier()` (Line 58) guarantees that the input text has been distributed. Each thread then counts the number of letter occurrences in its assigned fragment.

```
59    // Privatize the pointer
60    int myNumOcc = 0;
61    char *myText = (char *) (upcxx::global_ptr<char>) p[myId];
62
63    // Check whether it is really local
64    if(!((upcxx::global_ptr<char>) p[myId]).is_local())
65      std::cout << "Thread " << myId << " not accessing local memory"
66                << std::endl;
67
68    // Find the local occurrences
69    for(int i=0; i<blockFactor; i++)
70      if(myText[i] == l)
71        myNumOcc++;
72
73    // Put the local occurrences accessible to all threads
74    upcxx::shared_array<int> occs(numT);
75    occs[myId] = myNumOcc;
76
```

```
77    // All threads must have put accessible the local occurrences
78    upcxx::barrier();
79
80    if(!myId){
81      int numOcc = myNumOcc;
82      for(int i=1; i<numT; i++)
83        numOcc += occs[i];
84
85      std::cout << "Letter " << l << " found " << numOcc
86                << " in the text " << std::endl;
87    }
88
89    // Deallocate the local memory
90    upcxx::deallocate<char>(p[myId]);
91
92    // Terminate UPC++
93    upcxx::finalize();
94    return 0;
95  }
```

Listing 10.8: Main computation of parallel letter counting.

The counting performed by each thread is shown in Listing 10.8. We use the privatization approach explained in the previous section to improve performance. Each thread only accesses the fragment of text stored in its local memory using the standard C++ pointer `myText`. Note that we must explicitly indicate the cast to convert a global to a standard pointer, as shown in Line 61. We know that traditional C++ pointers only access local memory, so we do not need to take care of data affinity in the loop between Lines 69 and 71.

We use a shared array of one integer per thread (see Line 74) to save the partial results of all threads. Once one thread has finished its local counting, it puts its partial result into the position `myId`, which is always in local memory because the block factor of the array is 1 (pure cyclic distribution). The `barrier()` of Line 78 guarantees that all threads have finished their work and all partial results are written, so Thread 0 can add all partial results (loop of Lines 82–83) and print the letter counting. The program finishes with the memory deallocation (Line 90) and UPC++ finalization (Line 93).

Listing 10.8 introduces an additional method provided by global pointers of special interest for UPC++ programmers. Concretely, the method `is_local()` is seen in Line 64, which indicates whether the address of a global pointer is local to the thread. It can be used to guarantee that only local accesses are performed.

This version of the program for parallel letter counting makes use of the `copy()` method for blocking bulk copies. Nevertheless, we can also use non-blocking copies that let the CPU continue working while the network completes the remote copy. Thus we can overlap computation and data copies to reduce the overall runtime. The syntax of the method for non-blocking data copies is:

- `upcxx::async_copy<T> (T *src, T *dst, size_t count, upcxx::event *e);`

where `src` and `dst` are the pointers (of type `T`) to source and destination addresses, respectively. `count` specifies the number of elements to copy and the last parameter is a pointer to a structure of type `event` which is provided by UPC++ to synchronize computation to previous events such as non-blocking data copies or asynchronous functions (which we study in Section 10.6). This structure provides two synchronization methods:

- `wait()` blocks thread execution until the event has finished.
- `test()` returns 1 if the task has been completed or 0 if it is still being processed. It is equivalent to the method `async_try()`.

We will use non-blocking copies and events in our parallel letter counting to optimize the distribution of the text among the different parts of the global address space. Instead of reading the input text at once, Thread 0 will read it fragment-by-fragment. Furthermore, it will overlap copying of the current fragment with the reading of the next fragment from the input file. The use of `async_copy()` and `event` for this purpose is shown in Lines 49–62 in Listing 10.9. As the copy of the last fragment is not overlapped to any next reading, it is performed with the blocking `copy()` (see Line 65).

```
1  int main (int argc, char *argv[]){
2    // Initialize UPC++
3    upcxx::init(&argc, &argv);
4
5    int numT = upcxx::ranks();
6    int myId = upcxx::myrank();
7
8    if(argc < 3){
9      // Only the first process prints the output message
10     if(!MYTHREAD)
11       std::cout << "ERROR: The syntax of is ./letter l n"
12                 << std::endl;
13     exit(1);
14   }
15
16   char l = *argv[1];
17   int n = atoi(argv[2]);
18
19   if(n < 0){
20     // Only the first process prints the output message
21     if(!myId)
22       std::cout <<"ERROR: 'n' must be higher than 0" << std::endl;
23     exit(1);
24   }
25
26   if(n%numT){
27     // Only the first process prints the output message
28     if(!myId)
29       std::cout << "ERROR: 'n' must multiple of the number of
30                 processes" << std::endl;
31     exit(1);
32   }
33
34   // Create the array of global pointers
```

```
35  upcxx::shared_array<upcxx::global_ptr<char>> p(numT);
36
37  // Each thread allocates the memory of its subspace
38  int blockFactor = n/numT;
39  p[myId] = upcxx::allocate(myId, blockFactor*sizeof(char));
40
41  // Thread 0 reads the text and copies the fragments
42  if(!myId){
43    char *text = new char[blockFactor];
44    char *text2 = new char[blockFactor];
45    upcxx::event e;
46
47    readText(blockFactor, text);
48
49    for(int i=0; i<numT-1; i++){
50      upcxx::async_copy<char>
51                      (text, p[i], blockFactor, &e);
52
53      // Overlap the copy with reading the next fragment
54      // We cannot use text for the next fragment before it is sent
55      readText(blockFactor, text2);
56      char *aux = text;
57      text = text2;
58      text2 = aux;
59
60      // The previous copy must have finished to reuse its buffer
61      e.wait();
62    }
63
64    // The last copy does not overlap
65    upcxx::copy<char>(text, p[numT-1], blockFactor);
66
67    delete text;
68    delete text2;
69  }
70
71  // Threads must wait until Thread 0 has copied all the fragments
72  upcxx::barrier();
```

Listing 10.9: Initialization of structures for the parallel letter counting with non-blocking data copies.

Once the data distribution has been optimized, we will focus now on the summation of the final result. Instead of forcing Thread 0 to traverse a shared array (with $numT - 1$ remote accesses), we can use the collective routines provided by UPC++. As already learned in Section 9.5, collectives are routines that implement common patterns of data copies that involve all threads. The main advantages of collectives are increased usability (as we do not need to implement the patterns ourselves) and performance improvements (as the provided implementations are usually efficient, especially if optimized for specific architectures [6,11]). The available UPC++ collectives are:

- `barrier()`: All threads block execution until all threads have reached that point of the code.

- `bcast(void *src, void *dst, size_t nbytes, uint32_t root)`. The data of size `nbytes` stored at `src` in Thread `root` is replicated to the memory address pointed by `dst` in all threads.
- `gather(void *src, void *dst, size_t nbytes, uint32_t root)`. The data of size `nbytes` stored at `src` in each thread is aggregated in the space addressed by `dst` of Thread `root`.
- `allgather(void *src, void *dst, size_t nbytes)`. The same as `gather()` but the output is replicated for all threads.
- `alltoall(void *src, void *dst, size_t nbytes)`. The `nbytes` stored at `src` in all threads are scattered among all threads and the outputs are written into the memory pointed by `dst`.
- `reduce<T>(T *src, T *dst, size_t count, uint32_t root, upcxx_op_t op, upcxx_datatype_t dt)` performs the operation `op` to the `count` elements of type `dt` stored at `src` in all threads. The output is written in the memory pointed by `dst` of Thread `root`. The available UPC++ operations and datatypes are specified in `upcxx_types.h`.

Line 89 of Listing 10.10 shows how to use the `reduce()` collective so that Thread 0 can obtain the total number of occurrences of the letter in the text.

```
73   // Privatize the pointer
74   int myNumOcc = 0;
75   char *myText = (char *) (upcxx::global_ptr<char>) p[myId];
76
77   // Check whether it is really local
78   if(!((upcxx::global_ptr<char>) p[myId]).is_local())
79     std::cout << "Thread " << myId << " not accessing local memory"
80               << std::endl;
81
82   // Find the local occurrences
83   for(int i=0; i<blockFactor; i++)
84         if(myText[i] == l)
85      myNumOcc++;
86
87   // Reduce number of occurrences
88   int numOcc;
89   upcxx::reduce(&myNumOcc, &numOcc, 1, 0, UPCXX_SUM, UPCXX_INT);
90
91   if(!myId)
92     std::cout << "Letter " << l << " found " << numOcc
93               << " in the text " << std::endl;
94
95   // Deallocate the local memory
96   upcxx::deallocate<char>(p[myId]);
97
98   // Terminate UPC++
99   upcxx::finalize();
100  return 0;
101 }
```

Listing 10.10: Main computation in the parallel letter counting with reduced collective to obtain the final results.

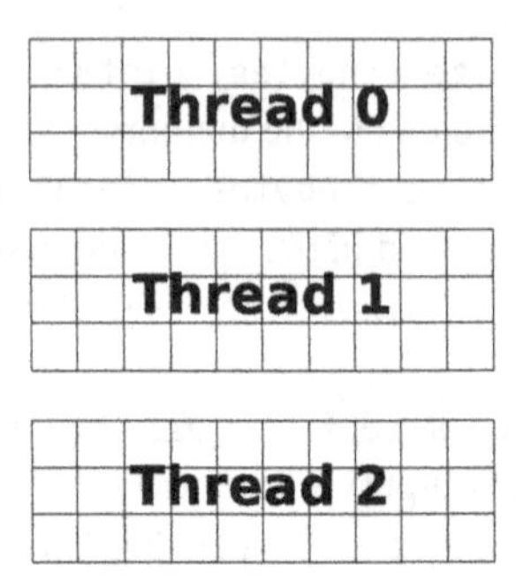

FIGURE 10.7

Example of block distribution by rows of an image using three threads.

10.5 LOCKS (IMAGE HISTOGRAMMING)

Synchronization of accesses to global memory is important in UPC++ in order to avoid race conditions (i.e., two threads modifying the same element at the same time). We have explained `barrier()` and `event` as basic UPC++ features to synchronize thread computation. However, how can we assure that only one thread is reading/writing certain variables? There is a mechanism in UPC++ inherited from OpenMP: the `shared_lock`. A `shared_lock` is a structure that contains one bit that determines its current state, i.e., whether it is being used (locked) or not (unlocked), as well as an integer to indicate the thread holder in case it is locked. When one thread acquires the lock, no other thread can enter the subsequent code section until the lock has been released. Thus UPC++ guarantees that a critical section will be performed by only one thread at a time. In particular, UPC++ `shared_lock` provides the following methods for synchronization:

- `void lock()`. The issuing thread tries to acquire the lock. If the lock is open, the thread acquires it and becomes the new owner. Otherwise, the thread waits until the lock is available.
- `void unlock()`. The issuing thread releases the lock and makes it available for other threads.
- `int trylock()`. The thread tries to acquire the lock but does not wait until it is available. It returns 1 upon success and 0 otherwise.
- `int islocked()` returns 1 if the thread is the holder and 0 otherwise, but does not try to acquire the lock.

We will use UPC++-based image histogramming as a parallel application example where locks are useful. This program reads a grayscale image (2D matrix of integers in the range 0 to 255) and counts the number of pixels in the image for each gray level. Histogramming is commonly used in image-processing applications. Our program receives as input through the command line the dimensions of the image and employs a block distribution by rows as shown in Fig. 10.7. For simplicity we assume that the number of rows is a multiple of the number of threads.

The whole program except the functions to read the input and print the output (not shown looking for simplicity) is listed in Listing 10.11. It starts by declaring the `shared_lock` that will be used later (Line 3), initializing UPC++ (Line 7), checking the parameters obtained through command line

(Lines 12 to 45) and distributing the input image among threads (Lines 47 to 85). The data distribution is performed with non-blocking copies, as described in the previous section for the optimized parallel letter counting code.

```
1   #include <upcxx.h>
2
3   upcxx::shared_lock l;
4
5   int main (int argc, char *argv[]){
6     // Initialize UPC++
7     upcxx::init(&argc, &argv);
8
9     int numT = upcxx::ranks();
10    int myId = upcxx::myrank();
11
12    if(argc < 3){
13      // Only the first thread prints the output message
14      if(!MYTHREAD)
15        std::cout << "ERROR: The syntax of the program is
16                     ./histo rows cols" << std::endl;
17      exit(1);
18    }
19
20    int rows = atoi(argv[1]);
21    int cols = atoi(argv[2]);
22
23    if(rows < 0){
24      // Only the first thread prints the output message
25      if(!myId)
26        std::cout << "ERROR: 'rows' must be higher than 0"
27                  << std::endl;
28      exit(1);
29    }
30
31    if(cols < 0){
32      // Only the first thread prints the output message
33      if(!myId)
34        std::cout << "ERROR: 'cols' must be higher than 0"
35                  << std::endl;
36      exit(1);
37    }
38
39    if(rows%numT){
40      // Only the first thread prints the output message
41      if(!myId)
42        std::cout << "ERROR: 'n' must multiple of the number
43                     of threads" << std::endl;
44      exit(1);
45    }
46
47    // Create the array of global pointers
48    upcxx::shared_array<upcxx::global_ptr<int>> p(numT);
```

```
49
50  // Each thread allocates the memory of its subspace
51  int blockRows = rows/numT;
52  p[myId] = upcxx::allocate(myId, blockRows*cols*sizeof(int));
53
54  // Thread 0 reads the image and copies the fragments
55  if(!myId){
56    int *block = new int[blockRows*cols];
57    int *block2 = new int[blockRows*cols];
58    upcxx::event e;
59
60    readImage(blockRows, cols, block);
61
62    for(int i=0; i<numT-1; i++){
63      upcxx::async_copy<int>(block, p[i], blockRows*cols, &e);
64
65      // Overlap the copy with reading the next fragment
66      // We cannot use "block" for the next fragment because
67      // it has not been sent yet
68      readImage(blockRows, cols, block2);
69
70      // The previous copy must have finished to reuse its buffer
71      e.wait();
72      int *aux = block;
73      block = block2;
74      block2 = aux;
75    }
76
77    // The last copy does not overlap
78    upcxx::copy<int>(block, p[numT-1], blockRows*cols);
79
80    delete block;
81    delete block2;
82  }
83
84  // Threads must wait until Thread 0 has copied the fragments
85  upcxx::barrier();
86
87  // Privatize the pointer
88  int *myImage = (int *) (upcxx::global_ptr<int>) p[myId];
89
90  // Check whether it is really local
91  if(!((upcxx::global_ptr<int>) p[myId]).is_local())
92    std::cout << "Thread " << myId << " not accessing
93                local memory" << std::endl;
94
95  // Declare the histogram
96  upcxx::shared_array<int> histogram(256);
97  for(int i=myId; i<256; i+=numT)
98    histogram[i] = 0;
99
100 // Threads must wait until the histogram has been initialized
```

```
101    upcxx::barrier();
102
103    // Examine the local image
104    for(int i=0; i<blockRows*cols; i++){
105      // Close the lock to access the shared array
106      l.lock();
107
108      histogram[myImage[i]] = histogram[myImage[i]]+1;
109
110      // Open the lock again
111      l.unlock();
112    }
113
114    // All threads must have finished their local computation
115    upcxx::barrier();
116
117    if(!myId)
118      printHistogram(histogram);
119
120    // Deallocate the local memory
121    upcxx::deallocate<int>(p[myId]);
122
123    // Terminate UPC++
124    upcxx::finalize();
125    return 0;
126  }
```

Listing 10.11: Parallel UPC++ histogramming.

The proper parallel histogramming starts in Line 87, with the privatization of the distributed image so that each thread uses a standard C++ pointer `myImage` to access the local fragment of the input. After checking that the memory is really local with the `is_local()` method in Line 91, the shared array of 256 elements that holds the histogram is declared in Line 96 and initialized to 0 in the loop of Lines 97 and 98. The loop that traverses the partial image for histogramming is between Lines 104 and 112. Each thread examines its own pixels and increments the corresponding histogram index every time the corresponding pixel value is found. The problem is, however, that two of the threads may find the same pixel value locally and attempt to increment the same element of the histogram at the same time, which creates a race condition. Thus the result from one of those operations may be lost. Therefore the lock `l` is used to protect the update of the histogram, with the corresponding call to the `lock()` and `unlock()` methods before and after the variable update.

The `barrier()` in Line 101 is necessary so assure that the histogram has been completely initialized to 0 before starting to update its values. In this example we use a cyclic array (i.e. a block factor of 1) for the `histogram` array. As accesses depend on the values of the image, there is no block factor that guarantees obtaining higher percentage of local accesses for any image.

This implementation suffers from a simple but serious problem. There is only one lock and it has to be used by all threads, even if two threads are trying to update different entries in the histogram, despite the fact that these two operations could have been performed in parallel. To resolve this problem and enable a higher degree of parallelism, we could use as many locks as the number of entries in the histogram. Thus each histogram entry is controlled individually.

Listing 10.12 shows the modified part of the optimized program (initialization, parameter checking, and data distribution are similar to Listing 10.11). We should also declare a shared array of locks, accessible to all threads, as:

- `upcxx::shared_array<upcxx::shared_lock> locks`

```
100   // Initialize the locks
101   locks.init(256);
102   for(int i=myId; i<256; i+=numT)
103     new (locks[i].raw_ptr()) upcxx::shared_lock(myId);
104
105   // Threads must wait until all locks and
106   // histogram have been initialized
107   upcxx::barrier();
108
109   // Examine the local image
110   for(int i=0; i<blockRows*cols; i++){
111     // Close the lock to access the shared array
112     ((upcxx::shared_lock) locks[myImage[i]]).lock();
113
114     histogram[myImage[i]] = histogram[myImage[i]]+1;
115
116     // Open the lock again
117     ((upcxx::shared_lock) locks[myImage[i]]).unlock();
118   }
119
120   // All threads must have finished their local computation
121   upcxx::barrier();
122
123   if(!myId)
124     printHistogram(histogram);
125
126   // Deallocate the local memory
127   upcxx::deallocate<int>(p[myId]);
128
129   // Terminate UPC++
130   upcxx::finalize();
131   return 0;
132 }
```

Listing 10.12: Parallel UPC++ histogramming using several locks.

Therefore each thread blocks at Line 111 only if another thread is accessing the same histogram slot at the same time.

We also use this code to show how to dynamically allocate shared arrays. Up to now, we have always used shared arrays specifying the number of elements in compilation time (see, for instance, Line 96 of Listing 10.11). However, we can also only declare it, and allocate the memory when necessary with the `init()` method (see Line 101). Moreover, if the datatype of the shared array is a pointer, we can use the method `raw_pointer()` to obtain the standard C++ array just behind the UPC++ `shared_array` structure and then dynamically allocate the pointers, as is done in Line 103 for the `shared_lock` pointers.

10.6 REMOTE FUNCTION INVOCATION (MANDELBROT SETS)

Another important UPC++ feature studied in this book chapter is remote function invocation using the `async()` macro. This is a novelty compared to other PGAS languages such as UPC. Asynchronous remote tasks in UPC++ provide a powerful strategy for encapsulating bulk updates of remote matrix elements in a manner that can be executed in isolation. UPC++ permits the application programmer to explicitly specify bulk data movement and custom update logic, and then offload the latter for execution on the target (ensuring consistency guarantees, maximizing data locality, taking advantage of known optimizations, etc.). The user can start an asynchronous remote function invocation with the following syntax:

- `future<T> f = async(place)(function, args...)`

where `place` specifies the thread where `function` will be called using the parameters `args`. A UPC++ `async()` call optionally returns a `future` object, which can be used later to retrieve the return value of the remote function call by `future.get()`.

In addition, UPC++ provides a mechanism to specify dynamic dependencies among tasks: event-driven execution with the already presented `event` class (see Section 10.4). The user may register `async()` operations with an event to be signaled after task completion. An event can be signaled by one or more `async()` operations, and used as a precondition to launch following `async` operations. We now look at some examples:

- `async(place, event *e)(task1, args...)`
- `async_after(place, event *after)(task2, args...)`

In the first case `task1` will start as soon as the task pool advances (using the method `advance()`) and we can use the methods `wait()` and `test()` of the event `e` (see Section 10.4) to control if the task has finished. In the second line, we specify that `task2` must start once the previous tasks associated with event `after` are finished.

We study in this section the parallel computation of the Mandelbrot set. It is known that certain number $c = x + i \cdot y \in \mathbb{C}$ where $x, y \in \mathbb{R}$ is part of the Mandelbrot set if and only if the recursively defined complex-valued sequence $z_{n+1} = z_n^2 + c$ with initial condition $z_0 = 0$ converges i.e., $\lim_{n\to\infty} z_n$ exists (is a finite value). Without proof, one can show that every complex number c with squared radius $|c|^2 = x^2 + y^2 > 4$ generates a diverging sequence. Thus we can approximate the Mandelbrot set by the set of all complex numbers c whose radius $|c|$ is smaller than 2 after a fixed number of iterations. If we map each pixel of an image to a complex number (the coordinates x and y being the real and imaginary parts, respectively), we just have to check if it belongs to the set by iterating the formula, and color the pixel black if it does. Since the iteration may never terminate, we limit the number of iterations. Listing 10.13 shows the function that returns the number of iterations for a certain pixel

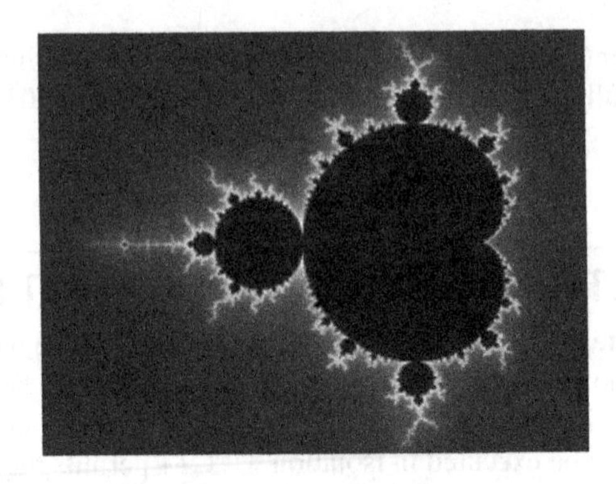

FIGURE 10.8

Mandelbrot set (black area) sampled with 1920 samples per axis.

(i, j) in an image of size $rows \times cols$ using $maxIter$ as maximum number of iterations. It returns 0 if the pixel does belong to the set. Fig. 10.8 shows an example.

```
1  int mandel(int i, int j, int rows, int cols, int maxIter){
2    float zReal = 0.0, zImag = 0.0, cReal, cImag, temp, lengthsq;
3
4    cReal = -2.0+j*4.0/rows;
5    cImag = 2.0-i*4.0/cols;
6    int k = 0;
7
8    do { // Iterate for pixel color
9      temp = zReal*zReal-zImag*zImag+cReal;
10     zImag = 2.0*zReal*zImag+cImag;
11     zReal = temp;
12     lengthsq = zReal*zReal+zImag*zImag;
13     k++;
14   } while (lengthsq<4.0 && k < maxIter);
15
16   if(k>=maxIter)
17     return 0;
18
19   return k;
20 }
```

Listing 10.13: Function to obtain the Mandelbrot value of a pixel.

The first approach that we present to implement the parallel drawing of these sets with UPC++ is straightforward: just divide the image pixels statically among threads, with pure block distribution by rows (the first $\frac{rows}{numT}$ rows computed by Thread 0, the next block of $\frac{rows}{numT}$ rows computed by Thread 1, etc.). It is shown in Listing 10.14. We start with the common initialization and parameter checking from the beginning to Line 53. Note that, as usual in this book, we ask the number or rows to be a multiple of the number of threads for simplicity.

```
1  #include <upcxx.h>
2
3  int main (int argc, char *argv[]){
4    // Initialize UPC++
5    upcxx::init(&argc, &argv);
6
7    int numT = upcxx::ranks();
8    int myId = upcxx::myrank();
9
10   if(argc < 4){
11     // Only the first process prints the output message
12     if(!MYTHREAD){
13       std::cout << "ERROR: The syntax of the program is ./mandel
14                     rows cols maxIter" << std::endl;
15     }
16     exit(1);
17   }
18
19   int rows = atoi(argv[1]);
20   int cols = atoi(argv[2]);
21   int maxIter = atoi(argv[3]);
22
23   if(rows < 0){
24     // Only the first process prints the output message
25     if(!myId)
26       std::cout << "ERROR: 'rows' must be higher than 0"
27                 << std::endl;
28     exit(1);
29   }
30
31   if(cols < 0){
32     // Only the first process prints the output message
33     if(!myId)
34       std::cout << "ERROR: 'cols' must be higher than 0"
35                 << std::endl;
36     exit(1);
37   }
38
39   if(maxIter < 0){
40     // Only the first process prints the output message
41     if(!myId)
42       std::cout << "ERROR: 'maxIter' must be higher than 0"
43                 << std::endl;
44     exit(1);
45   }
46
47   if(rows%numT){
48     // Only the first process prints the output message
49     if(!myId)
50       std::cout << "ERROR: 'n' must multiple of the number of
51                     processes" << std::endl;
```

```
52      exit(1);
53    }
54
55    // Output array
56    int blockRows = rows/numT;
57    int myImage[blockRows*cols];
58    upcxx::shared_var<upcxx::global_ptr<int>> outImage;
59
60    // Only the owner allocates the array to gather the output
61    if(!myId){
62      outImage.put(upcxx::allocate(0, rows*cols*sizeof(int)));
63    }
64
65    // To guarantee that memory is allocated
66    upcxx::barrier();
67
68    // Mandel computation of the block of rows
69    for(int i=0; i<blockRows; i++)
70      for(int j=0; j<cols; j++)
71        myImage[i*cols+j] = mandel(i+myId*blockRows, j, rows,
72                                   cols, maxIter);
73
74    // Copy the partial result
75    upcxx::copy<int>(myImage, (upcxx::global_ptr<int>)
76                     &(outImage.get())[myId*blockRows*cols],
77                     blockRows*cols);
78
79    // All threads must have finished their local computation
80    upcxx::barrier();
81
82    if(!myId){
83      printMandel((int *) outImage.get(), rows, cols);
84      // Deallocate the local memory
85      upcxx::deallocate<int>(outImage.get());
86    }
87
88    // Terminate UPC++
89    upcxx::finalize();
90    return 0;
91  }
```

Listing 10.14: Parallel generation of a Mandelbrot set with block distribution by rows.

The next step consists of creating a structure where threads can save the results for each pixel. Instead of using again a `shared_array`, we will show how to allocate the whole image in the subspace of only one thread (in this case Thread 0) and make it accessible to all threads. We already know that global pointers can be used to dynamically allocate memory in the desired subspace. However, if we directly declare `outImage` in our code, we will create a different instance of this global pointer. When Thread 0 allocates memory in Line 62, only the instance of this thread will point to the recently allocated space. This can be solved by declaring this pointer as a shared variable (shared_var) as in Line 58. There is only one instance of this variable shared among all threads and the modifications made by one thread are visible to others. Methods get() and put() are used to read and write the value of a shared variable. We must take into account that we must synchronize accesses to this shared variable in order to avoid race conditions. This is solved in this code with the barrier() in Line 66 to guarantee that memory has been allocated before other threads want to write in *outImage*.

Between Lines 69 and 72 each thread calculates the number of iterations for its associated pixels and initially saves the results in a local array *myImage*. Then, the partial images are copied to the corresponding position of *outImage* with one bulk copy (copy() method) per thread. Once all threads have copied their partial results (barrier() in Line 80), Thread 0 prints the output (Line 83) and frees the memory previously allocated in its subspace for *outImage* (Line 85).

In this approach all threads analyze the same number of rows. Since the computational cost of the rows varies, this could lead to an unbalanced workload. Looking at the function of Listing 10.13, we can see that points that do belong to the Mandelbrot set must repeat the execution *maxIter* times. Otherwise, only iterations to find the convergence level are needed. There exist scenarios where one thread analyzes more points that belong to the Mandelbrot set than others, and thus its workload would be lower. This leads to a lower utilization of parallel resources as some threads finish their computation (and are idle) while others are still working.

We solve this problem using a master-slave approach where Thread 0 (master) distributes the work among other threads (slaves). It initially assigns one row to each slave. Once a slave finishes its computation, the master assigns another row to it until the whole set has been computed. Listing 10.15 shows the master-slave variant of the parallel computation of Mandelbrot sets. The beginning of the code is similar to the previous version (initialization and parameter checking). The main difference is that we need to create a function that analyzes one row, that will be remotely invoked from the master in the slaves with async(). This function is mandelRow (Lines 9–23). Those arrays and variables that are used within this function must be declared globally outside the main() function. On the one hand, *outImage* is declared as in the previous version. On the other hand, we create a shared array busyTh in Line 7 to identify which thread is busy. Once one thread finishes the computation of the assigned row, it indicates that it is idle (Line 22).

```
1  #include <upcxx.h>
2
3  // Output array
4  upcxx::shared_var<upcxx::global_ptr<int>> outImage;
5
6  // Array to know the busy threads
7  upcxx::shared_array<bool> busyTh;
8
9  void mandelRow(int iterRow, int th, int rows, int cols,
10                int maxIter){
11   int rowRes[cols];
12
13   for(int j=0; j<cols; j++){
14     rowRes[j] = mandel(iterRow, j, rows, cols, maxIter);
15   }
16
```

```
17      // Copy the partial result
18      upcxx::copy<int>(rowRes, (upcxx::global_ptr<int>)
19                       &(outImage.get())[iterRow*cols],
20                       cols);
21
22      busyTh[th] = false;
23  }
24
25  int main (int argc, char *argv[]){
26      // Initialize UPC++
27      upcxx::init(&argc, &argv);
28
29      int numT = upcxx::ranks();
30      int myId = upcxx::myrank();
31
32      if(numT == 1){
33        std::cout << "ERROR: More than 1 thread is required for this
34                      master-slave approach" << std::endl;
35        exit(1);
36      }
37
38      if(argc < 4){
39        // Only the first process prints the output message
40        if(!MYTHREAD){
41          std::cout << "ERROR: The syntax of the program is
42                        ./mandel rows cols maxIter" << std::endl;
43        }
44        exit(1);
45      }
46
47      int rows = atoi(argv[1]);
48      int cols = atoi(argv[2]);
49      int maxIter = atoi(argv[3]);
50
51      if(rows < 0){
52        // Only the first process prints the output message
53        if(!myId)
54          std::cout << "ERROR: 'rows' must be higher than 0"
55                    << std::endl;
56        exit(1);
57      }
58
59      if(cols < 0){
60        // Only the first process prints the output message
61        if(!myId)
62          std::cout << "ERROR: 'cols' must be higher than 0"
63                    << std::endl;
64        exit(1);
65      }
66
67      if(maxIter < 0){
68        // Only the first process prints the output message
```

```
69      if(!myId)
70        std::cout << "ERROR: 'maxIter' must be higher than 0"
71                  << std::endl;
72      exit(1);
73    }
74
75    if(rows%numT){
76      // Only the first process prints the output message
77      if(!myId)
78        std::cout << "ERROR: 'n' must multiple of the number of
79                     processes" << std::endl;
80      exit(1);
81    }
82
83    // Initialize the lazy array
84    // All elements with affinity to Thread 0
85    busyTh.init(numT);
86    busyTh[myId] = false;
87
88    // To guarantee that busyTh is initialized
89    upcxx::barrier();
90
91    // Thread 0 is the master
92    if(!myId){
93      outImage.put(upcxx::allocate(0, rows*cols*sizeof(int)));
94      int nextTh = 1;
95
96      // While there are more rows
97      for(int i=0; i<rows; i++){
98        // Check whether any thread has finished
99        while(busyTh[nextTh]){
100         nextTh++;
101         if(nextTh == numT){
102           nextTh = 1;
103         }
104       }
105       busyTh[nextTh] = true;
106
107       upcxx::async(nextTh)(mandelRow, i, nextTh, rows, cols,
108                            maxIter);
109       upcxx::advance();
110     }
111
112     // Wait for the last row of each thread
113     upcxx::async_wait();
114
115     printMandel((int *) outImage.get(), rows, cols);
116     // Deallocate the local memory
117     upcxx::deallocate<int>(outImage.get());
118   }
119
120   // Terminate UPC++
```

```
121     upcxx::finalize();
122     return 0;
123 }
```

Listing 10.15: Parallel generation of Mandelbrot sets using a master-slave approach.

The main function is modified from Line 83. First, `busyTh` is initialized indicating that all threads are idle (Lines 85 and 86) and Thread 0 (master) allocates memory into its subspace for the output image (Line 93). Then, the master assigns the different rows in the loop starting at Line 97. This loop starts checking which is the first idle slave (Lines 99–104). Once we have selected the thread that will compute the row, the master indicates that it is busy in the `busyTh` array (Line 105), creates the remote task (Line 107), and advances in the task pool so that task starts (Line 109). Finally, when all tasks are performed, it writes the output (Line 115) and frees memory (Line 117).

10.7 ADDITIONAL EXERCISES

1. Let A be a matrix of dimensions $m \times n$ and x a vector of length n. Calculate in parallel using UPC++ the vector y of length m resulting from $y = A \times x$. The values of the result vector must be: $y_i = \sum_{j=0..n} A_{i,j} \times x_j$.
   a. Distribute both the rows of the matrix A and the elements of x among the different parts of the global memory space. Apply a suitable distribution to y in order to minimize remote accesses.
   b. Distribute the rows of the matrix A among the different parts of the global memory and replicate the vector x. Apply a suitable distribution to y in order to minimize remote accesses.
   c. Distribute the columns of the matrix A among the different parts of the global memory. Apply a suitable distribution to both vectors x and y in order to minimize remote accesses. You may need a reduction in order to obtain good performance results.

   Measure and discuss the performance of the three versions.
2. Sorting an array of elements is a common task in several applications. There exist several algorithms to sort a list, with their advantages and drawbacks. One of the most widely used and studied is merge-sort, which follows a divide-and-conquer approach with the following steps:
   a. Divide the unsorted list of length n in $\frac{n}{p}$ fragments of length p.
   b. Order the fragments independently.
   c. Merge the fragments by pairs in order to have sorted fragments of size $2 \times p$. Perform the merge iteratively until you obtain the final sorted list.

   Fig. 10.9 illustrates an example of a parallel UPC++ version of the merge-sort algorithm mapping the different fragments of the list to different parts of the global memory. Each fragment must be sorted by certain threads in order to minimize remote memory accesses.
3. Several bioinformatics algorithms work with DNA reads, i.e., a sequence of bases which can have four values: adenine, guanine, cytosine, and thymine. These values are usually represented with letters 'A,' 'C,' 'G,' and 'T,' respectively. Develop a parallel UPC++ algorithm that counts the number of appearances of each base on a DNA sequence. Your code can be based on the parallel letter counting presented in Section 10.4. You should only iterate once over the whole input sequence (one-pass algorithm).

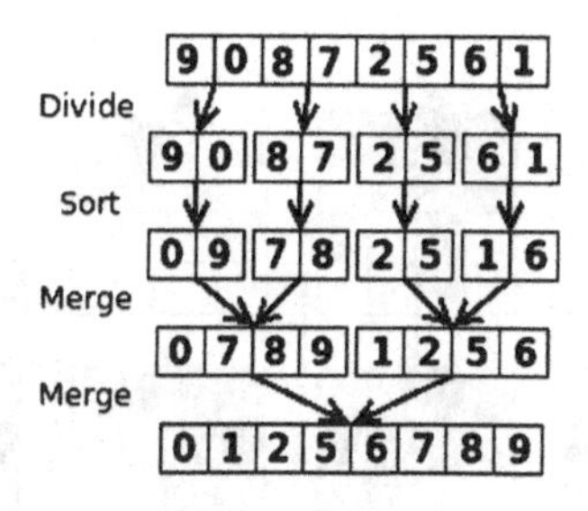

FIGURE 10.9

Example of merge sort over a list with eight (n) elements using fragments of two (p) elements.

4. The DNA to RNA transcription consists in translating a DNA sequence to RNA to create a protein from the core of a cell to the cytoplasm. RNA sequence has the same length as the DNA one but with the following translations: A to U; C to G; G to C; and T to A. Develop a UPC++ where each thread reads a fragment of a very long DNA sequence and all of them collaborate to translate the sequence to the corresponding RNA sequence. Each thread only works over the fragment read by it.
5. The N-Queens problem is a classic problem in computer science, due to its interesting computational behavior, depth-first searching and backtracking, and the fact that its processing time grows at a nonpolynomial (NP) rate. Thus, as the problem size grows, the execution time grows at a much more dramatic pace. Based on the chess game, there are numerous variations of the problem description. We ask you to develop a parallel UPC++ program that seeks to find all solutions to the problem of placing N queens on an $N \times N$ chessboard such that no queen can capture the other. As the queen is the most versatile chess piece and it can be moved in horizontal, vertical, and diagonal directions, this implies that no two queens can be placed on the same row, column, or along the same diagonal.
   Fig. 10.10 shows the computational flow in the N-Queens problem using a simple example of a 4×4 chessboard. The algorithm starts by placing the first queen in the first column of the first row, then shading all boxes along the same row, column, or diagonal. Then we attempt to place the second queen in the second row in the first open position. It is clear after placing the second queen that no more queens can be placed on the board; therefore, this path was abandoned and we had to backtrack to where a queen was placed in the second column of the first row. The algorithm continues, and at the end it is apparent that only two valid solutions, in which all four queens can be safely placed, exist.
   As the problem size increases, the number of iterations required to search for all possible ways that the N-Queens can coexist on the same board grows dramatically. Luckily, N-Queens lends itself to parallelism. This is because N-Queens is a tree-search problem in which the subtrees are independent of one another and therefore can be distributed across a number of threads.
   Develop a parallel UPC++ program to solve the N-Queens problem for an arbitrary chessboard dimension (indicated through command line). Distribute the elements of the first row among the UPC++ threads so that each thread performs the sequential problem for the case where a queen has been placed in the first row and the assigned columns. You should represent the constraints (gray

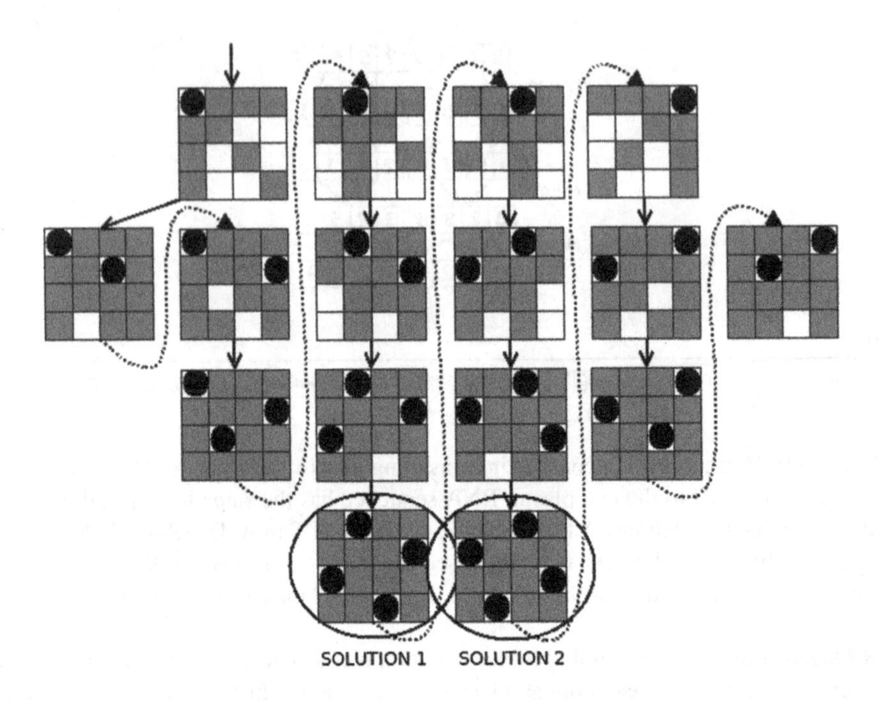

FIGURE 10.10

Example of the sequential solution for the N-Queens problem with a chessboard of dimensions 4 × 4. Black circles represent the queens. White and gray squares represent legal and illegal position for a new queen, respectively. Forward movements are represented by continuous arrows, while backtracking movements are represented by spotted arrows.

squares) present in the current state. You can use a data structure to represent them in a binary format (0 for legal positions and 1 for illegal positions).

6. Develop a UPC++ program that simulates the classical problem known as *dining philosophers* [3], where each thread represents a philosopher. Fig. 10.11 illustrates the situation of this problem. A number of philosophers sit at a dining table eating short meals and thinking, in-between with one fork between every two persons. However, each philosopher needs two forks to be able to eat. Each philosopher thinks for a random period of time until he or she gets hungry, then attempts to grab two forks, one from the left and one from the right, to start eating the meal. If the person succeeds, he or she starts eating. If the person does not get any fork, he or she continues to try. If the person gets only one fork, he or she puts it back. The program finishes when all philosophers have eaten a certain number of times (this number can be specified through command line).
7. Current next generation sequencing technologies often generate duplicated or near-duplicated reads that (depending on the application scenario) do not provide any interesting biological information but increase memory requirements and computational time of downstream analysis.

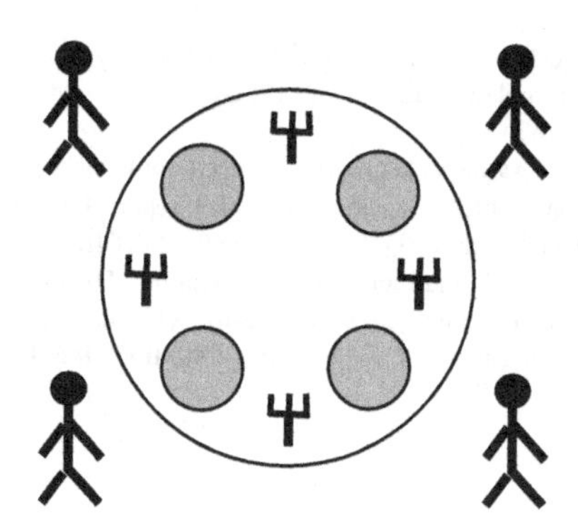

FIGURE 10.11

Illustration of the philosophers problem for four threads.

a. Implement a UPC++ parallel program that will receive as input a set of DNA sequences (strings formed by the alphabet A, C, G, T) and will provide the same set but removing the duplicate sequences.
b. Extend the previous code to discard near-duplicate sequences too. It will receive through command line a number of allowed mismatches. Two sequences are considered near-duplicated (and, thus, one of them must be discarded) if the number of DNA bases (i.e., characters of the string) that are different is less or equal than the number of allowed mismatches.

8. UPC++ can be used together with CUDA in order to exploit several GPUs placed in the same or different nodes of a cluster. Develop a code that integrates UPC++ and CUDA in order to complete the following matrix multiplication on several GPUs: $C = A \cdot B + C$, where $m \times k$, $k \times n$, and $m \times n$ are the dimensions of matrices A, B, and C. Note that m, n, and k do not need to be equal.

REFERENCES

[1] Christian Bell, et al., Optimizing bandwidth limited problems using one-sided communication and overlap, in: Procs. 20th IEEE Intl. Parallel and Distributed Processing Symp. (IPDPS'06), Rhodes Island, Greece, 2006.
[2] Cray Chapel Group, The Chapel parallel programming language, http://chapel.cray.com/ (visited on 08/20/2016).
[3] Edsger W. Dijkstra, Hierarchical ordering of sequential processes, Acta Informatica 1 (1971) 115–138.
[4] GCC Developers, CAF wiki, https://gcc.gnu.org/wiki/Coarray (visited on 08/20/2016).
[5] Jorge González-Domínguez, et al., Design and performance issues of Cholesky and LU solvers using UPCBLAS, in: Procs. 10th Intl. Symp. on Parallel and Distributed Processing with Applications (ISPA'12), Leganés, Spain, 2012.
[6] Amith R. Mamidala, et al., MPI collectives on modern multicore clusters: performance optimizations and communication characteristics, in: Procs. 8th Intl. Symp. on Cluster Computing and the Grid (CCGRID'08), Lyon, France, 2008.
[7] Rajesh Nishtala, et al., Tuning collective communication for partitioned global address space programming models, Parallel Computing 37 (9) (2011) 576–591.
[8] Project Kenai, Project Fortress, https://projectfortress.java.net/ (visited on 08/20/2016).
[9] Carlos Teijeiro, et al., Evaluation of UPC programmability using classroom studies, in: Procs. 3rd Partitioned Global Address Symposium (PGAS'09), Ashburn, VI, USA, 2009.

[10] Titanium Developers, Titanium project webpage, http://titanium.cs.berkeley.edu/ (visited on 08/20/2016).
[11] Bibo Tu, et al., Performance analysis and optimization of MPI collective operations on multi-core clusters, The Journal of Supercomputing 60 (1) (2012) 141–162.
[12] UPC++ Community, UPC++ wiki, https://bitbucket.org/upcxx/upcxx/wiki/Home (visited on 08/20/2016).
[13] UPC Consortium, UPC Language and Library Specifications, v1.3, report, Lawrence Berkeley National Laboratory, Nov. 2013, http://upc.lbl.gov/publications/upc-spec-1.3.pdf (visited on 08/20/2016).
[14] David E. Culler, et al., Introduction to UPC and Language Specification, IDA Center for Computing Sciences, 1999.
[15] X10 Developers, X10: performance and productivity at scale, http://x10-lang.org/ (visited on 08/20/2016).
[16] Yili Zheng, et al., UPC++: a PGAS extension for C++, in: Procs. 28th IEEE Intl. Parallel and Distributed Processing Symp. (IPDPS'14), Phoenix, AR, USA, 2014.

Index